NH

WORK, ORGANIZATIONS, AND TECHNOLOGICAL CHANGE

NATO CONFERENCE SERIES

I   Ecology
II   Systems Science
III Human Factors
IV Marine Sciences
V   Air—Sea Interactions
VI Materials Science

II SYSTEMS SCIENCE

WORK, ORGANIZATIONS, AND TECHNOLOGICAL CHANGE

Edited by

Gerhard Mensch

Case Western Reserve University
Cleveland, Ohio

and

Richard J. Niehaus

Office of the Assistant Secretary of the Navy
(Manpower and Reserve Affairs)
Washington, D.C.

Published in cooperation with NATO Scientific Affairs Division

PLENUM PRESS · NEW YORK AND LONDON

Library of Congress Cataloging in Publication Data

Main entry under title:

Work, organizations, and technological change.

(NATO conference series. II. Systems science; v. 11) "Proceedings of a NATO
Conference on Work, Organizations, and Technological Change, held June 14–19,
1981, in Garmisch-Partenkirchen, Federal Republic of Germany"—T.p. verso.
"Sponsored by the Special Panel on Systems Sciences of the NATO Scientific
Affairs Division"—Pref.
Includes bibliographical references and index.
1. Technological innovations—Congresses. 2. Machinery in industry—Congresses.
3. Manpower planning—Congresses. I. Mensch, Gerhard. II. Niehaus, Richard J.
III. NATO Conference on Work, Organizations, and Technological Change (1981:
Garmisch-Partenkirchen, Germany) IV. North Atlantic Treaty Organization. Division
of Scientific Affairs. Special Panel on Systems Sciences. V. Series.
HD45.W67   658.3   82-3751
ISBN 0-306-40993-3   AACR2

Proceedings of a NATO Conference on Work, Organizations, and
Technological Change, held June 14–19, 1981, in
Garmisch-Partenkirchen, Federal Republic of Germany

© 1982 Plenum Press, New York
A Division of Plenum Publishing Corporation
233 Spring Street, New York, N.Y. 10013

Printed in the United States of America

NA

PREFACE

This volume is the proceedings of the Symposium entitled, "Work, Organizations and Technological Change" which was held in Garmisch-Partenkirchen, West Germany, 14-19 June 1981. The meeting was sponsored by the Special Panel on Systems Sciences of the NATO Scientific Affairs Division.

In proposing this meeting the Symposium Directors built upon several preceding NATO conferences in the general area of personnel systems, manpower modelling, and organization. The most recent NATO Conference, entitled "Manpower Planning and Organization Design," was held in Stresa, Italy in 1977. That meeting was organized to foster research on the interrelationships between programmatic approaches to personnel planning within organizations and behavioral science approachs to organization design. From that context of corporate planning the total internal organizational perspective was the MACRO view, and the selection, assignment, care and feeding of the people was the MICRO view. Conceptually, this meant that an integrated approach was needed if all the dimensions of such problems within private and public organizations were to come out correctly.

As with any meeting of that kind, the Stresa conference left many dangling ends suggesting that now even a more macro view was needed. Also during the intervening four years, the constrained views of corporate manpower planning and organization design have given way to the broader concept of human resource planning. Here, not only are internal organizational relationships important, but also more of the organization's relationships with the external world are felt to be needed to be included. Notably such dimensions as the impact of external labor market developments and the advance of technology need to be added. In this instance the relationships to the external world becomes the MACRO view and internal manpower planning and organization design the MICRO view. The purpose of the Garmisch meeting was to explore this broader view using the organization as the focal point.

In putting together the program, the Symposium Directors
started with the larger picture, and then emphasized the important
parts of the problem followed by a mixture of broad and specialized approaches to parts or all of the problem. Also, discussion
was encouraged both during the formal sessions as well as during
the social events. This served to develop better linkages between
seemingly unrelated parts and to surface parts of the problem not
completely covered by the prepared presentations. Perhaps the
most important of these added dimensions was the need to include
the interrelationships between human value system and technical
human resource planning systems.

The organization of the symposium had to take into account
the development of the meeting topic coupled with balance among
contributions from different countries and research institutions.
The selection of papers was difficult with so many worthwhile
papers from many countries with a limited space on the program.
The difficulty was resolved by allocating the papers to one or two
types of presentations: plenary sessions for papers that appeared
to fit into one of the five Symposium themes: and parallel
presentations of special interest topics. In the event, following
the keynote paper by Dean George Kosmetsky, Graduate School of
Business, University of Texas at Austin, twenty-one papers were
presented in the plenary sessions and eleven in the parallel
sessions for a total of thirty-three papers. The session themes,
together with the session chairperson(s) are listed below:

SESSION	CHAIRPERSON
•Technological Change and Human Resources	J.A. Sheridan (USA)
•Labor Market Impacts on Organizational Planning	W.L. Price (Canada)
•Methods and Studies (Parallel Sessions)	R.E. Boynton (USA); M. Rowe (USA); and C. Gaimon (USA)
•Social Impact on Personnel Policy Analysis	D. Sadowski (Germany)
•Technological Change and Work Organization	P. Miret (France)
•Quality of Organization Life	R. Pearson (UK)

The Symposium Directors would like to thank Dean Kozmetsky and
the Session Chairpersons for a job well done.

The proceedings follow the Symposium program to some extent. However, the order of individual papers has been changed to produce a balanced publication geared towards the requirements of the general reader. Also, because of page limitations, a number of papers were included in the form of one page abstracts. The full text of those papers can be obtained by communication with the authors.

The formal and informal discussions at the Symposium shaped the final structure of the proceedings. This was coupled with the desire to move from the general to the specific with case examples in between. The Symposium Directors are thankful to Dr. Lisl Klein of the Tavistock Institute, London for striking a strong note for adding the dimension of human value systems beyond the needs of organizational technocrats. With the exception of the brief discussion of the final session provided below, the issues are left to the summaries at the beginning of each section and the papers themselves.

In the final summary session, an attempt was made to find what consensus was present in the diversity of topics discussed. The discussion started with the issue that new technology would displace workers in a way that would also change the underlying interrelationships between the work place and society. It was pointed out that the issue of technological change was not new, but that new technology might have outstripped the graduate required to manage the desirable direction of innovation and maintain the necessary rate of innovation in a competitive world. Again, the theme arose that there was a need to do research on joint issues of technological development and cultural development. This brought about a spirited discussion of the problem of who determines which cultural values are important in democratic societies. Furthermore, from the institutionalist point of view, it was suggested that with very large private and public organizations existing in all sectors of industrial societies, there is no choice but to include the effects of regulated technological change in human resources planning. However, there are not absolute solutions. The process is open and moving, at an uneven pace. Some organizations, even large ones, are being pushed by rapid series of technological innovations, whereas in other organizations it takes years to introduce technological change. It was suggested that the best way may be to look at controllable systems and use them to develop policies that can guide the future, but that view was questioned by reason of rationality and legitimacy. The divided opinions indicated that additional study is needed, particularly on how to systematically place value systems into the human resources planning process either when both are being driven by technological change, or when both have become the major inertial factors in creating useful affirmative technological developments.

Thanks are in order for all who helped to contribute to the success of the meeting. Special mention is for the officials of Garmisch-Partenkirchen and the Kongresshaus for the facilities and smooth running of all the logistics of the meeting. Secretarial help was provided by Ms. Annie Knott and Ms. Loretta Orrock of the Office of the Assistant Secretary of the U.S. Navy (Manpower and Reserve Affairs), and by Mr. Margery Sperling of the Weatherhead School of Management of Case Western Reserve University. Also, help in reviewing the proceedings was provided by Dr. Carol Schreiber, General Electric Company, and Mr. Ed. Bres, OASN(M&RA). In addition, there were many others, including our wives, who had considerable part in the social program and the orchestratian of the informal international meetings that took place in and about Garmisch-Partenkirchen.

Finally, we would be remiss if we did not acknowledge the role of the NATO Scientific Affairs Division in supporting the Symposium. We owe a special debt to Dr. B. A. Bayraktar and the members of the Systems Sciences Panel who provided counsel and advice during the early stages of the Symposium preparation.

Gerhard O. Mensch

Richard J. Niehaus

CONTENTS

CONTENTS

NA

SECTION 1

KEYNOTE ADDRESS

Professor George Kozmetsky, Dean of the Graduate School of
Business Administration, University of Texas at Austin, sees
that we are in the midst of chronic and often unplanned technological change where the value of the new technologies is difficult to assess and the need to adapt to change is vital.

The allocation of human and technological resources takes
place within an institutional setting. The personnel requirements for the technologies of the 80s will create a major
societal displacement. Millions of workers who now do only
repetitive jobs will need to acquire new skills. According to
Dean Kozmetsky, technological change is likely to have three
dramatic impacts on the general work force. First, machines will
take over repetitive work tasks at an accelerated rate. Second,
elementary decision making will become part of the mechanized
process. Finally, human resources will be freed to supervise
more complex machinery and assume a wider decision-making role.
Kozmetsky asks: How should we approach these developments?

6210
9120
U.S.

PERSPECTIVES ON THE HUMAN POTENTIAL

IN TECHNOLOGICAL CHANGE

George Kozmetsky
Dean
College of Business Administration
Graduate School of Business
The University of Texas at Austin

It seems to me that this NATO Symposium on "Work, Organizations and Technological Change" can be characterized as follows: First, technological change is the key influencing factor for all sessions; second, technology per se is not a primary focus; and third, your overall emphasis is on entity goal formulation which evaluates the effectiveness of incentives and advancements that enhance the efficiency and quality of life.

The forthcoming symposium sessions will examine technological change, considering its utilization by type of institution ranging from military to process controlled, batch controlled, automated and mechanized services institutions. In most cases the papers to be presented will analyze technology's impacts on the workforce, manpower planning and training. There are a number of provocative presentations that identify developments that are taking place and some that extend the current methodologies for manpower planning, control, and audits. These methods papers will encompass technological change in terms of institutional manpower planning and control, productivity, social impacts and restructuring of work organization.

3

You will be addressing a multiplicity of problems that require
unique solutions for each of the nations and institutions represented at this conference. We will all learn from each other. We
will be able hopefully to take away from these presentations and
discussions relevant knowledge that is applicable to the solution
of issues that face our own countries as well as our individual
organizations.

In a broader context, this symposia is unique. The underlying theme deals with two very significant resources -- human and
technological. This is the first meeting in which I have been
privileged to participate where technological and human resources
are linked. Most meetings would view these resources as separate
and distinct.

The importance of this linkage is that it provides for each
of us a way to deepen our scientific understanding of the process
by which these resources are jointly developed, allocated, and
utilized to meet institutional needs and selective missions. This
linkage also permits us to determine individual worker objectives.
to evaluate the choice mechanisms for selecting the alternative
uses of technological and human resources, and finally to audit
the effectiveness and efficiency of the transformation of these
resources in meeting institutional goals. The sum total of the
institutional goals will in a large measure determine the societal
goals.

For the next five days, we will be involved with the over-
riding issue that faces us individually and collectively: How do
we organize our human and technological resources between as well
as within institutions while maintaining a high quality of life
environment for most individuals?

I would like to focus my remarks on two broad themes. First,
I will review technological change for the 1980s from a personal
perspective. Second, I will discuss the role of professional
schools in the education of human capital in the midst of conscious technological change.

The Consequences of Technological Developments

Any solutions for the problems that we will be addressing
this week depend upon an understanding of the consequences of
technological change. Our perception of those consequences will
affect not only the way in which we link human and technological
resources but also the manner in which we approach the decision-
making process.

The identification of technology as a resource places yet another burden on institutional leaders who are responsible for making complex decisions about the distribution and allocation of resources. Each group within our respective organizations perceives and values technology in a variety of ways. Some groups see no need for it. Others see technology as a means of saving society or rescuing a firm's future whenever it's in trouble. However, technology exists and has profound effects. It has the capability of improving or worsening any situation.

There are two broad consequences that may result because of the importance of current and impending technological developments.

These consequences are based on the following assumptions:

1. Technology is a body of knowledge.

2. Knowledge is wealth and power.

3. Whoever controls technology controls other resources. Hence, one may say that technology is a "master" resource.

Therefore, the perceived consequences are centered around the control of technology as a resource.

The first consequence is to have this control highly centralized.

The second consequence is to have this control widely distributed.

In the highly centralized case, wealth and power could be aggregated in the hands of a small corps d'elite. This aggregation could lead to a rigidly bureaucratic organizational structure. Generally, bureaucratic organizations tend to lead to alienation and inflexibility.

The widely distributed consequence could produce an integrated yet pluralistic society. In the case of a firm, it could broaden the base of participation and decision making at all levels of organization including supervision over robotics and semi-automated functions. Technology gives us the means to distribute knowledge, and therefore power, down to a very local level. With appropriate checks and balances, this same technology could give us the means to reduce the centralization of power and to increase our responsiveness to needs through a more effective structural as well as organizational basis.

The choice between the two consequences and their associated spectrum of alternatives exists. Some technologists have voiced a strong opinion that if trends of the 1960s and 70s continue this choice will no longer be available and that a trend towards centralization will be irreversible.

I believe that we need to depart from moves toward centralization. This departure requires a determined, sustained and a managerial cross-disciplinary effort. This effort must be well planned, and range from small amounts of basic research to a large amount of development, implementation, production, and marketing. To achieve a more pluralistic society as well as increased participation in a firm, initial applications must be carefully selected to integrate technological change with work and organizational needs.

Society to date has placed the responsibility for solving such problems in the hands of key decision makers working through existing institutions. These institutions are characterized by a more or less coherent group of individuals, resources and values related at least in part to an organizational structure and a formal process of decision and communication.

Alternative Generation for Decision Making

There is now a real question in my mind whether the past loci of organizational decision making can be maintained. We are in the midst of chronic and often unplanned technological change where the value of the new technologies is difficult to assess and the need to adapt to change is vital. To determine the continued viability of the past system of work and organization, we need to ask several specific questions of our decision makers:

(1) Do they have the time to spare from the current short-run problems and pressures generated by their institutions?

(2) Do the operational rules of their institutions permit them to concentrate on anything but the short run?

(3) Do they have the motivation necessary to tackle problems whose time-frame extends beyond their tenure of office or position?

(4) Do they have the current means of distribution of resources (especially technological) to provide sufficient resources to meet the demands?

(5)   Do the current problem solvers have the appropriate conceptual apparatus to perceive the wider problems facing their firms and all their constituencies?

(6)   Do they have the techniques, including hardware and software and communication, to structure solution models and procedures?

(7)   Do they have the information technology required to gather the data for problem formulation, problem solution and monioring of results in a timely and effective fashion?

(8)   Do they have the appropriate organizational structure mechanisms both to solve the problem and to take the necessary implementation steps?

These questions ought to be the boundaries upon which we extend the scientific and methodological borders of knowledge to incorporate technological change for work and organization.

What is required in addition to the answers to these questions is an examination of the premises underlying the responses. We can then ascertain those factors which will permit managerial and worker adaptivity.

There will be a critical need in the 1980s for individuals with talents to:

(1)   perceive and formulate for solutions those problems facing our various institutions as well as our society (problem formulators);

(2)   relate the formulated problems into a unified usable conceptual model that identifies the desired structure of society and institutions (modelers);

(3)   translate these problem formulations into understandable opportunities for execution and implementation by the policy makers of institutions (problem solvers or decision makers);

(4)   provide a means for the problem solvers or decision makers to validate the acceptability and efficiency of their solutions (comprehensive auditors).

Today there is inadequate attention to alternative generation
for decision making. Too often problem formulators are working
on predicting future problems without reference to present
realities, available resources, political realities, or even the
current range of possible solutions. Too often problem solvers
are working on current problems without considering adequately the
consequence of their actions in the future. Like nuclear reactions, decisions can have unseen yet far-reaching repercussions.
If the leaders of our institutions are to manage technological
change and not simply be engulfed by it, they need to anticipate
the chain reactions their decisions may set in motion. And this,
perhaps, is the major point: important choices lie before us, but
they cannot be made in any monolithic way. We are entering upon a
period of critical experimentation, and it is too tempting to
assume that we will find "the best way" for there is probably no
best way at all; only better ways for a while, for different
groups, and for different areas.

The management of technological change then becomes the
management of the creation and application of knowledge. The
problem for managers of all policy-making centers at every level
of decision making, in reality, becomes the problem of auditing
and selecting those elements of technology that will enable
institutions to achieve their goals within a wider context of
quality of objectives and/or within a framework of national policies.

The Implications of Technological Changes

The potentials of technology for the 1980s have critical
implications for all the independent and sovereign nations that
are a part of NATO or cooperate with it. If we are to advance
the quality of life in all our countries, we must understand the
implications that technological changes will bring socially,
politically, economically and to the individual. In the 1970s,
most decision makers focused technological changes on short run
objectives that were concerned with the bottom line or designed
to meet the competition. Longer term capital investments were
severely constrained by inflation and capital availability.

The technologies of the next decade will provide for our
new growth industries. They will influence changes in opinions,
attitudes, concerns, life styles and cultural values, and affect
our defense postures and strategies.

By innovating and then applying technology in anticipation of
these developments, we can make the 1980s the age of conscious
technological change. Through the judicious application of
technology, we can continue to improve our economic, socio-political
and cultural conditions. We need not settle for lower quality of
life or standard of living. The potential of the emerging technologies for the 1980s makes it possible for us to do more than
dream about reaching beyond the status quo.

Without question, technology, tempered with the lessons of
the 60s and 70s, promises not only stable future growth, but
also a more productive and fulfilled workforce. All of us who
have been a part of technology's science, its management or
support know it is not always predictable, nor does it always
achieve precisely what we had hoped. But when treated with
respect, concern, common sense, understanding and general
concensus, technology does deliver a fair share of its promises
to all mankind.

Five Perspectives on Technology

How do we assess these promises? It is my contention that
only by viewing technology from several "emerging perspectives"
can we hope to deal adequately with the dramatic changes in work
activities and organizational structures that are bound to come
in the next two decades. I believe that the technologies for
the 1980s must be viewed as a national and a world resource, as
a generator of wealth, as the means to increase productivity and
international trade, as an area for assessment of public and
private risk taking, and last as an influencing factor of changes
in the organization, education and training of the workforce.

First, technology is not simply an engineering "thing," a
gadget or even a "process." It is a national resource. Unlike
natural and human resources, it is not consumed in the process
of use. Rather, like a catalyst, it can be a stimulant, or it
can be self-generating as in fusion. The use of technology
actually creates more technology. There is, however, no
systematic means for allocating or evaluating technology as a
resource. Today, we cannot tell whether the allocation function
is efficient or inefficient; or even if it is working and meeting
the objectives for which it was created let alone the needs of
society or our individual institutions.

Second, technology is a type of wealth; one which we do not
yet know how to measure for economic purposes. Since we cannot
measure technology as an economic resource, our ability to determine its full value is limited. Wealth, of course, is a means of
attaining economic, social and cultural status for individuals
as well as a way of achieving institutional objectives and the
general welfare of society. It is recognized as perhaps the single
most important variable in determining an individual's, corporation's, nation's, state's or locality's alternatives. Wealth is
the key to power in any political system. Its ownership, control,
and transformation determine not only the structure of society
but also the success or failure of the system itself. To ensure
the continued strength of our economic and political systems in
terms of our current institutions, we need to develop ways of
estimating the real wealth, that is, the productive wealth,
provided by new and improved technologies.

Third, technological innovation is a prime factor in
stimulating our productive capacity and insuring healthy
competitive international trade. We must, therefore, adopt
more viable national policies for technological development.
Thus far, few if any nations, have effectively formulated such
critical policies. We can bring about significant short-term
as well as long-term gains by encouraging the establishment of
an effective working relationship between government, business
and universities regionally, nationally and internationally in
order to strengthen and broaden our R & D environment. Government
can be a very important stimulant to technological development.
By developing policies that encourage academic and industrial
research, strengthen the educational structure, promote a
positive business and economic climate, and identify special areas
of concern, government can be a positive force in technological
growth.

Fourth, the relationship between the public and private
sectors is changing. Each nation's technological needs are
linked with other nations' commerce in a number of important
ways. In no case is one market wholly insulated from what happens
in other countries. This is a fundamental political and economic
fact. Shortages can result in a crisis, a comfort, or a boon
depending on the circumstances. Some shortages can produce a
crisis of national consequences. To reduce the consequences of
the crisis puts us in the arena of potential "public risk" taking.
When such an event threatens national security either economically
or militarily, there is a "public risk." Investments in some
critical technological areas may be so large that the private
sector alone is unable to assume the potential risks. Petroleum
import dependence, for example, is both a factor of national
economic and technological importance as well as a threat to
national security. It is appropriate to consider the risk

involved as a possible public venture to be implemented as well as
supplemented by our private firms.

Fifth, we need effective technological management training
at all levels in both the private and public sectors in operations
as well as research. This is in addition to scientific and
engineering education and training. Over the next decade, a variety
of changes in technology are likely to occur that will significantly
affect all institutions. Any predictable developments in technology,
however, can only come about as a result of the actions of public
and private institutions. Thus changes in technology become matters
of what we decide rather than merely events to be predicted.
Conjecture, of course, will play a part; but the emphasis should
be on the transformation of the managerial component. Conjecture
is a matter for experts; management is a matter for managers;
the former is sterile without the latter. Management of change
depends on how we manage the creation and application of technology
and its acceptance as well as constructive contributions by the
workforce. Technology management education and training is central
to this function and responsibility.

Renewed Enterprise System

The allocation of human and technological resources takes
place within an institutional setting. Technological change is
occurring with such velocity that it is difficult, if not impossible
for individuals and institutions to respond. As change in the
environment accelerates, more novel first-time problems arise.
Traditional institutional forms may prove inadequate, and
innovative thinking may be required if we are to keep pace.

In the 1950s and 60s, the federal government in the United
States financed research and development through its focus on
military preparedness and need. In the 1970s, the government's
emphasis shifted. Medical, environmental, and human delivery
systems gained prominence as our government directed funds toward
federal programs aimed at broadening social welfare. Currently
there is strong support for a return to greater allocation of
money for defense, research and development expenses. The linkages
of such efforts to more non-defense applications have not yet been
addressed but need attention.

Technology from 1950-1970 was generally applied to the
advancement and marketing of a particular segment of an industry
such as integrated circuits for new generations of large scale
computers. In my view, the technologies of the 1980s will not
be solely used in this manner. A "clustering" pattern will be the
"hallmark" of the 1980s-1990s period and will be based on
prioritized economic and social needs. It will complicate the

organizational structure as well as increase the complexity of
training the workforce. In the process, however, it could well
lay the groundwork for the developments of a renewed enterprise
system. To accomplish this, the private or public sector or both
will have to develop a new, integrated incentive program that will
include such diverse elements as expanding perquisites, stock
options, security systems, pensions and bonuses in addition to
attractive salaries. The allocation of human capital is an
institutional process that is profoundly affected by technological
change. The objective is to ensure adequate manpower supply in
institutions at the proper time.

Individual and institutional goals are complex and sometimes
contradictory. Consequently, the motivation of specialists, who
are always in short supply, is changing. Specialists will tend
to be more impatient; they will be in positions to demand
immediate gratification. Unlike the others in the workforce who
are willing to take risks and seek to optimize the economic
resources of the organization, specialists will be more apt to
move out of the organization to pursue their own objectives.
Consequently, this independence and increased mobility among
skilled specialists may present a real manpower shortage in the
years ahead.

The personnel requirements for the technologies of the 80s
will create a major societal displacement. Millions of workers
who now do only repetitive jobs will need to acquire new skills.
Technological change is likely to have three dramatic impacts on
the general workforce. First, machines will take over repetitive
work tasks at an accelerated rate. Second, elementary decision
making will become part of the mechanized process. And third,
human resources will be freed to supervise more complex machinery
and assume a wider decision making role. How should we approach
these developments? Required training and upgrading must be
related to appropriate incentives.

Meeting Individual and Organizational Needs

Conscious technological change presents us with a different
kind of challenge--to organize work in such a way that the
individual is improved and has the appropriate incentive to keep
updated. This will require us to fulfill the needs of the
individual while meeting the needs of the organization. Organizational leadership will have to facilitate the adaptive process
and thereby reduce resistance to change. What responsible management should seek from their workforce is not simply passive
acceptance of new ways but active collaboration. More active
participation will mean increased flexibility in the workplace,
the advent of the "electronic cottage," and the customization
of products and services.

To be successful within our technological society, institutions must attract men and women who accept responsibility, who handle even larger tasks, and who adapt swiftly to changed circumstances and the new technologies. This type of workforce will be less pre-programmed and react faster to changes in the environment. They will tend to be complex, individualistic and proud of the ways in which they differ from other people. Workers who seek meaning, who question authority, who want to exercise discretion, or who demand that their work be socially responsible, may be regarded as troublemakers in some institutions, but our techno-logical society cannot run without them.

In the past, we worried about providing too much leisure time for fear that workers would not know what to do with it. At the same time, we were providing the requisite capital to build the infrastructure for recreation and leisure. Currently, people are demanding more time to be utilized by their individual prefer-ences. They are even willing to buy larger segments of their own time. In the future, we will need to become more aware of the trade-off between individual time and remuneration, and then more effectively balance the two with appropriate incentives.

Comprehensive or full-scope audits of all phases of manage-ment's responsibilities and activities will become a reality in the 1980s. These audits will have to deal with long-range plans, manner of governance, the institution's contributions to society's general welfare, and management effectiveness. All of these deal with the allocation of human and technical resources. In this context, the essence of management in the 1980s must be foresight, calculated action and accountability.

Technology in Managerial Education

I would like to conclude my paper with some remarks on my particular area of expertise, that is, the management of a professional school. As dean of one of the important business schools in the United States, I am constantly reassessing what our role as an educational institution ought to be in our complex and rapidly changing technological society.

Effective manpower planning is directly related to education and training. Training focuses on specific jobs and skills and prepares one to cope with change. Education focuses on adequate support systems and research, and prepares one to manage change. In both cases, experience plays a crucial role, and organizational management must find ways to provide the necessary experience. In addition, management should determine on a timely basis what the opinions, attitudes and concerns of the workforce are, and then take appropriate action to provide adequate training and education.

Higher education has been exemplary in education of
scientists and technologists. It must continue to meet this
responsibility. However, I have been asking myself what will
the technologies for the 1980s require of professional schools
in their task of preparing students for managing technology. The
education process must, of course, be cross-disciplinary. It will,
therefore, be necessary to broaden the scope of management education
at the master's level. We will need to include many existing as
well as some new academic disciplines including science, engineering, computing, business, law and public affairs. Managerial
education must include technology as a necessary function with
the same importance as marketing, finance, production, accounting
and human resources.

The job of the professional school is essentially twofold.
On one hand, it must import problems and develop them into issues
suitable for research aimed at solutions. On the other hand, it
must export and communicate results. This importing and exporting
process should be geared around several strategic research areas
within the overall context of "Work, Organizations and Technological
Change":

-- Managing productivity, efficiency and competitiveness

-- Management of innovation, research and development

-- Managing international business

-- Managing strategic relationships and social responsibilities

-- Managing human resources including manpower, compensation
   rewards, and working life

-- Entrepreneurship, business development, and capital
   venturing

-- Financial management and markets including capital
   requirements and investment decisions

-- Management of information and management information
   systems

-- Managing the environment in the interests of quality of
   life

-- Development of new growth industries

-- Assessment of technology as a national and world resource

-- Evaluation of technology as a value resource

-- Analysis of public risk

-- Assessment of the influence of technology on organization, education and training

Responsibilities of Professional Schools

If professional schools are to develop curriculums that will meet the needs of our changing technological society, they must integrate three responsibilities.

First, they must emphasize situational and strategic management and control. For too long, managers of firms have paid little attention to the scientists and engineers in their organizations, while the scientists and engineers have ignored the issues and needs of managers. What we educators need to do is to redesign the educational process in our management programs. Professional schools must provide the crucial link between basic sciences and management practice. That link can be provided by a new core of professionals who demonstrate genuine entrepreneurial skills. These highly qualified Techno-Socio managers will be able to deal effectively with this age of conscious technological change. They will be able to recognize, understand and implement the fundamentals of both the physical and social sciences, and apply them to social needs as well as specific issues on national agendas.

Second, professional schools must focus on comprehensive audits. We need to work to create an appropriate system for evaluation that provides real accountability. In addition to education for the management of technological change, the academic community has a responsibility to do the research required to develop an acceptable body of knowledge in the area of compliance and comprehensive audits. It is important for our academic community to be the leader in this area of concern.

Third, professional schools must expand our state of scientific knowledge by developing methodologies and techniques to deal with complex problems involving our technological and social evolution. An example of such a problem is the allocation of world resources to meet the needs and secure the rights of all humans. Other macroengineering problems include concerns about pollution, scarcity, social welfare, energy and public risk. The management of such large systems requires that the leaders of our institutions exhibit the abilities to conceptualize multiple objectives, to interact with other economic, social and cultural institutions, and to operate within a highly dynamic environment. Research in education should, therefore, be directed toward establishing a way in which man and society will reconstruct the world based on the technological change that is taking place.

What I am advocating is an increased scope for professional schools if they are to meet perhaps the most demanding challenge of the next century namely, to manage our human resources in a technologically based society in such a way that we actually anticipate change and not just accept or react to it. We must develop the ability to organize work in such a way that for most individuals -- regardless of race, color, creed or intelligence quotient -- work becomes more complementary to human needs and thus more fulfilling and self-satisfying.

SECTION 2

TECHNOLOGICAL CHANGE AND HUMAN RESOURCES

Technological developments do not happen at an even pace,
Mensch distinguishes between several types of technological in-
novation(basic innovation, improvement innovation, and pseudo
innovation). He offers a scenario with an increasing market pull
for sophisticated replacement (improvement) technology, and an
evolving structural readiness of capital market for another push
of basic innovations. Contrary to the belief of many that a
high degree of technologization will induce the economy to create
more non-technological products and services, Mensch sees the
market mechanism reacting to the abundance of some technology
with even stronger demand for other, different technology. Hence,
it is not the availability of accumulated capital per se, but the
ability of the population and work force to accept and develop
better technologies that becomes the bottleneck in the process
of the creation of wealth, employment and income.

Schreiber describes an example of best practice for corporate human resource planning in the world of the large industrial corporations. The General Electric system that she
describes is directed towards emerging risks and opportunities
for human resource planning. This approach, as part of a shift
in emphasis from financial restrictions on growth and development towards human resource restrictions, recognizes the fact
that selective shortages of some types of human resources evolve
in the midst of a general over-supply of labor.

Within the military sector, West perceives a similar shift
in emphasis. In the United States and other NATO countries,
technological superiority has been seen as a way to achieve the
qualitative superiority in armament needed to balance quantitative disadvantages. The need for better technology is now
increasingly driven by human resource considerations. The
acceptance and further development of technology is a way of
compensating for the quantitative and qualitative manpower
shortages anticipated for the years to come.

THE CO-EVOLUTION OF TECHNOLOGY AND WORK-ORGANIZATION

Gerhard O. Mensch

Weatherhead School of Management

Case Western Reserve University, Cleveland, Ohio USA

ABSTRACT

According to Benjamin Franklin, "Man is a tool-making animal." Had he lived today, Franklin might have said that man is a technology maker, and that he is very successful at that. Skeptics, on the other end, think that this success has generated over-industrialization, over-technologication, and man has become a technology taker--nolens volens.

This paper argues that the market process seems not to follow the skeptic's value judgment. Instead, the market forces react to the fact of over-supply of some types of technology with even stronger demand for technology; for different technologies, though. Inter-industrial, inter-regional, and inter-national competition propels the re-direction and acceleration of technological change, and in the future as in the past, the individual and organized human capabilities for technology making and technology taking will determine the course of events in the co-evolution of technology and society.

In discussing the universal tendencies and the temporal deviations from this development, a distinction between different types of technological innovation is in place. As these different sub-trends in the rate and direction of product and process innovations are pre-programmed by developments in the life styles and work organizations of the population, they in turn impact upon the way of life and the development of new forms of work organization.

INTRODUCTION: THE CO-EVOLUTION OF TECHNOLOGY
               AND WORK ORGANIZATION

     This article is chiefly addressed to managers in industrial
corporations and public service delivery organizations, and to
those professionals who assist these managers in making strategic
decisions in a world of intensifying competition, accelerating
technological change, and increasing uncertainty about future
economic developments. Those turbulances call for organizational
creativity, and for corporate leadership capable of coping with the
technological developments that originate within and act upon the
corporate world by way of product and process innovations.

     The objective I have in mind for this audience is to help them
in becoming more effective in innovations management, both in the
creation of higher-valued new products and in the increasing of
productivity, reliability, etc., of on-going productions. I do
this by pointing at some ongoing technological trends of major
proportions. In 1980, the intensified technological competition
pushed the number of bankruptcies in private business to an all-
time high. Thus, if for no other reasons than for economic
stability alone, technological innovations must be dealt with more
effectively, both in the sense of "technology making" and
"technology taking." With respect to the change, firms must learn
to advance from plain struggle for survival to a state of stable
well-being. As a rule, industrial corporations must learn not only
to live with changes, but to thrive on it.

     The point I am trying to make is this: Contrary to many
peoples' belief that the industrial nations are heading into a
POST-INDUSTRIAL SOCIETY, on the assumption that we are over-
industrialized, our lives are over-technologized, and under-
developed in terms of non-technical services and public goods, I
see much evidence for us heading for even more industrialization,
which will, however, be based on different technologies. I also
see us heading for more industrial organization in the service
sector, and an even stronger dependence on complex and "high"
technologies than at present will be the likely result. My point
is that this development is driven by the forces of competition,
but it is restricted in scope and pace by human resources, and
the ability of people to organize themselves creatively. This
point is not much more than the application to problems of work
organization of one general principle, the Principle of
Co-Evolution of Technology and Social Formation. Technology and
technological change have a profound influence on the social
structure and its restructuring, and vice versa. Social
conditioning and preprogramming of technology and technological

change must also be taken as a fact.[1] Just as organization is embodied in "human capital", technology is embodied in "physical capital", and if one changes, so does the other, thereby acting as a restriction or as an incentive for change (Figure 1).

Hence, the general Principle of Co-Evolution of Technology and Social Formation can be applied to a host of substantive areas of work: Work in production, administration, distribution, transportation, research and development, banking and finance, exchange and defense, etc.

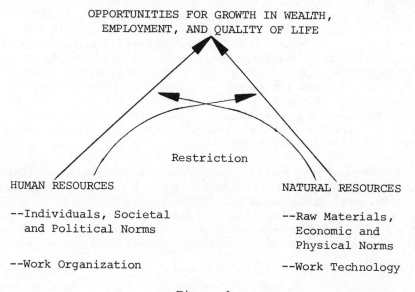

OPPORTUNITIES FOR GROWTH IN WEALTH,
EMPLOYMENT, AND QUALITY OF LIFE

Restriction

HUMAN RESOURCES

--Individuals, Societal
  and Political Norms

--Work Organization

NATURAL RESOURCES

--Raw Materials,
  Economic and
  Physical Norms

--Work Technology

Figure 1

[1]G. O. Mensch, Social Preprogramming of Technological Change, forthcoming in Technological Forecasting and Social Change (extended version of German original in: Tijdschrift voor de Studie van de Verlichting, 8(1980/81), 127-141.

This is a fairly general principle. It leads us to understand
phenomena that are somewhat paradoxical: For example, is it not
somewhat paradoxical that the market mechanism works in such a way
that it reacts to an over-supply of some kinds of technology with
an even stronger demand for new, alternative technology? Such
seemingly paradoxical behavior is by no means restricted to the
realm of market competition; in markets at home or in world
markets. International power competition obviously works in the
same paradoxical way: The acceleration of the arms race is an
increasing function of the sum of the stocks of arms accumulated
within the rivaling nations or political blocks, and it is not a
decreasing function of that sum, as one' intution would suggest.

THE PRIME MOVER IS COMPETITION

For the purpose of gaining a better understanding of the
co-evolutionary interplay between organizational design and
technology making or technology taking, we have to extend the
scope of the notion of competition in order to grasp the nature of
the driving forces behind both technological and organizational
change. Without an understanding of the extended scope of
competition there is no firm grasp of the changing intensity of
competition, and the technological risks and opportunities
involved. I am certainly not propagating Social Darwinism, but
suggest that we face some Social-Darwinistic realities as a
preliminary step in the direction of becoming master instead of
being servant of technological change.

One of the supply-side puzzles that is currently perplexing
economists is the paradoxical fact that at a time of increasing
affluence, during the 1960s and 1970s, the intensity of competition has increased rather than decreased. Many economists would
have expected economic competitiveness to decrease with affluence.
The fact is that economic competition has gotten stiffer nearly
everywhere:

- International competition in world markets: Given that many
  Third World countries have learned how to make many modern
  industrial products at nearly the same level of quality,
  pressure on many lines of mass production has developed in
  the advanced industrial nations; and that pressure is likely
  to grow as more countries enter into these industrial
  sectors, such as synthetic fibres, shipbuilding, and
  machinery.

- Inter-Regional competition: In this time and age, the
  increases of market effectiveness of the leading

corporations are, of course, achieved at the cost of erosion
of less effective business in other parts of the nation and
the world. In turn, these places fight back with various
policy instruments, such as trade barriers, differential tax
incentives, and active policies of attracting firms from
outside the region to move into and settle in this region.

· Inter-Industrial competition: For practical purposes of
corporate strategy formulation, and product/market planning,
the old SIC classification system is nearly useless.
Industrial products compete not only against their kind in a
given market (intra-sectorially), but also indirectly with a
whole range of substitutes in other sectors (inter-
sectorially). While it may be that the large corporation is
able to decrease the degree of (monopolistic) competition
intra-sectorially, it is usually unable to control for inter-
sectorial competition. Thus, if we have a merger wave as
over the last 12 years, competition is not only intensifying
gradually, but in quantum jumps. If the big manufacturers
of whole systems such as aeroplanes, communication or power
networks, ships or computers change the design or the
materials base, then a seemingly minor change in the whole
system will create a major change in the business potential
of a great number of small and medium large firms which are
the suppliers of the large systems manufacturer. That is,
even a marginal change in the demand structure or consumption
pattern of the population will cause a sweeping restructuring
motion of the supply-side of the industrial economies.

Such restructuring of the supply-side always requires an
enormous amount of human energy, organizational creativity, and
longer-term oriented corporate leadership. Even a superficial
look at the current change in intensity of competition in industry
makes us aware of the fact that the ability of organizations and
individuals to cope with this intensification has become the
bottle-neck factor in the creation of wealth and well-being in the
future. Consequently, we are currently undergoing a major
transition: <u>Certain human resources become decisively scarce
whereas many capital goods become obsolescent</u>. We call this
hyper-inflation. This point gets completely lost in modern
monetary and Keynesian economics, when one speaks of the need of
the level of supply to have to adjust to the level of "inflated"
demand. Competition intensifies when the structure of the demand
changes faster than the supply side is able to adjust to it
structurally. Then, under intensified competition, the adjustment
pressure on the human resources of the corporations increases even
further, and not the availability of capital but the ability of
people in work organizations to cope with the rate and direction

of change becomes more and more the upper bound to the feasible
rate of change. This is exactly the aspect which corporate
managers in the western countries envy in the Japanese system of
decision making.

Up to this point, we have only looked at horizontal
competition, that is competition between units on the same level,
such as between firms in one market, between sectors, regions, or
between rivaling nations. International competition, intersectoral
competition, interregional competition, these are horizontal
concepts. In addition to this, there is vertical competition
between hierarchical units. It has important intervening
influences on the intensity of horizontal competition. Vertical
competition reinforces hierarchical behavior such as top-down
delegation of tasks and bottom-up delegation of problems. Notably
in times when things go badly we observe on any level of the
organization a tendency for passing more problems on up to the next
higher level, and for reducing the number of problems that would
have been passed down. At times of relatively mild horizontal
competition, the usual behavior of passing the buck vertically
works quite well, and does no immediate harm. In the longer run,
though, it lowers productivity and adaptability to change. Thus,
if horizontal competition intensifies by vertical competition in
corporations, and among corporations in bureaucratically regulated,
industrially concentrated sectors, then the going gets really
rough, as in present times. Then this vertical pattern becomes
outright dysfunctional: economic systems become frozen, locked-in.
For example, a government speaker may say that the country is
ungovernable, or a company top executive may explain corporate
failure by saying the firm has become unmanageable (Fortune
Magazine, October 5, 1981). What makes such a situation so
critical is this: If horizontal competition intensifies, as in
many "mature" industries today, the intra-firm or inter-firm
vertical competition can drive the system into a vicious circle
of inefficiency, inflexibility, confusion, and waste. How many of
us, who are engaged in intra-firm competition, have not experienced
such vicious circles in our daily work? It acts on us in the way
of a double bind: While the deadline for output approaches, you
cannot get the support needed from your superior, and cannot get the
job done by your subordinate either. It locks you in. In an
analoguous manner, your department may be locked in, your firm,
your city, your region, your country, your race, or your whole
world may be locked-in into a gloomy development. Many young
people these days feel that the world is locked-in into a gloomy
technological development. They protest the rate of technology
taking, and they object against the direction in which technology
making is going.

GROWING DEMAND FOR PROCESS INNOVATION

Innovations research reveals that despite the high degree of
technologization already achieved, there will be a huge demand for
more technology. First of all, this market-pull is for more and/or
better process innovations that incorporate new, productivity
enhancing, more sophisticated replacement technology. Most
industrial corporations are temporarily locked-in to the
technological state of the art prevailing in the branch of industry
they are presently in. But over time, as competition intensifies,
they will try to escape by developing alternatives: new products
to sell in different markets, but first of all, new processes to
more cheaply manufacture the existing products with. Innovation
researchers vary in their overall estimates, and their estimates
vary from industry to industry; on the whole, they estimate that
about 80 percent of the industrial products and processes now sold
in the markets will be phased-out by 1990 and replaced by some
alternatives. And of 75 percent of the products and processes to
drop from the market, it is presently not known by what they will
become replaced over the years. Hence, about 60 percent of the
present industrial produce will be replaced during this decade by
something still to be developed or to be specified. To a large
extent, this increasing demand for sophisticated replacement
technology is the dynamical consequence of a backlog that resulted
from a relatively low rate of significant innovation during the
past decade. Yes, I mean it. Contrary to common belief, the past
decade was one of relatively little innovation and of relatively
slow structural change in industry. If supply-side change speeds
up to the long-term trend level, it will induce a much higher rate
of innovation.

However, the increasing demand for replacement technology in
the industrial sector of virtually all modern industrial societies
will not easily find its supply. The supply side of the economy
has been lagging in response to this growing demand. Why? My
first point is that with respect to this faster supply-side
adjustment induced by the growing demand for replacement technology,
the human resources were and will remain the obvious and foremost
inert factor that slows the restructuring process down.

In the past, the slow-down was partly due to awkward
management, partly to organized labor. Most obvious is the slow-
down in industries and in regions with a high degree of unioniza-
tion. One seldom finds strong labor leaders and strong corporate
leaders in one place, as one tends to balance the other. Where we
find the strongest labor leaders, there we also find most of the
compromising-type corporate leaders. These are the leaders who would
not only compromise on productivity and cost but also on product
quality and sales (customer satisfaction), thus getting the

corporation into decline. Note, then, that protection of office
and factory workers against a high rate of productivity enhancing
progress innovations is a blessing only in the short run. It may
easily turn into what economists call the English Disease. History
teaches that economic crisis develops if the slow-down of
rationalization ("technology taking" in production and office work)
is accompanied by a low rate of demand-stimulating product
innovations ("technology making"). This is the reason why I think
that the market forces will generate a higher rate of technological
change in the years to come, whether we like it or not, because we
opt for it by way of our day-to-day purchasing decisions.

TOWARD A NEW CLUSTER OF BASIC INNOVATIONS

     The next point I am trying to make is an extension of the
first one. That portion of human resources which is most needed
for making and taking of the new replacement technology, will
become scarce during the 1980s not only because it is limited in
supply while the demand for it is growing. Inventive genius,
organized creativity, and technical entrepreneurship never were and
probably never will be available in abundance. However, scientific
evidence leads me to think that this human resource will become
particularly scarce because the growing demand for replacement
technology will grow faster at a time when the world economy
develops additional demand for yet another type of new technology:
final demand augmenting technology. This is a consequence of long-
term regularities in the historical pattern of ebb and high tide in
the flow of basic innovations, a pattern regularity which I expect
will extend into the future.

     It has become widely accepted to equate today's economic
stagnation in most industrial countries with a lack of basic
innovations[2] in previous years. That is, slowdown in demand in
many industrial sectors is not due to a Keynesian under-consumption
and/or under-investment (over-saving), but to a Schumpeterian
under-demand for existing industrial products (which are available
in relatively ample supply) and under-supply for alternative
products (that would be in relatively high demand if they were now
available in the markets). On the aggregate, my diagnosis of the

_____

     [2]G. Mensch, Stalemate in Technology: Innovations Overcome the
Depression. Cambridge, Mass., Ballinger, 1979 (German original
Frankfurt: Umschau, 1975).

present economic distress is structuralistic: The small number of
today's growing industries does not offset the stagnation and
decline in the large number of today's mature industries. During
the past decade, there have been too few basic innovations that
created new branches of industry. Again the point is that a past
backlog of basic innovations will create a cluster of basic
innovations in the future. For an illustration of the long-term
fluctuation in the frequency of basic innovations, see Figure 2.

Basic innovations are defined as creating new markets (infant
industries) and revolutionizing existing (mature) industries. By
contrast, improvement innovations are defined as subsequent
product or process innovations that occur in series within the
branches of industry newly created or revolutionized by basic
innovations. The beforementioned type of horizontal and vertical
competition often involve the implementation of product or process
innovations in existing branches; it may be denoted as "<u>actual</u>"
<u>competition</u>, as opposed to "<u>potential</u>" <u>competition</u> that operates
roundaboutish by way of basic innovations. Potential competition,
then, refers to all activities of invention, research, development,
planning involved in preparing for basic innovation. Needless to
say, these activities demand highly qualified personnel.

Thus, at a time of intensified actual competition, which
requires courage, endurance, creativity and leadership in great
measure, potential competition also comes in with ever larger
appetite for the very finest of human talent. It is safe to predict that this intersection of actual competition and potential
competition will generate a shortage of the most valuable and rare
part of "human capital" in the future. This is my <u>second point</u>.

This is not the whole story, though. The scarcity has
dysfunctional effects: We should expect that in addition to the
evolving shortage in quantity of talent relative to the evolving
needs, the competition between actual and potential competition
generates inefficiencies in the utilization of the available
talent. This is easily understood as a consequence of conflicting
priorities. On the individual level, it can be illustrated by the
example of a boxer. While he is fighting in the ring, he cannot
possibly attend to the question of what he might like for supper
tomorrow. Conflicting priorities between the needs of actual and
potential competition generate lagged responses not only on the
individual level but on any aggregate level. As an example, think
of the competing units as countries such as the U.S., Japan,
West Germany and Russia, that are rivaling for superiority in some
military and civilian technologies such as biotechnical instruments,
microprocessors, robots, nuclear equipment, etc. The intensified
actual competition weakens the efforts allocated for the search of

less competitive alternatives, and the lack of such alternatives,
in turn, reinforces the actual competition in locked-in subsystems.
As another example, consider the rivalry between regions such as
the manufacturing belt and the sunbelt of the United States. They
compete for high incidence of start-up and relocation of industrial
corporations, and over the efforts to induce firms from the outside
to move into this region the efforts to develop the endogenous
potential of the region is being largely forgotten. The same
principle holds true for rivalry between divisions and departments
within the corporation. The conflict is over more investment funds
and less payback to the mother corporation. In our innovations
research work in industrial corporations, we see in the great
majority of cases that the allocation of funds and talent is
dominated by the short-term needs of actual competition. So far,
so good. But we also see that in the majority of cases the short-
term demands crowd out the longer-term needs, and firms lose the
ability to allocate the right kind of people and funds into
research, development, and planning of basic innovation
opportunities.

     Consequently, the firm, industry, region, country, or
international community such as the European Market countries or
OECD countries run into a temporary problem of too few commercially
viable new industries compensating for the many stagnant industries
that release capital and labor instead of absorbing it. After
years of scientific research and reflection on the causes of
industrial stagnation, decline in productivity growth, and economic
crisis, the reason just given seems the most profound. Lack of
creativity and leadership in our private and public institutions
generated an intensified competition for the most valuable
innovations that come about, and while the intensification of
competition absorbs the greater part of the creative leadership
talent with an unsatiable hunger, search effort for more valuable
alternatives is being further diminished. Such a period in
economic development I call a "Stalemate in Technology", when many
of the old technologies do not provide enough employment
opportunities for manpower and stock of money, whereas too few
really desirable technologies are emerging as a basis for fast-
growing industries. However, history teaches that such periods of
deep recession or even depression are but phases in the economic
evolution, and subject to transition.

THE PROCESS OF CREATIVE DESTRUCTION
IN ACTUAL PERSPECTIVE

     In order to characterize the phase transition from
"Technological Stalemate" to recovery and economic growth, I shall

draw on the evolutionary concept of Schumpeter, the great economist, whose ideas are but slowly making inroads into the economist's profession.

Joseph A. Schumpeter characterized the economic process as a "Process of Creative Destruction." Technological innovation and changes in industrial organization, he maintained, are designed by rivaling industrial firms as to beat the competitors. As a consequence, Schumpeterian competition (as opposed to perfect competition) disrupts the economic equilibrium. However, the economic gains to the corporation that accrue from the innovations do not balance with the economic losses borne by the corporation's competitors. In general, the social benefits and the social dis-benefits of creative activities are considered unequal in size, and unevenly distributed over members of the population. Sometimes the overall advantages overcompensate for the overall costs, sometimes the disbenefits exceed the benefits. Over time, at some times the creative benefits that outweigh the destructive side effects, and other times the other way around.

The problem of lack of demand-pull for a wide variety of modern technical products, and the problem of supply-side inertia in view of slow development of better products, services, and service delivery systems, have to be seen in this context of temporary cost-benefit inequality and temporal unevenness in the appropriation of benefits and costs. We must recognize these inertia factors in the markets for consumer goods (under-consumption) and in the work places (under-investment) as unintended consequences of prior creative achievements. Not less creativity but redirected creativity is thus needed. Reduction in creativity results in missed employment opportunities and greater unemployment risks in the next years. What matters is the direction of creativity because both employment opportunities and unemployment risks evolve by the same economic process.

Especially with respect to medium-term developments in most industrialized countries, which are troubled simultaneously by the Scylla of unemployment in most branches of once prospering industries, and by the Charybdis of inflation (stagflation), the question of economic policy cannot safely be left in the sterile dichotomy of "capital-oriented" (a la Thatcher and Reagan) or "labor-oriented" (a la Mitterand). According to our basic hypothesis, employment, income and wealth are being created by a co-evolutionary process of technological change and organizational achievement, yet the flow of benefits and costs may at any short-time interval be so unbalanced that social resistance against a short-term net cost may prevent the attainment of the greater but later benefits. Unfortunately, such policy topics are not found on the agenda of government economic policy makers in any of the

Western countries. Hence, it is not unsafe to assume that over the
next years, just as in economic history, it will again be the
market mechanism which largely determines the specific path of
development in technology and industrial organization. By the same
token, the pattern of transition may follow the same basic
organizing principles as it did in the long-term economic history
of industrial evolution since the Industrial Revolution. That
means the "technology-making" influence of governments will still
be relatively small: Government priorities will reflect market
realities, and technology policy is supportive rather than
directive. Of course, this assumption is only probabilistic in
nature. If we assume, however, that governments act mainly as
technology-takers, then the beforementioned long-term perspective
may be further extended for the medium-term purposes at hand.

In economic history, we have observed four major clusters of
basic innovations. The data is given in my Stalemate of Technology,
Chapter 4, Tables 4.1-4.4, and is being depicted in Figure 2.
According to the data, the four clusters occurred

    around 1775 (Industrial Revolution)

    around 1825 (1825 $\pm$ 18 years)

    around 1886 (1886 $\pm$ 11 years)

    around 1935 (1935 $\pm$ 8 years)

These clusters are depicted in Figure 2. By way of non-linear
trend extrapolation, subject to the condition stipulated in the
above probabilistic assumption, we can derive a conditional
projection into the future: The next cluster of basic innovations
may be expected around 1989, with a standard deviation of $\pm$ 5 years.
Especially noteworthy is the socio-economic climate prevailing at
the times of clusters of basic innovations. Each time, the new
basic technologies came about during a period of economic
depression: "The Depression" around 1935, "The Great Depression"
around 1886, "The Time of Poperism" around 1825, and the "pre-
revolutionary" misery that induced the Industrial Revolution.[3] Are
we to conclude from this coincidence that depression is a necessary
condition for basic innovation clusters? I think not. I rather
think that a set of desirable basic innovations can be achieved
without the economy first suffering from a general depression.

_____

[3]G.O. Mensch (with H. Freudenberger), "Bruno Study." A Contribution
   to the Political Economy of Social Innovation, Illustrated by the
Cluster of Innovations of the Industrial Revolution in the Bruno
Region, Goetinger, 1975).

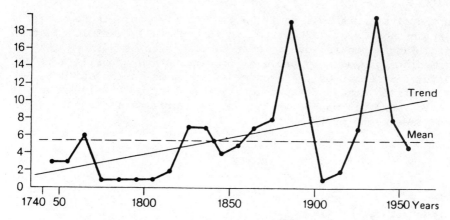

Figure 2.   Frequency of Basic Innovations in 22 Ten-Year Periods
            (1740-1960).   Source: Fig. 4-1 in Stalemate, p. 130.

Historically, clusters of basic innovation coincided with
transition from depressions to recoveries. Although it may be
oversimplification to say that the expansionary forces of the
cluster of basic innovations created the income and employment
effects of economic recovery at those times, it is certainly true
that the new opportunities ended the depressed investment attitude
of the capital market in those times. While the output effects of
the newly created industries may not be strong in the aggregate for
some number of years, the swing in the investment propensities by
capital owners induced by the innovative opportunities occur rather
suddenly. Figure 3 reveals that at the end of the depression years,
the mix of basic innovations is no longer biased for rationalization but heavily biased for expansionary types of new technology,
which come mostly in the form of new industrial products that
establish a new branch of industry. This change in the composition
of basic innovation mirrors the swing in the investor's propensities at those times.

In order to further characterize the transition from
depressions to growth, by means of the biases just mentioned, we
should take note of the fact that the balance in the Process of
Creative Destruction was positive (strong expansionary bias) in the
recovery, and negative (strong rationalization bias) before and
during the depressions; see Table 1. Figure 3, summarizing the
information given in Table 1, also indicates that prior to the
expansionary pushes (high E/R Ratios) there existed periods of low
E/R Ratios which stand for strong rationalization biases. A cluster of basic innovations is initially composed mostly of radical
process innovations, which are designed to cut costs even deeper;
but the mix of innovations shifts to an expansionary bias only

when the cluster is well in its way. Here is where the analogy to
the Process of Creative Destruction is especially useful. Cost
cutting (and squeezing-out of cost factors such as workers and
intermediate products by process innovations) is actually a very
powerful depressant of the economy if it is not accompanied by
equally strong doses of expansionary investment. A well-known
historical example may serve to further clarify this aspect,
which leads me to my third point.

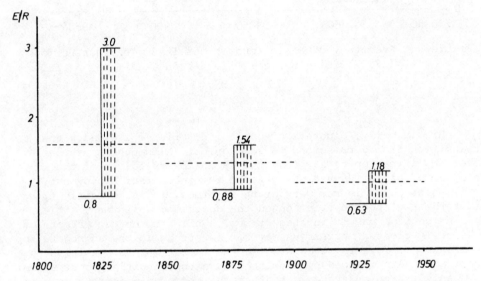

Figure 3.   Jumps of the E/R-Ratio at the Times of Clusters of
             Basic Innovations Before and After 1825, 1876 and 1925.
             Legend: Data see Table 1 (Tables 4.1-4.4 in Stalemate)

        From the human resource perspective, my third point is a
reflection on the tendency to drive creative skills and leadership
talent away from the organization at times of strong rationalization biases, given the cost-squeezing efforts and competitive
processes predominating at such times. As a consequence,
accumulated human capital in many industrial firms and public
service corporations is going to waste during the phase prior to
increasing need for exactly this resource. The Process of
Creative Destruction works especially destructively on the
creative fragments of the human resource base. Unfortunately, our
accounting systems are not designed to measure the actual and
potential contributions of the creative parts of human capital.

Hence, firms have no warning system that provides them with an early assessment of the extent of depreciation of human capital, which in times of economic depression may far exceed the depreciation of physical capital.

If technological change is accelerating and changing directions in nearly the manner we anticipated above, then my points one, two and three taken together suggest a theory of economic crisis that is a variant of the classical labor theory of value, and hence of considerable significance to human resource planners. Corporations suffer, and so does the economy if the creative parts of the human resource base are being allocated predominantly to the purposes of cost saving and fighting actual competition, because that temporarily diminishes the allocation targeted for expansion and potential competition. Hence, the disproportionality in the allocation of creative skills and leadership talent may strengthen the depressing factors and weaken the anti-depressant forces for still some years to come.

My third point hinges upon the shift in bias in technological development at times of clusters of basic innovations, such as around 1826, which is the date of the opening of the Stockton-Darlington route, which marks the basic innovation in the railroad industry.

Before railroad construction in Britian took off in and after 1825 and created huge new investment and employment opportunities, which resulted in a super boom that lasted through the middle of the 19th century, the steam engine was actually introduced shortly after 1800 as a stationary rationalization device in numerous factories. There, during the 1810s and 1820s, it replaced workers by the thousands, and drove many non-competitive firms into bankruptcy. Rationalization, if it is not accompanied by expansion, acts as a depressant. Since the early basic innovations in a cluster tend to reinforce the prevaling rationalization bias, and the expansionary basic innovations in a cluster come only later, we learn from history that it is the lack of the anti-depressant expansionary basic innovation we must gard against.

It may be useful to reflect the pattern in the rate and direction of technological change in view of the standard definition of the production function, which is the economist's description of work organizations in terms of its inputs and output. Whereas an ordinary process innovation is defined as a shift of the production function (representing an intra-plant rationalization measure that may save labor, energy, materials, etc.), a basic process innovation is defined as setting-up of a new product function (representing a new inter-plant industrial

TABLE 1

SHIFTING EXPANSIONARY AND RATIONALIZATION
BIASES: THE E/R RATIO BETWEEN
1800-1950

	PERIOD	BASIC INNOVAT. N	RATIO OF EXPANSIONARY TO RATIONALIZING BASIC INNOVATIONS		
			PERIOD	TWO PERIODS	
1	1800-1850	21	1800-1850 13 E : 8 R	before 1825 4 E : 5 R	after 1825 9 E : 3 R
2	1850-1900 Electrotechnical Industry	22	1850-1900 11 E : 11 R	before 1886 3 E : 3 R	after 1886 8 E : 8 R
3	1850-1900 Chemical Industry	28	1850-1900 17 E : 11 R	before 1886 5 E : 6 R	after 1886 12 E : 5 R
4	1850-1900 Total of 3 and 4	50	1850-1900 28 E : 22 R	before 1886 8 E : 9 R	after 1886 20 E : 13 R
5	1900-1950 (Stoikov)	50 (50)	1900-1950 25 E : 25 R (25 E : 25 R)	before 1929 5 E : 8 R (4 E : 11 R)	after 1929 20 E : 17 R (21 E : 14 R)

Legend: All 121 basic innovations published in
         G. O. Mensch, Stalemate in Technology,
         Tables 4.1-4.4, have been classified as
         E-type ("expansionary") or R-type
         ("rationalizing") by a team of
         12 colleagues at the International
         Institute of Management, Berlin, to whom
         I am very grateful for their help.

NOTE: A very similar observation of the step-up of the E/R Ratio
       is found in the work of Vladimir Stoikov who distinguished
       "product-adding" (E) versus "product replacing" (R) innova-
       tions, and who also classified the 50 innovations in row 5
       of Table 1 (in parenthesis). See Stoikov, Vladimir, The
       Classification of Inventions: A Sample Study, in: The
       Southern Economic Journal, 29 (1962/63), pp. 15-20; also: A
       Note on Product-Adding versus Product-Replacing Innovations,
       in: Kyklos, XVI (1963), pp. 138-140.

organization with multiple cost savings magnifying the rationalization effect). That is what the steam engine was used for
initially after 1800, and that is what today the microprocessor is
mainly used for. Micro-electronics make their inroads today into
a wide range of rationalization devices such as microcomputers
that reduce waste, change-over time of machine tools, transaction
costs in banks, inventories in supermarkets, and clerical work in
offices. Microprocessors guide robots that drill, punch or spray
faster, cheaper, and more precisely than men. Initially, micro-
processors are implemented for rationalization mainly. However,
studies conducted by and for the large electrotechnical corporations all reveal a common belief of experts in the electronics
field that in the medium-term future, microprocessors will find
application in a vast array of new, mostly yet unknown product,
thus generating an expansionary bias in all microprocessor applications later on. This may be so, and this may indeed later
create a demand-pull for these new products. But we should be
aware of the possibility that before it comes to that expansionary
drive, a deepening of the depressing rationalization bias must be
expected for the next couple of years without it being compensated
for as yet by an equally strong, anti-depressant expansionary
force.

    Thus, just as in the example of the steam engine, which
became implemented as an intra-plant and inter-plant rationalization device at a time of increasing competitive pressure in the
major industries of the 1800-1815 period, reinforcing the
prevailing rationalization bias with a quantum jump in efficiency of work organization, the microprocessor reinforces the rationalization bias that today dominates in all major industries
And just as the steam engine became the major expansionary force
only after 1825, it is expected that the expansionary forces of
the microprocessor, and of other new technologies now emerging,
will not unfold their balancing effect but later. The
Schumpeterian Process of Creative Destruction still is in its most
risky phase.

                              CONCLUSION

    According to Benjamin Franklin, "Man is a tool-making
animal."

    Had Franklin lived today, he might have said that man is a
technology-maker. Or, upon second thoughts on the high degree of
technologization in modern times as compared to his times, he
rather might have said that man is a technology-taker.

At any rate, given the infrequency[4] of invention and technological innovation throughout the ages up to the Industrial Revolution, the acceleration of technological change at Franklin's time was recognized by many enlightened citizens of Western countries as an opportunity to overcome poverty. A contemporary of Franklin, Jeremy Bentham hailed the gifts of innovation in the Foreword of his Fragments on Government (1776), thereby giving an alternative theory (utility theory) of prosperity to the one given in the same year by Adam Smith in the Wealth of Nations (efficiency theory). That chasm in economic thought has broadened during the 200 years of industrialization, neither to the advantage of better utility theory of innovation nor to the advantage of better efficiency theory of innovation. Whereas we are just starting to develop a reasonable efficiency theory of technological change, based on neo-Schumpeterian concepts, we are still far remote from but primitive knowledge on the utility of technological change. Thus, while technological changes have become a very frequent phenomenon in our times, and most people have become technology-takers, a utility theory of technological innovation is wanting.

Utilized in production, consumption, and distribution, in foreign relations and national defense, technology is omni-present, and the rate of introduction of new or different product and process technology has increased to a point where technology-taking has become of great disutility to many people. When a list of 40 negative items such as "disease", "divorce", "accident", etc. was offered to a group of individuals for rating, the word "change" was picked up by the majority as the most irritating item. It seems as if the more frequent but minor disturbances add up to more anxiety than the perception of less frequent but greater disasters.

In any case, in modern times just as in the past, the ability of populations to produce new technology and absorb it has always been a scarce factor. Even in situations of apparent abundance (such as today's information overflow) the scarcity of the truly valuable item is being felt. For example, among the many, many

---

[4]Infrequency is, of course, a relative term: The more we learn for example about the Middle Ages, the more we see invention and innovation increase in numbers and in impact. An excellent series of articles on Innovation and Growth in the Middle Ages (15th and 16th centuries) and Wildwest in Europa (14th century) have been published by Wolfgang von Stromer, Freie Universitaet Berlin.

pseudo innovations (with low utility) one has to look hard for the
few high utility improvement innovations. Similarly, the relative
abundance of employable people coincides with at least a temporary
 shortly of innovative skill--or skill in the right places in
organizational slots and leadership positions. The main conclusion from this analysis of technology-making and technology-taking
is that the human capital (and not the physical capital) has become
the main determinant of attainable wealth of nations.

Hence, at a time when unemployment is high and is
threatening to grow, the leadership talent for creating creative
organizations is as scarce as the set of workable hypotheses that
make sense both in the framework of a trusted utility theory of
innovation and in the framework of the market. This lack of
practical theory projects itself also into economic policy.

"The United States today has no conscious manpower policy
specifically designed to strenghten the environment for
technological innovation and to respond to the needs of workers in
a technologically changing economy."[5]

Clearly, the lack of conscious government innovation policy
and scarcity of corporate innovation strategy is largely the
result of the lack of a good utility theory, just as bad policy
and strategy making is mostly the result of wrong theory. "The
ideas of economists and political philosophers, both when they are
right and when they are wrong, are more powerful than is commonly
understood. Indeed the world is ruled by little else. Practical
men, who believe themselves to be quite exempt from any intellectual influences, are usually the slaves of some defunct
economist" (J. M. Keynes, last page in his General Theory). Next
to no practical guidance is afforded to the public policy maker
and corporate strategy planner by most of the available theories
of innovation. If it is a demand-pull theory, it is useless for
specifying the last term in the equation:

$$demand = purchasing\ power \cdot need.$$

If it is a supply-side theory, it comes as un-handy as
Oswald Spengler's Decline of the West. If it takes off from
Marx's notion of monopoly capitalism, it postulates a general

---

[5]M.I.T. Center for Policy Alternatives, Report to the Office
of Technology Assessment, Congress of the United States, on
Government Involvement in the Innovation Process, 1978.

vanishing of investment opportunities (Hansen), or the disappearance of the supportive social climate for entrepreneurial innovation
(Schumpeter), as government regulators (bureaucrats) and corporate
administrators (technocrats) crowd-out the individual risk taker
while modern technology evolves from single-invention type implement to large-system type regiment (Galbraith).

Without additional input from a convincing utility theory of
innovation, the co-evolution of technology and work organization
is likely to proceed as a Schumpeterian Process of Creative
Destruction. The scenario is that market competition will push
for more cost-squeezing rationalization (process innovation) of
work organizations during the next five or six years, without
enough anti-depressant, expansionary forces offsetting the
depressive tendencies. Expansionary basic innovations will come
later only if we do not learn to cope with the specific human
resource shortages that hinder and delay basic innovation now, and
that will restrict the utilization of the innovative potential
later. The time has come for a reconsideration of human resource
priorities, and for new managerial techniques for improving the
human capital investment and utilization process, as this has
become the constraining factor in the creation of employment,
income, and economic stability.

USING DEMOGRAPHIC AND TECHNOLOGICAL FORECASTS FOR HUMAN RESOURCE
PLANNING

Carol T. Schreiber

General Electric Company
Corporate Employee Relations Operation
Fairfield, CT 06431

ABSTRACT

Human resource planners have begun to recognize the complex interdependence of internal and external social, political, economic, technological and demographic factors and their impact on planning for the human resource contribution to business development. This paper documents one industry's approach to analysis and forecasting of environmental trends and events, and the integration of these analyses into a human resource planning process.

INTRODUCTION

As the decade of the 1980s unfolds, more of us look forward to the turn of the century, to the year 2000--now less than 20 years away. Until recently, a mention of the benchmark years of 1984 or 2001 brought visions colored by science fiction, fantasy. Today, 1984 represents tomorrow, and 2001 is next week--in a time frame marked by accelerated change and high-pitched complexity, a discussion of the future is neither the novelist's sole province, nor an academic exercise. It is a business necessity--a critical key to organizational survival, the basis for today's decisions for tomorrow's outcomes.

Writers of popular management literature such as Peter Drucker (1980) agree with recent scholarly works on organizational effectiveness (Cameron and Whetten, in press) that today's changing environment poses new challenges for all organizations. These authors concur in their recommendations that information about the shifting environment be linked to organizational planning. They

agree also that all organizations should question their internally
shared assumptions about policies, programs, practices, business
direction, in the context of a changing social and technological
environment.

As Drucker has proposed, in these times, today's leaders and
managers are compelled to attend to current realities and their
meaning for the future. Thus, while organizations and businesses
can be <u>understood</u> by looking backwards, they can only be sustained
and developed by rethinking past assumptions in the light of current information, and <u>thinking forward</u>. Thinking forward means
beginning to rely on assumptions that tomorrow will be different
from today. Since they recognize that we can no longer anticipate
an extension of today, many organizations are working to pull themselves away from past assumptions to reconsideration of present information relevant to the futurity of today's human resource
decisions.

Human resource planning professionals in the corporate world,
then, are concerned with the analysis, understanding and utility
of information available now, and its application to those of today's
human resource decisions with implications for tomorrow. Since this
task is new to all of us, and each has undertaken the work differently, we have not yet had the time to demonstrate which of our
approaches works more effectively for our own organization, let
alone for other institutions. There is no generic prescription.
Without the test of time, though, it is useful to describe our current efforts, exchange ideas about method and approach--to learn.

In this spirit, I shall describe some features of an <u>approach</u>
to human resource planning. My description will emphasize the
linkage of environmental analysis and forecasting with analysis
of key business strategies and the identification of human resource
implications. Additionally, I shall delineate some of the most
relevant implications to emerge from our recent analyses, high-
lighting the impact of technological and demographic forecasts.

THE HUMAN RESOURCE PALNNING PROCESS

Human resource planning for industry today involves more
than the consideration of quantity/mix of people. It encompasses
a range of internal/external environments which impact personnel
and/or business strategies positively or negatively. Underlying
this approach, we have developed a process for integrating consideration of business strategy with consideration of the internal/
external environment.

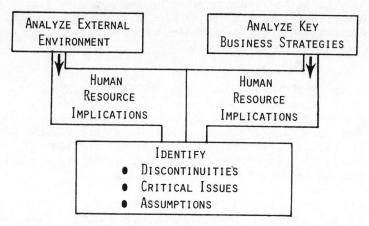

Figure 1.   The Human Resource Planning Process

This process begins with a parallel approach to the identification of human resource issues. (See Figure 1.) For example,
one major piece of work focuses on review of key business strategies;
with thorough analysis of business strategic plans, to elicit human
resource issues presented both implicitly and explicitly in those
documents. To carry out this task, staff teams study and analyze
strategic plans for their human resource implications--some obvious
and some not so obvious. The involvement of broad mix of human
resource professionals in analysis of the business plans spreads
both responsibility and learning among these professionals, many
of whom are functional specialists in such areas as benefits, compensation, management education and professional recruiting. Expo-
sure to business plans has been enlightening and useful to participants, and has offered unusual and unanticipated insights to the
review process.

Paralleling work with the business plans is a comparable
analysis of the human resource environment, documented in the
Human Resources Environmental Scan. The two analyses generate sets
of human resource problems, which in turn, are identified and
prioritized as human resource issues of strategic significance if
they are: pervasive across the diverse Company businesses; indicated as important by the environmental analysis; have noticeable
impact on corporate strategic thrusts; and are assessed by decision
makers as having high level corporate risk. From this identification and prioritization come specific products--human resource
issues of strategic significance, which are documented and reported
to top management; and the next year's human resource planning
challenges, which are incorporated into the strategic planning

guidelines for the following year. The third product of the combined
analyses are the human resource planning assumptions, statements
about highly probable trends and events, pertaining to human resources, which are disseminated throughout the Company for consideration in specific business planning.

HUMAN RESOURCE ENVIRONMENTAL ANALYSIS AND FORECASTING

    The environmental analysis and forecasting process is particu-
larly salient to a consideration of the use of present data for
future decisions. For this discussion, I will consider its methods
and then introduce some of its findings--examples of substantive
contributions to the planning process itself.

    For the planning work we do, the human resource environmental
scanning process has been conducted for four years; during the last
two of which a major volume has been produced. In 1979-1980, the
first Human Resource Environmental Scan was published, a 60-page
volume; and in 1980-1981, a companion Data Source Book was produced,
as a revision, update and supplement to the original document.

    The objectives of the human resource environmental scanning
process are fourfold:

    1. To systematically examine those external conditions and
        forces with implications for management of the Company's
        human resources;

    2. To identify changes in external and internal conditions
        with implications for the Company's human resource
        management;

    3. To establish priorities among these implications by time
        period, based on probability, importance and ability to
        manage; and

    4. To distill from these priorities critical assumptions
        about the changing environment as a base for human resource
        strategic planning.

To reach the Scan's objectives, a specific logic path is followed:

    First, we identify and document current and future trends in
the key causal conditions in the environment--five conditions
(e.g., politics, economics, demographics, attitudes, technology)
identified as central features of the human resource environment.

We then develop from these conditions a base case scenario for the near-term time period, which is analytically linked to: current and future trends in the external work force environment and our work force situation. From the combined analysis of these sets of conditions, we derive human resource implications, which are then assessed for criticality and importance; and conclude by combining the environmental implications with implications of the business plans to develop human resource planning assumptions for the next planning period.

To start the scanning process, a first challenge involves selecting and documenting analyses and forecasts about the relevant external environment. As the environment has become increasingly complex and subject to accelerated change, we use many people and informational resources to understand the issues. For instance, as resources to track key driver conditions, we use inputs from economists on economy and energy; technology specialists, and our own staff expertise on demographics, politics and attitudes. The actual impact of these environmental factors on human resource programs and practices is profoundly effected by many expected and unexpected events from the domestic legislative/regulatory arena. To keep abreast of these events, public issues analysts monitor the employer/employee legislative/regulatory situation regularly. Our Scan Appendix lists a summary analysis and forecast of 20 domestic employer/employee issues.

Our effort to capture the complicated human resource environment emphasizes three interconnected sets of trends, events and issues from which human resource implications are derived.

First among these three sets are the key drivers, identified earlier--those pervasive conditions from which we initially developed the base case scenario for the 1980s. These key drivers include: economics, demographics, technology, politics, attitudes.

From the combined analysis of key drivers--trends and projections--emerged a domestic base case scenario for the near future.

The second set of environmental conditions we consider are more specific features of the immediate work place environment, which have separate and interdependent effects in relation to key drivers. Those include:

- General retirement trends;
- Private pension trends;
- General employer/employee legislation and regulation;
- General work force attitudes;
- Education: enrollment trends and projections;
- Health and safety trends;
- Management education trends.

Information and background on these topics is provided by
Corporate Employee Relations functional experts, internal organiza-
tion/manpower consultants, medical officers and corporate legal
staff. These experts offer analysis and insight on internal conditions as well, including:

- Company demographic trends;
- Retirement trends;
- Employee benefit trends;
- EEO;
- Employee attitudes;
- Human resource practices: hiring, appraisals, promotions;
- Organization and management;
- Organization experiments;
- Education programs/benefits.

In the logic flow of the Scan analysis, both the base case
scenario and the key drivers are linked singly and in combination
with those features of the internal and external work force environment to generate human resource implications.

One example of our systematic analysis of a key driver condition
is illustrated by the approach to demographic analysis, based on
available census data.

Overall, we have forecasted that U.S. labor force growth in the
1980s will be slower than the 1970s, i.e., 2.2% vs 1.6%. Still,
age mix and participation rate will not cause an "across-the-board"
shortage. Labor force growth in the 1980s will be greater than in
the 1930s, 1940s and 1950s. Although the growth of working age
(16+) population is decreasing, the contribution of rising participation is larger than before. For example, female labor force expansion continues to exceed male by better than 2:1 margin. Thus,
in the 1980s we anticipate that the number of employed persons will
increase by 16.7 million and the portion of the population that is
employed will climb.

In regard to age and sex composition, we have predicted that
the U.S. domestic population will be characterized by an aging
population undergoing regional shifts, with a work force changing
in sex and race composition, with scattered entry-level shortages
forecasted (20-24 year olds will decline from 9.1% of population in
1980 to 8.5% in 1985 to 7.1% in 1990). By 1985, the 25-40 age group
will make up 32% of the total U.S. population.

Along with the changing age composition of the U.S. population
and labor force, we have witnessed a shift in the sex composition of
the labor force. Proportionally, more women are now working in the
U.S. than in previous decades. Today, women comprise 42.4% of the

total civilian labor force; up from 38.2% in 1970; 29.6% in 1950.
Of all women of working age, 51% are in the labor force. Most of
the record gain during the 1970s in women's labor force participation
occurred among women under 35 years of age. Since 1960, the proportion of women aged 25 to 34 who worked grew from 36% to 64%.

In the 1980s, due to the projected increase of females in the
labor force from 50% to 56% and the concomitant changing sex composition of the work force, we anticipate an increase in multi-earner,
dual-career families and an increased percentage of labor force participants who are members of multi-earner families.

In addition to demographic analysis, consideration of educational enrollment is central. U.S. educational enrollment trends
have shown changes since 1965, mostly responding to downward demographic trends, which are partially offset by substantial increases
among women and over 25 year olds enrolling in two-year and part-time
college studies.

The number of high school graduates peaked in 1979, will gradually decline until 1985 and then will resume a gradual increasing
trend. The proportion of high school graduates enrolling in degree-
granting colleges has leveled off between 63% and 66%. The significant growth of college students receiving baccalaureate degrees will
slow down and level off in the latter half of the 1980s. Advanced
degrees are expected to continue to grow faster than Bachelors
through the 1980s.

Enrollments have risen, particularly among women and adults
over age 25, in vocational and technical schools, two-year colleges
and part-time continuing education programs. Much of this growth
is attributable to significant increases in government funding and
emphasis. The upward trends are expected to continue.

Regarding critical technical skills, the supply of engineers
graduating in key traditional disciplines as well as computer/elec-
tronics/manufacturing education have been increasing. Considerable
emphasis is being placed on new manufacturing technologies by many
engineering schools. Women engineering graduates are increasing
dramatically. The supply of people graduating with baccalaureate
degrees in engineering technology (with applications orientation)
has increased, although it is expected to peak and level off during
the late 1980s. In the 1980s, overall demand for technical graduates is expected to increase and heavy competition will persist.
Supply of newer technologies will probably lag demand.

ANALYTIC APPROACH

Independently, each set of key driver analyses yields large
quantities of data. Once we combine documented key environmental

trends and events which pertain to human resources, we need an approach to guide the analysis and assessment of the vast array of
issues identified by our Scan. To which of these issues/implications
should the Company pay attention?

Our assessment process begins once the major trends, events and
issues have been identified. We then compile all their associated
human resource implications and assess their criticality for three
time periods--near-term, mid-term and long-term. The importance
of an implication is based on an assessment of probability vs impact
of it happening. The format we use for this assessment is displayed
in Figure 2.

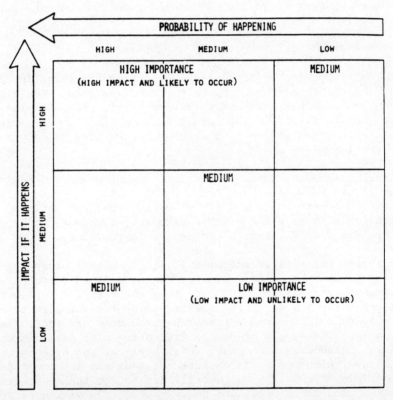

Figure 2.   Assessment screen for importance of human resource
            implications.

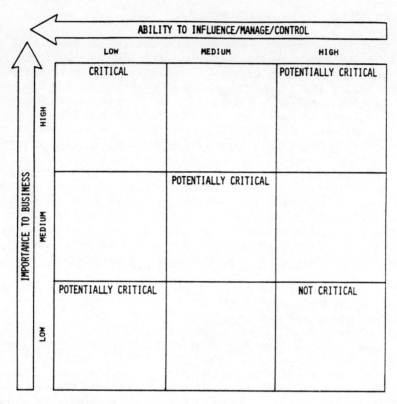

Figure 3.   Assessment screen for criticality of human resource
            implications.

    The second feature for analysis emphasizes criticality, which
pertains to importance vs ability to manage.   The format used for
this analysis is portrayed in Figure 3.

    Our overall assessment highlights those human resource implica-
tions which will have impact on the corporation and which require
management attention and/or action sooner or later at the corporate/
sector and/or Strategic Business Unit (SBU) level.   To assure inclusion of the broadest range of judgments in our assessment of importance and criticality, we've asked for the viewpoints of corporate
staff on criticality and importance, in addition to viewpoints of
key operating managers.

HUMAN RESOURCE PLANNING ASSUMPTIONS

Once the assessment process is complete, we review the human
resource implications identified as probable and important for the
near-term and mid-term periods. From this set, we identify those
which could qualify as human resource assumptions for the upcoming
five-year time period. Human resource planning assumptions are
based on those trends/events which will most probably occur, if
present patterns continue; will be critical to our overall human
resources/strategic planning; will require managerial attention.

Because our company is so differentiated by business location,
business strategy, each assumption can have different meaning and
usefulness to our planners. We do not expect each assumption to
impact uniformly on each part of the business; but we do expect
strategic business planners and human resource professionals to be
alert to the assumptions as a base-line and to be prepared for plan-
ning/action necessary in their specific environment.

Among the human resource planning assumptions derived from our
1980-1981 Human Resource Environmental Scan were two directly tied
to demographics and technology.

We stated that the supply of skilled people from traditional
sources will decline in the 1980-1985 period. Driving this assumption are these conditions: the rapidly changing shape and supply of
the U.S. work force causes a shift in skill mix and selective skills
shortage. U.S. labor force growth in the 1980s will be slower than
in the 1970s (1.6% vs 2.2%), although the shrinking 16-24 domestic
age group could be supplemented by more female, 65+, immigrant
labor force participation. This new mix could ease some shortages.

Practical implications of this assumption are different for
GE's different businesses. Clearly, SBU skill demands will determine the criticality of skill supply shortages; and the stress on
SBU's will vary by the intensity of business growth/technology
strategies, as well as local regional population, labor force and
education demographics. Additionally, some circumstances will offer
the opportunity to competitively position a business, through competitive pay, use of over-65's and selective retraining.

A second major human resource planning assumption asserts that
the continued infusion of new technologies in design, planning and
production will alter the structure of work and size of the work
force. This assumption is driven by our observations that:

● Microprocessor and information-processing technology are
  poised for new breakthroughs.

- New technology will continue its introduction in traditional electrical/mechanical processes, products and methods of product design and process control.
- New information-processing technology will continue to heighten office and administrative automation, providing increased information capability.
- Advances in robotic training will create new capabilities and applications (e.g., complex assembly work, medical diagnostics).
  - By 1990, half of workers in factories will be white-collar engineers/technicians, programming, maintaining robots/ microprocessors (Society of Manufacturing Engineers prediction).
- Automated factories will link technological innovation in design, process planning, requirement planning, control and production.
  - Broad applications in U.S. due to comparative age, obso- lescence of manufacturing facilities.

Implications of this assumption are:

- Substantive and immediate impact on human resources will continue until systems become totally integrated.
- Increased vertical integration of work from design to pro- duction to distribution will stress traditional organiza- tion, structure, job design and job ownership.
- Skill mix changes will occur in the work force, changing work classification, hastening dislocation, new hires, retraining.
- Attitudes regarding acceptance of change can impact the timing, payback and cost of effectiveness of new technology investments and applications.
- As programmed expenditures for productivity are increased, new technology applications will change employment needs. Similarly, there is potential for new technologies to alter skill needs of business, especially re shortage of critical skills.
- Innovative approaches for recruiting/retention of scarce skills will be necessary. Varying work life styles will need to be accommodated.

Each year, the set of human resource planning assumptions is revised, updated and extended, based on current and anticipated circumstances. Thus, in 1979, four assumptions were written; in 1980, two more were added; and in 1981, four more were added.

Strategic consideration of these assumptions is implemented by a broad array of planners and managers, including corporate strategic

planners; Corporate Employee Relations professionals, sector strategic planners in the development of their annual strategic plans;
Strategic Business Unit planners and human resource planners for use
in the next year's business plans.

The assumptions and the Environmental Scan are reviewed and
used throughout the corporation. Once the assumptions have been
disseminated, the process is initiated again. The matrix organization is reconvened to produce sequels to the Scan, either as updates
or revisions. We have also developed a course module and workbook
for our human resource planning course to train human resource professionals from headquarters and operations in skills of environmental analysis and forecasting. In this way, we have sought to
disseminate analytic skills as well as substantive resources to
strategic situations throughout our varied business organizations.

CONCLUSION

In today's complex, changing business environment, decisions
based on independent trend projections such as skill supply forecasts become far more complicated in their identification and analysis. Today, we have recognized the multiple and differentiated impacts of national and international economic and demographic, technological development trends and political events on business decisions and directions. To encompass these factors in an analytic formulation, to consider these factors in the broader planning process,
and finally to integrate these considerations into decisions about
present and future human resources, pose a challenge for all of us.

My discussion has presented one approach to this challenge, an
approach which is subject to regular revision and refinement. No
matter what the revisions and refinements, though, some conclusions
are apparent: that labor supply and technological development are
rapidly becoming interdependent facets of our human resource decisions and business plans. The interdependence of skill mix and
technological application requires regular, almost constant, monitoring, within the specific framework of each business environment,
to assess current and impending problems or opportunities. For
those human resource planners who document and understand the complex relationship between technological development and skill mix,
contributions to business development can be significant. In the
1980s, then, human resource planners have the opportunity to extend
their own analyses into the future and by doing so, to inform
management decisions for tomorrow's business outcomes.

REFERENCES

Cameron, Kim S. and Whetten, David A. (Editors), 1982, "Organiza-
    tional Effectiveness: A Comparison of Multiple Models,"
    Academic Press, New York.
Drucker, Peter, 1980, "Managing in Turbulent Times," Harper & Row,
    New York.

MANAGING U. S. ARMY TECHNOLOGICAL CHANGE: CONSTRAINTS AND

OPPORTUNITIES FROM THE HUMAN RESOURCE PERSPECTIVE

Harry M. West, III

Deputy Director, Army Manpower, Programs and Budget
Headquarters, Department of the Army
Washington, D.C.   20310

ABSTRACT

Technological superiority is essential to the national security of
the United States and NATO.   Three elements are important to
maximize the technological superiority of the United States Army:
(1) Equipment reliability and maintainability, (2) Doctrine which
includes strategy, tactics and the recognition of political
realities, and (3) Human resources.   This paper focuses on the
human resources element since it provides the greatest opportunity
for expanding or constraining the military benefits of technology.
Conclusions reached in this paper suggest that while technological
change is essential to meet military superiority requirements, it
must be equally used to reduce the demands placed on human resources
in the Army of the future.

INTRODUCTION

Amid the discussions of the value and importance of
technological change to the U.S. Army comes the clashing of two
viewpoints.   The first deals with the proponent of technological
change as an essential ingredient to military qualitative
superiority when faced with numerical adversarial advantages.   The
second deals with the practical matters of providing the human
resources to support a technological based military structure,
especially in the All Volunteer Force (AVF) environment.   Too many
questions are raised and remain unanswered between these two
viewpoints.   This situation suggests much more than simply divergent
opinions.   In fact, when viewed from the human resource arena, it
would appear as though there is a breakdown in the interdependencies

53

between the proponents of these two viewpoints. This breakdown may
delay the technological change essential to overcome present United
States and NATO quantitative, and sometimes qualitative
technological disadvantages.

OVERVIEW AND PERSPECTIVE

     Managing the U.S. Army in transition to higher technological
based weapons systems is a complex issue. Factors such as
increasing and changing threats, strategic planning scenarios,
near-term readiness requirements, force modernization issues,
investment trade-offs in the resource management area and the
allocation and use of human resources must all be considered.
Indeed, all are critical factors in the success of the U.S. Army in
carrying out its commitments. This paper focuses on one of these
factors in the transition to a higher technological based Army--that
of human resources. In fact, it may be the single most important
"force multiplier," or conversely "force constrainer" of the
technology necessary to overcome the current Soviet Union and Warsaw
Pact quantitative military advantage.

     There are many perspectives and levels of detail from which to
view the problems faced by the U.S. Army in support of its
world-wide commitments. For our purposes at this conference, a
comparison of major fighting systems is sufficient. Figure 1
indicates a NATO vs. Warsaw Pact aggregate comparative advantage of
the European Center Region[1] equipment capabilities.

Major Weaponry	1970		1980		1980 Pact Advantage Over 1970's
	NATO	Pact	NATO	Pact	
Tanks	5,745	13,550	6,200	18,000	+ 151%
Armored Vehicles	13,000	14,400	14,400	18,000	+ 257%
Artillery Pieces	2,000	5,150	2,300	6,500	+ 133%
Anti-Tank Weapons	3,300	3,400	5,000	7,400	+2,40 %
Air Defense Systems	1,900	4,400	2,300	5,000	+ 108%
Aircraft	1,550	2,850	1,420	3,200	+ 137%

Source: Adapted from Dr. Phillip A. Karber, The BDM Corporation, "The European Arms Race, 1948-1980," Preliminary draft of June 9, 1980 as presented at the June 11-13, 1980 Arms Control Conference in Ebenhausen, West Germany.

     Figure 1.   NATO vs. Warsaw Pact Equipment Levels, 1970 vs.
                  1980--Center Region.

     There are other comparisons available which analyze strategic
nuclear and theater nuclear forces, land forces in the aggregate,
naval forces, air forces, mobility[2] and others. One aspect is clear
in all these comparisons, "...for ground and probably for tactical
air in the Center Region, the numbers favor the Pact, but NATO's

continuing qualitative advantage—not only its unfortunately
narrowing lead in technology—act to reduce the possibility that
NATO forces would find themselves overwhelmed, at least in the early
stages of a future war in Europe."[3] Technology and U.S./NATO
qualitative advantages have been the key ingredients in the defense
of the Western Alliance. Most will agree with Professor Morgenthau
that—

> "The fate of nations and of civilizations has often been
> determined by a differential in the technology of
> warfare for which the inferior side was unable to
> compensate in other ways."[4]

THE ISSUES

This paper analyzes the dependence placed on technological
based military superiority. Other resources apply as well, but in
terms of importance, human resources represent potentially the
greatest problem or opportunity in achieving the U.S./NATO
qualitative objective. Two issues are important. First,
dependencies on technology to support military superiority are
clear. Applications of technology and resulting technological
change are accepted as the most important aspect of the development
of long-term qualitative advantages for the United States and its
NATO allies. The second issue deals with the ability of the United
States and its NATO allies to take full advantage of technology to
enhance deterrence and, if required, to overcome numerical
disadvantages. Within this second issue, two sub-issues are
important: (1) The capacity of NATO and the Western Alliance
technological base upon which to develop and build additional
qualitative advantages, (2) The capability of the alliance to fully
utilize this qualitative advantage under the present system of
managing human resources. At the outset, basic indicators suggest
we do not. This position forms the major hypothesis of this paper
and it suggests three things:

1. Dependence on the qualitative military advantage will drive
the U.S. Army beyond its ability to attract and retain highly
qualified skilled personnel at some future period. Indeed, all
countries in the NATO Alliance are included. West Germany, for
example, will experience a decline in its present qualified
recruiting pool of 310,000 eligible 18 year olds to 177,000 in 1994.
Currently, West Germany inducts 200,000 per year to maintain its
military structure.[5]

2. Failure to recognize human resource constraints now will
have an adverse impact on future benefits expected from today's
investments in technology.

3. Opportunities exist to optimize the required U.S./NATO
qualitative military advantage if policies supporting human
resources are modified with sufficient lead-time to remove human
resource constraints.

This last point merits some additional discussion. If the
forecast of a human resource constraint is predicted for a given
future time period t(p), the actual solution must take place in t(p)
- t(i); where t(i) is the implementation time required for the
solution or resolution. This is based on the fact that a closed
military personnel management system (with little or no lateral
entry and inputs mostly from the bottom up) requires time to respond
to adjustments, especially qualitative adjustments based on
technological demands. Since the latter point is so critical, a
simplified model of this problem is presented in Figure 2.

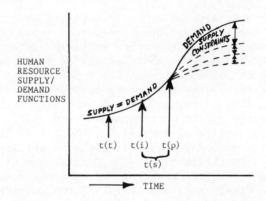

Figure 2. System effectiveness constraints

If t(p) (constraint problem) is not forecast by t(i) (minimum
lead-time to implement), then the t(p) (the problem) is slipped and
the reality of the human resource constraint negatively impacts on
designed technological benefits. T(t) (today) is shown before t(i)
which is an optimistic assumption incorporated in the hypothesis.
In other words, there is assumed to be either sufficient time or
other alternatives that could be used to solve the constraint
by t(p). These relationships will be addressed when a series of
non-traditional human resource policy options are discussed.

THE IMMEDIATE CONCERN

Due to the absolute U. S. Army and NATO dependence on
technological advantage, the U.S. Army has embarked on a major force
modernization program for the decade of the 1980's. Estimates today
suggests that most of our present day fighting systems will be
replaced during this period. This will impact on nearly all of our
present officer and enlisted personnel. Will the Army be able to

accommodate this transitional dimension? The present profile of
human resources suggests that there is already evidence of
qualitative and some quantitative problems today in the enlisted
career force. This is even before the additional challenges of
technology are present. While the active component of the U.S. Army
is achieving the accessions (recruits) needed to meet a constrained
end strength (filling authorized spaces with people), the
qualitative skill dimension is experiencing constraints. Thus,
today's Army is already constrained in achieving both the numbers
and quality desired for a 16 Division active peacetime force.
(NOTE: This author does not agree nor suggest that technology adds
complexity and thus increases human resource demands. Rather, this
has been the result of the recent past, but should not be an
accepted impact in the future.)

     Figure 3 indicates the relative distribution of selected age
groups in our population between 1960 and 1990. The enlisted
military personnel system of the active U. S. Army requires annual
accessions of 120,000 to 130,000 non-prior service males (136,000
FY-80 actuals), and 15,000 to 25,000 non-prior service females
(22,000 FY-80 actuals) to maintain its constrained military
structure during peacetime. Actually the Army quantitatively
achieved 101.2% of its assigned non-prior service male objective and
94.9% of its non-prior service female objective in fiscal year 1980.
The Army is achieving similar results for fiscal year 1981, but is
faced with drawing larger numbers from the 16-19 age group.

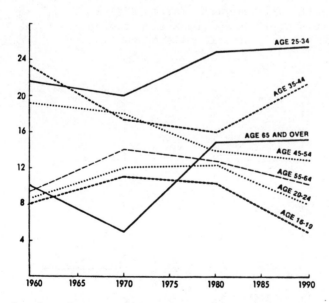

Figure 3. Age Groups as Percentage of Total Labor Force

Given the national constraints, we must assess the qualitatively aspect of this diminishing pool. This assessment reveals
that the active Army is experiencing qualitative difficulties in
recruiting High School Diploma Graduates (HSDG). Between fiscal
year 1976 and 1979, the Army's HSDG recruiting fell from 105,543 to
82,843. This coincided with a number of factors including the
removal of the GI Bill, reduced recruiting resources and an erosion
of entry pay. However, the Army reversed the downward trend in
fiscal year 1980 with 85,825 HSDG accessions. For 1981, the
objective is 96,000 HSDG's plus an additional 25,000 non-High School
Diploma Graduates which compares favorably to the 120,000 to 130,000
level projected for the 1980's. Of particular interest in the
reversal in the downward trend of HSDG's is the importance of the
appeal of technology itself to the potential recruit of the Army
today. "... the Army's fiscal year 1981 advertising plan was
revamped to include high technology opportunities for prospective
enlistees."[6] At the mid point of fiscal year 1981, the active Army
is achieving both its quantitative as well as its Congressionally
mandated qualitative HSDG recruiting goals. The critical dilemma
that faces the Army today is the rapidly accelerating increase in
the variety of technological complexity which requires larger
numbers of higher skilled personnel.

U. S. ARMY HUMAN RESOURCES TODAY: AN OVERVIEW

A brief manpower review of where the U. S. Army stands today
will serve as a reference for future discussions. In gross terms,
Figure 4 provides a perspective of where we have been since the
inception of the all volunteer force (AVF) in 1973. The concentration in this paper is on the Active component of the Total Force
and more specifically on the enlisted portion of that component.

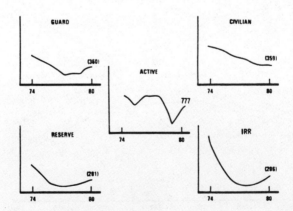

Figure 4.   Where We've Been (000).   Source: "U.S. Army FY 80
            Personnel-Posture Overview."

The five components shown comprise the total Army human
resources with the exception of the Inactive National Guard, the
Standby Reserve and Retirees. The civilian component is of course
all civilian. The National Guard, U. S. Army Reserve, Individual
Ready Reserve and the Active Army are all made up of Commissioned
Officers, Warrant Officers and Enlisted. Figure 5 displays detailed
information on the enlisted portion of the Active Army. The data
shown are the most current available. Actual data are indicated by
solid lines, while requested strength is shown as broken lines and
reflects the Army's budget request before the Congress today.

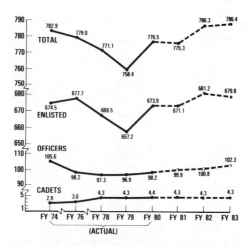

Figure 5.   Active Army Manpower Program.   End Strength (000).

The projected enlisted strengths shown in Figure 5 on the
broken line are calculated to be achievable.   The Army uses a
computer based enlisted inventory projection system entitled
ELIM-COMPLIP to determine their attainability. This system
forecasts strength, gains and losses over a seven-year period based
on both historical time-series data and projected effects of changes
in policy and other conditions.[7] This system is comprised of an
Enlisted Loss Inventory Model (ELIM) and a Computation of Manpower
Programs Using Linear Programming (COMPLIP). The current fourth
generation of the ELIM-COMPLIP system is producing strength
forecasts that are accurate within 0.5% for the total enlisted
strength of the active Army for 12 months beyond the latest actual
strength data.[8]

MARGINAL ANALYSIS OF TECHNOLOGICAL CHANGES

Before leaving the enlisted segment of the active force, let's

examine some of the enlisted qualitative trends of the Army. Figure
6 indicates that the Test Score Categories (I, II, III and IV)
derived from entrance examinations are all moving in the wrong
direction to support increased technological specialization,
especially in the total enlisted force. The Career Force provides a
somewhat more favorable trend with the exception of Category IV
increases. It should be noted that the enlisted career force grew
by approximately 75,000 between 1974 and 1980.

Figure 6.   Active Army Enlisted Inventory by Test Score Category.
            - Total Enlisted Force 1971/1975/1980 -

TEST SCORE CAT	% OF 71 INVENTORY	% OF 75 INVENTORY	% OF 80 INVENTORY	% CHANGE 71-80
I	4.7	3.2	3.0	-35.0
II	24.4	23.9	20.1	-16.9
III	38.1	46.1	36.2	-4.9
IV	20.6	13.5	31.5	+52.8
		-CAREER FORCE-		
I	3.5	4.0	3.8	
II	17.4	22.5	23.0	+32.2
III	33.2	38.6	40.9	+23.5
IV	15.5	15.5	19.7	+27.4

(NOTE: Test Scores renormed 1976-1980)

        Using this base of enlisted qualitative data, we can begin to
assess the impact of three additional variables that are likely to
further constrain the qualitative levels of the enlisted personnel
supplies in the future. These are (1) the decreasing propensity to
enlist in the U. S. Army (2) the decreasing size of the national
pool of 18-year olds, (3) The lessening of a national commitment.
Given that the propensity to enlist is decreasing, the expressed
commitment has decreased and the pool itself is decreasing, the
outlook is less than optimistic. Combined with the data presented
earlier on the decreases in accessions with high school diplomas, it
is difficult to anticipate much elasticity in the ability of the
enlisted force to accommodate the anticipated technological
complexity for operators and maintainers.

New Systems and Technological Complexity: Forecasting Human
Resource Demands An Indication of High Technology Emphasis

        The U. S. Army is experiencing a modernization program that
will rapidly increase during the decade of the 1980's. During the
period FY 82-86, the Army has $44 billion programmed for modernized
items and will introduce 47 major new fighting systems. In effect,
this involves the replacement of almost all fighting systems. As an
indication of the emphasis placed on technological change, the Army

has designated the 9th Infantry Division (about 17,000 soldiers) at
Fort Lewis, Washington as its High Technology Test Bed (HTTB). The
HTTB will be used to transition the 9th Infantry Division (9ID) into
a technologically advanced, combat ready, prototype infantry force
capable of rapid and strategic deployment. Accelerated infusion of
new weapon systems and available high technology will provide a test
of improved operational capabilities. At present however, the 9ID
HTTB does not provide a source of data on technological based human
resource demands. Even preliminary human resource information will
not be available until early 1982, and then only on an incremental
basis as new technological advancements are implemented.

Are Our Weapons Too Complex?

     To some nations, technology offers the opportunity to achieve
military strength without massive defense expenditures. Recently
Britain's Defense Minister, John Nott indicated that "I shall be
considering in the coming months...in consultation with our allies,
how technological and other changes can help us fulfill the same
basic roles more effectively without the massive increase in real
defense expenditures which the escalation of equipment costs might
otherwise seem to imply."[9] Technologically advanced weapon systems,
therefore, are sought not only to counter military advantages
achieved by an adversary but to do so at reduced costs.

     This approach has caused much controversy based on one very
fundamental issue, that of complexity. That is, can the weapons
system be maintained and operated effectively by the military in the
way the engineers designed the system? Obviously with the
dependence on technological gains for qualitative battlefield
superiority, the researchers and the engineers seek maximum
technological advances. This raises the question, have we exceeded
the human capacity for complexity norms in search of the maximum
technological superiority?

     In an attempt to cast a proper perspective on an answer to this
question, the Honorable William J. Perry, former Under Secretary of
the Defense for Research and Engineering indicated that the
application of advanced technologies are essential to gain the
"...performance edge critical to our forces in the face of the
numerically superior opponent."[10] He, of course, went on to
indicate that technology can be used in many ways, one of which is
to reduce the need for human resources. Indeed, amid the debate
between the two schools, this important issue is often overlooked,
but is fundamental if both are to achieve their fullest objectives.

     Two recent articles have discussed the opposite sides of this
issue. In one, Franklin C. Spinney,[11] a Department of Defense

analyst is quoted as saying, "The across-the-board thrust towards
ever-increasing technological complexity just is not working. We
need to change our way of doing business...Our strategy of pursuing
ever increasing technical complexity and sophistication has made
high technology solutions and combat readiness mutually
exclusive."[12] His main objection to the focus on technology is that
we as a nation have evolved into an unsupportable position of having
blind faith in the use of military technology. Contesting this
viewpoint, Perry is quoted and contends that one cannot answer the
question of complexity except in comparative terms, that is, in
comparison to the Soviet threat. He indicates that while they have
made vast technological gains, it has not been without their own
added cost and complexity. In his words: "Our trend has been
toward simplifying technological complexity. Their trend has been
in exactly the opposite direction...The Soviet T64, which is the
high technology Russian tank, is as complex as any tank we have
built in terms of the gadgets that are on it. The night vision
devices, the laser systems, the computer control firing systems...We
still think our XM1 will out perform the T64 or any other Soviet
Tank."[13]

A recent study by the United States General Accounting Office
(GAO) indicates that"...the demand for high performance has forced
designers to incorporate new technology into (weapons) systems often
before its reliability has been fully assessed...When this happens,
the cost in field repairs and low system readiness rates can be
high."[14] The report goes on to conclude that military planners are
mesmerized by high technology. Fallows also suggests that
technology may not be a singular answer to quantitative
disadvantages. He contends that a common thread running through all
the arguments supporting high technology is the idea that the United
States has no other choice. This, he suggests, leads to the
conclusion that technology will save us. He believes this to be
unfortunate "...because the concept is wrong."[15] Rather Fallows
recommends a more balanced mix of conventional and high technology
weapons as opposed to striving for qualitative superiority to
counteract quantitative disadvantages.

So the debate goes on but one point seems abundantly clear.
Facing numerical disadvantages, the United States is being forced to
utilize technology, sometimes in large advances, to develop
qualitative military superiority. In the midst of this debate is
always the human element. It is characterized as indicated in
Figure 7, by Mr. William T. Coulter, in his Washington Post
editorial artist's conception of 22 Feb 81.

FIGURE 7

An unfair characterization, of course, but is there a time in
the Army's future when such a situation might take place? Is it
now? These are difficult questions to answer. As Baker indicates,
"The Army is moving swiftly to ensure that the human resource
requirements of emerging systems do not exceed the projected
availability and skill levels of its personnel."[16] Today, however,
the Army has embarked on a major force modernization program that in
all likelihood may exceed the human resource management capabilities
of the system. The human resource controls indicated by Baker may
be insufficient to manage the human resource element relative to the
accelerated Army requirements for technological change. This is
especially significant when one considers the adverse impact imposed
by national demographics[17] and recent "propensity to enlist"
survey[18] findings.

The search for technological solutions to reverse the trend and
tip the military balance back toward NATO has imposed greater
pressure on human resources. Kerwin and Blanchard have directly
addressed the issue of equipment complexity and today's soldier. In
their words--

> "The U. S. Army has a major man/machine interface problem.
> There are not enough qualified people to perform the
> tasks required to effectively operate, support and
> maintain current Army systems...Increasing weapon
> complexity, the large number of new systems being
> developed, insufficient formal school training, a
> declining manpower pool, disproportionate numbers
> of CAT III B and CAT IV personnel, recruiting and

retention problems and unit turbulence-all will
contrive to strain the already overburdened personnel,
training, and development communities."[19]

A major recommendation by Kerwin and Blanchard is that the
personnel developer should be a co-equal in the Army's materiel
acquisition process. However, when there is as much emphasis on
technology as there is today to achieve military superiority, there
is generally no room for co-equality. It is often suggested that -
"There is always a human resource solution; don't you realize we are
developing a new technology here to overcome a Soviet advantage?"

Is There a Way to Resolve this Issue?

The Army Science Board meeting on 16-17 March 1981 confronted
this challenge directly by discussing an issue entitled, "Human
Resources to Support a Technologically Based Army". In summary
their issue was: (1) Change the design constraints on hardware for
the 1990's; (2) Seek solutions to expand the supplies of human
resources; and (3) Use technology to maximize the current known
relationships between manpower and equipment so as to reduce the
human demand while achieving mission responsiveness."[20]

MUCH NEEDS TO BE ACCOMPLISHED RAPIDLY

Baker and Shields depict an integrated and interdependent U. S.
Army manpower system. However, they suggest that this display,
while dynamic, is not an instantaneous, responsive system. Rather
they suggest, it is much like a predictor display found in advanced
systems.[21] The Army's manpower system does in fact have entropy as
well as reduced reaction time to dynamics. While only selective
human resource constraints are evident today, the trends of the
national pool, propensity to enlist, quality of the new Army
accessions and the demand for technological change all suggest
action today. In order to preclude human resources from
constraining technological advancements in the U. S. Army, three
major initiatives should be sequentially undertaken to permit the
Army's manpower system to react. In summary form they are:

1. Using empirical, estimated or delphi data, the Army must
determine its requirements for skills and numbers of enlisted
personnel to fully support its force modernization program.

2. Using model supported projections for the near-term
(mid-1980's), estimate the attainability of the requirements
indicated in the first step. For the longer term (late eighties to
early nineties), use extrapolations of modeled data, using at least

three scenarios which are sensitive to:

> Personnel policies
> National economic projections
> National sociological indicators

3. If the cross-over point between Step 1 and 2 is recognized
and is before t(p) (Figure 2), then adequate time for an implemented
solution is appropriate. If between t(i) and t(p), then
extraordinary means must be applied to solve an embedded problem.

If this process is employed, it is anticipated that in a future
time period, human resources will become a constraint on
technological solutions to U.S. Army objectives to gain qualitative
military advantage. This will then signal a need for the highest
priority solutions before combat capabilities are in fact
constrained.

Complicating this issue is a national recognition of the
problem of an inadequate defense posture for the United States. As
Laird and Korb have indicated, "...our force structure is too small,
our modernization programs are insufficient and our operational
forces have severe readiness and sustainability problems."[22] From
this perspective, human resources become "problems." In addition,
they often get caught in the ground swell of technologically based
force modernization in the Army and are unable to compete relative
to the objectives being sought. Recent articles quoting Defense
Secretary Caspar W. Weinberger suggest that the Reagan
Administration may cause additional pressure by seeking major gains
in military personnel over the next five years, perhaps as many as
250,000 additional recruits.[23] It appears as though insufficient
importance and attention are afforded to the human resource element
and that there is insufficient institutionalization of a program to
include human resources in the technological equation. The issue of
projecting supplies and demands is data dependent but can be
supported by state-of-the-art techniques. The reality of human
resource planning relative to technological force modernization is
passive at best. If it were otherwise, the initiatives above would
be an institutionalized process and would use the tools summarized
by Price, Martel and Lewis.[24]

PRESENT TRENDS WILL NOT AVOID FUTURE PROBLEMS

The transition to increased technology, on the basis that it
is essential to national security, will likely not avoid human
resource problems but accelerate their arrival. Clearly, avoiding
such problems should take highest priority. Three weapon system
modernization architectures are possible that will reduce and in
combination avoid future human resource constraints:

1. Highly disciplined integrated force modernization strategic planning that includes human resources as a critical element in the strategic planning equation.

2. Organizational structures to support the results of the strategic planning analyses consistent with both qualitative and technologically based military objectives.

3. Design of future weapons systems to fit within the envelope of human resource constraints, or modify the constraints marginally or seek greater use of robotics to free humans.

A recent U. S. Navy publication by the Naval Research Advisory Committee confirms this pessimistic outlook and suggests similar changes to "...avoid the potentially disabling mismatches between these systems and the personnel who must operate and maintain them."[25] However, again a divergent viewpoint is heard when one analyzes the recent work of C. W. Taylor. He suggests that most of the National Security Forces (re-named to coincide with his study of the Army in the Year 2000) have optimized man-machine systems performance. He suggests that most jobs in the military have been automated. He also suggests that "...military operations have increased in complexity...and that the man-machine relationships have been optimized."[26]

POTENTIAL MANPOWER AND PERSONNEL NON-TRADITIONAL POLICY OPTIONS

Recognizing in some instances that the pressure for technologically based force modernization may be constrained by human resources, are there marginal human resource policy options that can support added complexity? The answer to this question is dependent on a number of issues but can be discussed relative to the following two categories:

Variable Policies:

- The continuation of the draft-free environment.
- The continuation of emphasizing "initial capital cost" acquisition vice "human capital life cycle cost" as control mechanisms in the Army's force modernization.

Fixed Constraints:

- The continued decline of the national pool of 18-20 year olds.
- The continued reliance on technology to maintain Army battlefield qualitative superiority while facing numerical disadvantages.

The variable policy factors provide the greatest opportunity to "incentivize " within the draft free environment as suggested by Coffey, Moskos and others. These options must be pursued to gain maximum opportunities from the All Volunteer Force. It is unlikely that, other than registration, the United States will change the basic features of the All Volunteer Force unless national security requires full or partial mobilization. Therefore, options to enhance the opportunity to provide greater numbers of higher skilled volunteers must be pursued. The second policy factor has much more opportunity to reduce equipment complexity, that is, to directly reduce the complexities required of the operator and maintainer. A major policy change in this arena would permit a "design to human resource constraints" factor in the already complex Army System Acquisition Review Council. The U. S. Army should project the companion cost of military compensation to overcome the initial capital investment constraints placed on Army weapons systems acquisition. The military compensation costs should also include retirement liabilities so that as the level of skills as well as the number of skilled operators and maintainers increases, capitalized human resource costs can be more accurately displayed for each system. This approach recognizes that the aggregate system costs, including human resource costs, will in all probability require a totally new emphasis. That is, initial capital investments in hardware will most likely result in decreasing operator and maintainer interface systems.

In the fixed category, the factors are somewhat more difficult to deal with since they are embedded realities. The United States pool of 18-20 year olds will continue to decline for the rest of the decade of the 1980's as indicated earlier. The primary opportunity to change this factor is to enlist greater portions of a declining pool and retain larger percentages of those already in the Army. While this is a complex issue, opportunities do exist to deal with these constraints, largely through selective application of compensation incentives. The second factor in the fixed category is equally more troublesome. Without relief from the other three factors, this may be the only variable that is left to accommodate the realities of the human resource constraints of the future. Recognizing the dependence on this factor to maintain the qualitative superiority required under today's national security posture, Army human resource managers should not permit this to become the slack variable.

Non-Traditional Human Resource Policy Options

What are non-traditional human resource policy options? Basically they are options that recognize the above fixed and policy variable categories and center on enhancing the use of technology to

achieve the greatest military qualitative advantage possible in view
of the predicted human resource constraints. They are briefly
described as:

1. Using the fixed category factors, construct the "zone of
the attainable" in terms of human resource constraints. That is,
using traditional supply and demand projections and developing
expected characteristics over the total life cycle of the systems.
In some cases, this will mean 30 years. Then, within the best
information available, allocate these resources, on an "ownership"
basis to a set of prioritized competing demands for these resources.

2. Assuming that the human resource pool is insufficient,
perhaps qualitatively during the mid-1980's and then quantitatively
during the late-1980's, use the opportunities available to extend
the constraints as far into the future as possible. This could be
accomplished up to some new level by:

- Requiring systems that exceed the constraints to pay a
"human resource tax" in the form of multi-year appropriation
increases to the Military Pay Army (MPA) appropriation or
others required to acquire, train and retain people.

- Changing the manner in which we traditionally manage the
closed military personnel system with primary entry from the bottom.
That is, to use lateral entry throughout the system where specific
constraints are evident.

- Recognizing the fact that constraints in most other critical
resources, including strategic minerals, are overcome by strategic
assessment, conversely human resources are taken as a given. This
organizational problem must be met directly with a totally new human
resource planning dimension just now being implemented in the
private sector. Implementing models of this type as suggested by
Niehaus, Shafritz, Biles and Holmberg[27] and others will place the
proper strategic emphasis on human resources to maintain the
designed U. S. strategic qualitative advantage.

- Substituting the maximum number of civilians for military in
support roles, thus permitting current military personnel to be
freed to be applied to higher priority military requirements.

- Utilizing selective volunteers from the Western Alliance to
serve in the U. S. Army. Just as the NATO Rationalization-Standard-
ization-Interoperability (RSI) program recognizes equipment
interdependencies, there is every reason to expect that the same
strategic value can be obtained from "shared" human resources.

THE CRITICAL ISSUES

To recognize the importance of removing human resource
constraints consistent with strategic objectives, a U. S.
Army-Soviet Army comparison is presented in Figure 8. It displays
quantitatively the deployment of 24 U. S. Divisons and 173 Soviet
Divisions. While the manpower size of the U. S. Divison is larger
(about 17,000) than a Soviet Division (about 10,000), the
quantitative advantages are obvious.

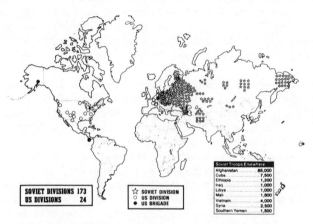

Figure 8.   U.S./Soviet Worldwide Deployment.   Source: "U.S. Army
            Overview FY 82"

Another quantitative comparison can be made using representative
fighting systems. Figure 9 indicates a comparison of primary
battlefield hardware systems. Again, the quantitative advantage is
obvious.

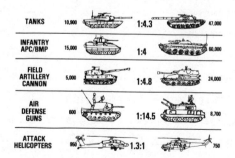

Figure 9.   U.S./USSR Quantitative Disparity.   Source: Collins, John M.
            "US-Soviet Military Balance, 1960-1980," 1980.

As noted at the beginning of this paper, the U. S. Army depends on a qualitative advantage for superiority. However, the Army has been unable to keep pace qualitatively with the Soviets in terms of modernized systems introduced.

> "A US expeditionary force dispatched to any of a dozen Third World countries would find itself matched or outclassed in weapon systems capability. Today's Army is equipped with many of the same systems it had 15 or more years ago. The need to replace that equipment-- to modernize--is of importance equal to any task facing us. Our goal is to attain qualitative equivalence by the middle of the decade and qualitative advantage before the next decade."[28]

To more completely indicate the potential impact on human resources in the future, as we gain equivalence and finally advantage, Figure 10 is a current assessment of the qualitative disparity. To overcome this disparity, the U. S. Army will rely on a major technologically supported modernization program. Technological superiority, such as the M-1 tank over the current Soviet T-72 and even the soon to be introduced T-80 is an absolute necessity. The same qualitative dimension is true for all Army objectives noted. Improved warheads for the TOW antitank guided missiles, laser homing capability for the 155mm cannon launched guided missile projectile, advanced attack helicopters with laser homing Hellfire missiles and others on the Army objectives list are technologically superior to the Soviet counterpart. All are essential to gain a qualitative advantage in view of quantitative disadvantages.

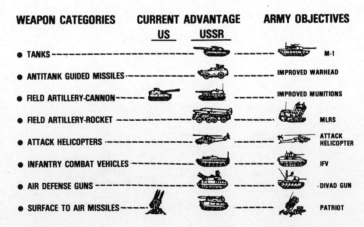

Figure 10.   Need for Modernization; Quality Issue.   Source: "U.S. Overview FY 82"

CONCLUDING COMMENTS FROM A HUMAN RESOURCE PERSPECTIVE

The challenges are great technologically. Equally challenging, and perhaps even more constraining or opportunistic, are the human resource issues, associated with technological change. If not properly managed, human resources can become a shackle on technological superiority. Conversely, they can become the factor to enhance technological superiority. There are ample opportunities to apply state-of-the-art techniques to the human resource planning models on this complex issue. Recognizing the issue is however the first step. If recognized too late, technological advantage and indeed Army resources could be wasted when the operator and maintainer are unable to support the technologically based system of the future. However, this need not be the case. In fact, the very bottom line of this issue suggests heavy front end capital investment in weapon systems to avoid human resource constraints. Emphasis on human resources in the qualitative equation can multiply the technological advantages of the U. S. Army of the future. Likewise, we may be able to reduce the total life cycle costs associated with a technologically based Army.

It has been said that "...the Soviets have very obviously committed themselves to the creation of a technology base second to none."[29] Since technology offers the greatest opportunity to compete successfully with the Soviets, we must ensure that human resources never become a constraint in this competition. The Chief of Staff of the United States Army has recognized the importance of this competition and indicates that "the resultant Army Modernization Program is the largest in peacetime history and is intended to achieve at least technological equivalence in fielded systems by 1985 and superiority by 1990."[30] Our challenge is to ensure that this competition is not constrained but enhanced by human resources.

REFERENCES

1. "The European Arms Race, 1948-1980", 1980 Arms Control Conference in Ebenhausen, West Germany.
2. Harold Brown, Secretary of Defense, "Department of Defense Annual Report, Fiscal Year 1982, " January 19, 1981. pp. 37-80.
3. Harold Brown, p. 76.
4. Professor Morgenthau, Politics Among Nations
5. Strategic Studies Institute, US Army War College, Futures Group, Periodic Report 4, December 1980, p. 12.
6. Lieutenant General Robert G. Yerks, "This Is Your Army-1981, Manning Overview", Department of the Army, March 1981, p. 3.
7. For more information see, Holz and Wroth, "Improving Strength Forecasts: Support for Army Manpower Management", INTERFACES, The Institute of Management Sciences, Vol. 10, No. 6, December 1980.
8. Holz and Wroth, p. 42.

9. Leonard Downie Jr, "Britain Costing Critical Eye on Costly, High
Technology Weapons, "The Washington Post, April 16, 1981, p. 34.
10. Senate Armed Services Committee testimony by the Honorable
William J. Perry, December 9, 1980.
11. See Spinney's "Defense Facts of Life", December 5, 1980.
12. George C. Wilson, "Are Our Weapons Too Complex?" The
Washington Post, February 22, 1981.
13. George C. Wilson, "Not If You Compare Them to the Soviets, "The
Washington Post, February 22, 1981, p. 23
14. The Comptroller General, Effectiveness of U. S. Forces Can Be
Increased Through Improved Weapon System Design, U.S. General
Accounting Office, January 29, 1981.
15. James Fallows, "America's High-Tech Weaponry," The Atlantic
Monthly, May 1981, p. 29
16. James D. Baker, "The Army's Match Game: Man to Machine",
Defense Management Journal. Second quarter, 1980, p. 24.
17. Richard L. Fernandez, "Forecasting Enlisted Supply:
Projections for 1979-1990," RAND, N-1297-MRAL, September 1980.
18. 1980 Youth Attitude Tracking Study
19. General Walter J. Kerwin and General George S. Blanchard, USA
(Retired), Man Machine Interface - A Growing Crisis, Discussion
Paper, Army Material Systems Analysis Activity, August 1980.
20. Army Science Board, 16-17 March 1981, San Antonio, Texas
21. James D. Baker and Joyce Shields, "Personnel Affordability: The
Army's Odyssey into the Year 2000," U. S. Army Research Institute,
CORS-TIMS-ORSA 1981 Joint National Meeting, May 1981.
22. Melvin E. Laird and Lawrence J. Korb, The Problems of Military
Readiness, The American Enterprise Inst., Wash, D.C., 1980, p. 27.
23. Richard Halloran, "Reagan Military Plan Envisions Up to 250,000
Additional Recruits", New York Times, April 26, 1981, p.1
24. W.L. Price, A. Martel & K.A. Lewis, "A Review of Mathematical
Models in Sciences, OMEGA Vol. 8, No. 6, pp. 639-645, March 1980.
25. Naval Research Advisory Committee, Office of the Assistant
Secretary of the Navy (Research, Engineering and Systems),
Man-Machine Technology in the Navy, December 1980, p. 4
26. C.W. Taylor, "Science Technology, and the Army 2000 (DRAFT),
Strategic Studies Institute, U. S. Army War College, Futures Group
Periodic Report #3, 1 November 1980. p. 33.
27. Richard J. Niehaus, Computer Assisted Human Resources Planning,
Wiley, 1979; George E. Biles and Steven R. Holmberg, Strategic Human
Resource Planning Horton 1980; Jay M. Shafritz, The Public Personnel
World: Reading on the Professional Practice, International
Personnel Management Association, 1977.
28. "US Army Overview, FY-82"
29. Jack Vorona, "The Soviet March Toward Technological
Superiority", Defense 80, Mar 1980, Government Printing Office.
30. General E. C. Meyer, "A Framework for Molding the Army of the
1980s into a Disciplined, Well-Trained Fighting Force", White Paper
1980.

SECTION 3

LABOR MARKET IMPACTS ON HUMAN RESOURCE PLANNING

One consequence of the widespread application of new technology is that organizations become more sensitive to labor market
restrictions. These labor market shortage are also partly the
consequence of technological change, suggesting a complex interrelationship that requires development of a new methodology for
human resource planning and analysis.

Atwater, Bres, Niehaus and Sheridan present a prototype of a
comprehensive supply-demand human resource planning system which
represents considerable progress, from the issue development stage
to the numerical results stage. As an example, projected annual
recruiting requirements for science and engineering technicians
in the U.S. Navy civilian work force between 1981 and 1987 have
been calculated based on different scenarios for technological
change in the economy. A further discussion of important labor
market externalities is provided by Pearson, who also identifies
additional information needs and sources of data for human
resource planners. Rowe and Silverman outline an integrated
system of models for determining the external and internal
feasibility of U.S. naval officer manpower plans that have been
proposed independently in various parts of the Navy. Techniques
such as goal programming become necessary to develop and assess
the benefits of navy-wide officer manpower plans in the presence
of joint constraints.

Further development of integrated model systems will proceed
along the lines of availability of powerful partial models.
Sadowski and Martel present partial models for special categories
of manpower or special manpower issues. Sadowski models corporate training investment decision making for apprentices under
German labor laws, regulations, and business practice. Martel
models total compensation of Canadian federal public service
employees in a quasi-market comparability framework, thus
accounting for the competitive relationship between the
governmental sector and the economy at large.

INTEGRATION OF TECHNOLOGICAL CHANGE

INTO HUMAN RESOURCES SUPPLY-DEMAND MODELS

Donald M. Atwater,* Edward Bres III**
Richard J. Niehaus, ** James A. Sheridan***

* Department of Economics
University of California, Los Angeles
Los Angeles, CA 90024, USA

** Office of Assistant Secretary of the Navy
(Manpower and Reserve Affairs) - ODASN (CPP/EEO)
Washington, D.C. 20350, USA

*** American Telephone and Telegraph Company
1776 On the Green, RM 48-4A18
Morristown, NJ 07960, USA

ABSTRACT

This paper exaimines the issues of integrating human resources factors external to an organization with the demands from within. A comprehensive system of models is used to move from issue development to numerical results. On the supply side reservation wage models are used which have both economic and technological change segments. On the demand side, constrained regression is used to project manpower flow rates which are then used as input to a standard goal programming human resources planning model. The prototype numerical example uses the U.S. Navy science and engineering technicians as the target job group in three alternative economic scenarios.

INTRODUCTION

The availability of integrated supply-demand human resources planning systems has been one of the most often stated desires of policy planners and managers since the mid-1960's. Until the last few years, most of the actual system development has been on the demand side. Comprehensive models for planning the internal labor markets are available in a wide variety and are in standardized use in many organizations. Models to estimate manpower supplies

at the level of skills required by an individual employer beyond
rudimentary census counts began to appear in the mid-1970's. This
paper provides a prototype system which integrates both
methodological developments.[1]

The system discussed in this paper uses U.S. Navy science and
engineering technicians as the target job group in the prototype
example. The crafts journeymen job group was also included as it
is an important internal feed to the Navy's supplies of science
and engineering technicians. In the study itself, these target
groups were also categorized by male and female, and minority
and non-minority so that the equal employment opportunity implications of the study could be included in more comprehensive
versions of the prototype. For sake of compactness most of the
data shown in this report is for males.[2]

The projections cover 1980-1985 using various historical data
in the 1970-1979 period. Three alternative economic scenarios
were used: a base (nominal technological and "average" economic
conditions) projection; a low private sector wage projection;
and a high private sector wage projection. The next section begins
the methodological development with a discussion of the reservation wage models.

ECONOMETRIC LABOR SUPPLY MODELING

The econometric labor supply modeling approach provides both
a new method for the development and analysis of economic incentives across multiple job sectors and occupations and an integrated perspective between macro economic factors and work decision behavior modeling at the micro level. As shown in the
theoretical methodology section and the application system description which follow, the approach is still developmental in
nature and is expected to undergo numerous enhancements in the
future.

The development of accurate and comparable economic incentive
measures (i.e. wages) is a principal concern in the approach.
The econometrics focus on disaggregating or unbundling the dynamics
of compensation into a sector component, an occupation component
and a job component. The sector component is linked to specific
macro economic conditions (i.e. unemployment, GNP, interest rates
and inflation) and to alternative views of technology and forecasted events. The occupation component is the core around which
public data on multiple job options can be tracked and projected
into the future periods. The interactions of the occupational employment and occupational wage process over time provide a basis
or foundation for integrating the sector and job specific information. A multi-tiered recursive modeling format is used. The

job component analysis is required to align the more aggregate
compensation information with special skill factors and for
work and leisure.

The economic incentives include both work and leisure wage
measures. The value of leisure time is calculated using a proven
and tested reservation wage modeling sequence.[3] The reservation
wage modeling takes into account sampling selectivity bias
modifiers and is structured to develop labor supply information
at the sector/occupation/job level for selected race or national
origin and sex groups. The three economic behavior decision
models compare reservation wages and market wage alternatives
which persons must choose between when comparisons of work
opportunities across jobs, occupations, and sectors are considered.

The economic labor supply modeling, as shown, is extremely
flexible in terms of the range of events and decisions which can
be examined. Wherever appropriate, cyclical conditions have been
included using residual regression techniques. Decision choices
have been analyzed using probit analysis to reflect dichotomous
behavior options. Constant dollar measures are used to avoid
errors in forecasting due to real wage tradeoffs. Most important-
ly a wide range of control variables are provided to examine
alternative views of the future. All of the econometric relationships are linked together in a balanced, dynamic structure which
is sensitive to change but, based on testing, is not prone to
instability.

THEORETICAL RESERVATION WAGE METHODOLOGY

In this study work participation decisions of individuals
in selected economic and technological environments are modeled.
The statistical model is summarized by the following three
equations:

(1) $\quad W_{0_{ij}} = \beta_0' \cdot X_{0_{ij}} + \varepsilon_{0_{ij}}$ $\qquad$ (Occupation (i) sector (j)
$\qquad\qquad\qquad\qquad\qquad\qquad\qquad$ Wage Offer Equation)

$\qquad W_{0_{ij}} \equiv \overline{W}_0 + DOW_{ij}$

(2) $\quad W_{r_{ij}} = \beta_1' \cdot X_{1_{ij}} + \varepsilon_{1_{ij}}$ $\qquad$ (Occupation (i) sector (j)
$\qquad\qquad\qquad\qquad\qquad\qquad\qquad$ Reservation Wage Equation)

(3) $\quad D_{ij} = \begin{cases} 1 \\ 0 \end{cases}$ if $\begin{matrix} W_{0_{ij}} \geq W_{r_{ij}} \\ W_{0_{ij}} < W_{r_{ij}} \end{matrix}$ $\qquad$ (Participation Equation)

The first equation is the occupation (i), sector (j) wage offer
equation. The wage offer is assumed to be determined by a set of
observable variables $(X_{0_{ij}})$ which include education, prior labor
market experience, age, occupation and a set of unobservable
variables unique to the individual, such as ability and the
quality of education. The effects of these unobservables are
captured by the disturbance term $\varepsilon_{0_{ij}}$. The occupation/sector
wage identity shows that the occupation/sector wage can also be
characterized as an average market wage $(\overline{W}_0)$ and a differential
occupational wage (DOW_{ij}). Equation (2) is the comparable
reservation wage equation. The observable determinants of the
reservation wage $(X_{1_{ij}})$ include the number of children by age
sub-groups in the home and the spouse's income (if any). The disturbance term $\varepsilon_{1_{ij}}$ reflects the effects of unobservables, such as
tastes for leisure and non-market productivity. The third
equation is the civilian participation equation. D_{ij} is a
dichotomous variable which indicates participation status (1 for
participation, 0 for non-participation). For persons in a
specified cohort a positive choice indicates a positive work or
participation likelihood under the given economic and technological environment.

An individual will be observed to be a participant in an
occupation and sector if the occupation/sector wage offer meets
or exceeds the reservation wage. That is if:

(4) $\quad W_{0_{ij}} \geqslant W_{r_{ij}}$

Using the equations (1) and (2) we may rewrite equation (4) as:

(5) $\quad \beta_0' \cdot X_{0_{ij}} + \varepsilon_{0_{ij}} \geqslant \beta_1' \cdot X_{1_{ij}} + \varepsilon_{1_{ij}}$

or equivalently as:

(6) $\quad \dfrac{\beta_0' \cdot X_{0_{ij}} - \beta_1' \cdot X_{1_{ij}}}{\sigma_{u_{ij}}} \geqslant \dfrac{\varepsilon_{1_{ij}} - \varepsilon_{0_{ij}}}{\sigma_{u_{ij}}}$, where $u_{ij} = \varepsilon_{1_{ij}} - \varepsilon_{0_{ij}}$

Under the assumption of the normality of $\varepsilon_{0_{ij}}$ and $\varepsilon_{1_{ij}}$, hence u_{ij},

the probability that a given individual participates in occupation
(i) and sector (j) is:

(7)
$$P_{ij} = \text{Prob} \left\{ \frac{\beta_0' X_{0_{ij}} - \beta_1' X_{1_{ij}}}{\sigma_{u_{ij}}} \geqslant u_{ij} \right\} = \int_{-\infty}^{\Theta_{ij}} \frac{1}{2\pi} e^{-\frac{1}{2}(t)^2} dt$$

where $\Theta_{ij} = \dfrac{\beta_0' \cdot X_{0_{ij}} - \beta_1' X_{1_{ij}}}{\sigma_{u_{ij}}}$ for persons in occupation (i)

and sector (j). The so-called probit likelihood function for
data consisting of n observations on individuals s of whom are
participants in the occupation (i) and sector (j) is:

(8)
$$L_{ij} = \prod_{k=1}^{s} P_{ij} \prod_{k=s+1}^{n} (1-F_{ij})$$

with the form of P given in equation (7). Probit analysis
generates maximum likelihood estimates of $\dfrac{\beta_0'}{\sigma_{u_{ij}}}$ and $\dfrac{\beta_1'}{\sigma_{u_{ij}}}$ for the

elements appearing only in either $X_{0_{ij}}$ or $X_{1_{ij}}$. If there are
elements which are common to both $X_{0_{ij}}$ and $X_{1_{ij}}$, then the probit
method will generate estimates of the difference between the coefficients in the occupation/sector wage equation and the
reservation wage equation divided by $\sigma_{u_{ij}}$. With probit we obtain

estimates of $\dfrac{1}{\sigma_{u_{ij}}}$. We can then solve for β_1' and derive a mean

reservation wage. Thereafter the variance of the reservation
wage distribution can be calculated.

APPLICATION WITHIN A SYSTEM FORMAT

The economic decision models which are based on specific
comparisons of reservation and market wage measures for current
jobs and selected alternative work options are the bridges
between theory and the real world. Using the theoretical relationships specified in the previous section, real information on
persons in the selected race or national origin and sex groups.
and information on jobs in defined sectors and occupations, the
reduced form and structural participation models were estimated.
The economic incentives were then calculated for the targeted
Navy civilian forces and their supply decisions were quantified.

The non-persistence choice model compares economic incentives
for current Navy Civilian jobs with the calculated reservation wages
for each race or national origin and sex subgroup. The percentage
of persons whose value of leisure time EXCEEDS their calculated
Navy Civilian job's wage is identified as the Technician (mid/senior)
non-persistence rate. The percentage is calculated as:

$$1 - \frac{WAGE_T - WAGE_R}{\sigma_r^2} = Index \Big|_{N(\overline{W}_r, \sigma_r^2)}$$

where $WAGE_T$ = Mean Technician (mid/senior) Wage

$WAGE_R$ = Mean Reservation Wage

σ_r^2 = Variance of Reservation Wage Distribution

$Index \Big|_{N(\overline{W}_r, \sigma_r^2)}$ = Cumulative Normal Distribution Index Value (Standardized).

For craft journeymen the migration model uses reservation
wages and economic incentives for both current (Craft Journeymen)
and alternative (Technicians: (mid/senior) jobs. A wage
availability comparison based on the differences in persistence
for current and alternative jobs is provided. In mathematical
terms the essential choice model calculation is:

$$\frac{(WAGE_T - WAGE_{RT})}{\sigma_r^2} - \frac{(WAGE_C - WAGE_{RC})}{\sigma_r^2} = \begin{array}{l} \leq 0: \text{ NO MIGRATION} \\ > 0: \text{ MIGRATION} \end{array}$$

where $WAGE_T$ = Mean Technician Wages (mid/senior)

$WAGE_{RT,RC}$ = Mean Reservation Wages

$WAGE_C$ = Mean Craft Journeymen Wage

σ_r^2 = Variance of Reservation Wage Distribution

The non-persistence choice does not take into account the
fact that current workers whose value of leisure is exceeded by
the economic incentive for their current job could leave if a
better paying job is found. Therefore the private sector loss
model uses the private sector wage distribution and identifies
the percentage of jobs with higher wages than current wage incentives. The number of private sector job openings are calculated
based on a non-persistence calculation for private sector technicians. In mathematical terms the choice model relies on:

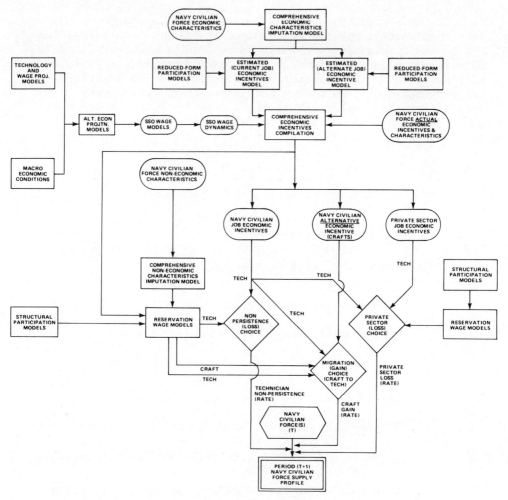

Figure 1.   Economic Labor Supply Modeling System

$$\frac{(WAGE_T - WAGE_{RT})}{\sigma_r} - \frac{(WAGE_C - WAGE_{RC})}{\sigma_r} = \begin{cases} \leq 0: \text{NO MIGRATION} \\ > 0: \text{MIGRATION} \end{cases}$$

where $WAGE_T$ = Mean Technician Wages (mid,senior)

$WAGE_{RT,RC}$ = Mean Reservation Wages

$WAGE_C$ = Mean Craft Journeymen Wage

σ_r^2 = Variance of Reservation Wage Distribution

The non-persistence choice does not take into account the fact that current workers whose value of leisure is exceeded by the economic incentive for their current job could leave if a better paying job were found. Therefore the private sector loss model uses the private sector wage distribution and identifies the percentage of jobs with higher wages than current wage incentives. The number of private sector job openings are calculated based on a non-persistence calculation for private sector technicians. In mathematical terms the choice model relies on:

$$1 - N \left(\frac{WAGE_T - WAGE_{TP}}{\sigma_T} \right) = PERWG$$

$$1 - N \left(\frac{WAGE_{TP} - WAGE_{RTP}}{\sigma_r} \right) = OPEN$$

where OPEN x PERWG = PRIVATE SECTOR LOSS

$WAGE_T$ = Mean Technicians Current Wage (Navy Civilian)

$WAGE_{TP}$ = Mean Private Sector Wage (Technician)

σ_T^2 = Variance of Private Sector Technician Wages

$WAGE_{TP}$ = Mean Private Sector Technician Wage

$WAGE_{RTP}$ = Mean Reservation Wage (Technicians Private Sector)

σ^2 = Variance Private Sector Technician's Reservation Wage Distribution

$N(.)$ = Cumulative (Standardized) Normal Distribution Function

The described econometric labor supply modeling system provided transition (loss/gain) information which was then linked to Navy civilian work force inventories for 1980 to yield forecasted Navy civilian force levels for 1981. The process was repeated for each year up to 1985. Since the resulting force levels reflected only supply considerations, they were subsequently integrated with demand factors. The demand factors and the integration process are described in the next sections.

LABOR SUPPLY ECONOMETRIC ESTIMATES

As indicated in the previous section, reservation wage availability rates were projected for three alternative views of technology and the U.S. economy in the period 1980-1985. The base projection assumes expected technological growth and average economic conditions. The high technology and low technology projections are based on economic assumptions leading to higher and lower relative private sector wages, respectively. These projections are illustrated by the data for senior level male technicians, Table 1. Such estimates were produced for each sex and minority/non-minority category in U.S. Navy's science and engineering technician occupation at the middle and senior levels. The full set of projections will be published in a subsequent report.

TABLE 1

Projection Reservation Wage Availability Rates
For Navy Senior Level, Male Technicians

YEAR	NON-PERSIS-TENCE	LOSS TO PRIVATE SECTOR	GAINS FROM CRAFTS-MEN	YEAR	NON-PERSIS-TENCE	LOSS TO PRIVATE SECTOR	GAINS FROM CRAFTS-MEN
HISTORICAL PERIOD				HIGH TECHNOLOGY			
72	.009	.054	.014				
73	.020	.054	.017	80	.024	.058	.021
74	.020	.061	.018	81	.026	.062	.022
75	.019	.070	.014	82	.028	.066	.021
76	.025	.089	.014	83	.030	.070	.020
77	.018	.093	.011	84	.033	.074	.018
78	.018	.063	.008	85	.035	.079	.017
79	.021	.061	.009				
BASE PROJECTION				LOW TECHNOLOGY			
80	.024	.065	.021	80	.024	.098	.010
81	.027	.075	.021	81	.027	.105	.016
82	.029	.084	.020	82	.031	.114	.012
83	.032	.093	.019	83	.035	.134	.013
84	.035	.102	.017	84	.040	.144	.009
85	.039	.111	.015	85	.048	.153	.006

APPLICATION OF LABOR SUPPLY ECONOMETRIC ESTIMATES

The econometric estimates for historical wage availability based transitions were first compared to actual transition data to evaluate the usefulness of projected estimates in predicting future transitions under the three alternative scenarios. Models were then developed to predict personnel flows based on the

numbers of people in the selected categories and estimated wage
availability rates for these flows. The projected flows and
resulting recruiting requirements for these categories were then
produced for the three alternative future scenarios using a standard manpower planning model with embedded personnel flow rates.[4]
Results showed increasing recruiting requirements to maintain
current strengths under all three scenarios. Embedding the projected flow rates in a manpower planning model allows consideration of additional constraining factors and managerial strategies
in further analysis of these scenarios.

FLOW RATE MODEL

     The population under consideration in this example consists
of nine job categories in the U.S. Navy civilian work force:
Entry, mid, senior, and supervisory levels for science and
engineering technicians; and apprentice, semi-skilled, journeyman,
leader, and supervisor levels for blue collar craftsmen. The
primary focus is on technicians and their response to technological change in the economy. Craftsmen are included as an
important internal source for movement into the technician
occupation.

     The wage availability rates produced by the econometric
analysis for flows from mid and senior level technicians to the
external economy and for flows from journeyman crafts to mid
and senior technician were used to attempt to predict observed
yearly flows over the period 1972-1980. Other factors included
the number in a category at the beginning of a year and the
number eligible for retirement in that category and year. A
least absolute value constrained regression approach was used
to allow inclusion of constraints on the coefficients and
predicted flows[5]. The predicted flow rates from this procedure were then fixed within a larger constrained regression
model used to estimate the remaining flow rates between job
categories.

     Inspection of the data and the results of preliminary models
showed that the observed flow rates did not move with sage availability rates from the same time period. Models with lagged wage
availability rates gave much better results. The data also
suggested models with leading wage availability rates. In such
models projected future wage availability is used to predict
current flows. These models were subsequently discarded as a
result of economic and informational feasibility considerations.
Several large industrial facilities operated by the U.S. Navy
were closed during the 1972-1974 period. These closings produced
turbulence in the U.S. Navy civilian work force that may explain
higher loss rates for those years. The fact that this data
supported the lead model was another reason to discard this model.

The 1972–1974 data were not used in the lagged model finally adopted. This model is:

$$f_{ijt} = m_{ij} \, p_{it} + b_{ij} \, w_{ij(t-1)} + c_{ij} \, w_{ij(t-2)} + e_{ijt}$$

where f_{ijt} = flow from job category i to j, time t

p_i = population in job category i, time t

w_{ijt} = number who are wage available for movement from category i to j, time t (including non-persisters, private sector losses, and internal flows)

e_{ijt} = error residual flow i to j, time t

and m_{ij}, b_{ij}, c_{ij} are parameters to be estimated

The parameters m_{ij}, b_{ij}, and c_{ij}, are further constrained to be between 0 and 1.

Coefficients were estimated separately for men and women with results given in Table 2 below:[6]

Table 2

Coefficients For Personnel Flow Model

Flow	$\hat{m}_{ij}$	$\hat{b}_{ij}$	$\hat{c}_{ij}$
Men			
Mid level technician loss	.069	.000	.312
Senior technician loss	.032	.000	.540
Craft journeyman to senior technician	.005	.062	.000
Women			
Mid level technician loss	.077	.000	.844
Senior technician loss	.014	.564	1.000
Craft journeyman to senior technician	.006	.294	.000

Coefficients could also be estimated for minority and non-minority categories.

The flow rates produced by this model were projected forward through 1987 under the three alternative scenarios. These rates are illustrated for male senior technicians in Figure 2. Actual and predicted flow rates for the historical period are also

LOSS RATE

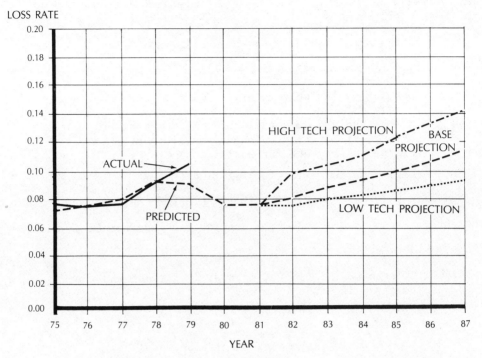

Figure 2. Actual and Projected Loss Rates; Senior Level (GS 9-12)
          Male Physical Science and Engineering Technicians,
          (Dept. of Navy).

indicated. In order to evaluate the projected wage availability
based flow rates in a larger context these flow rates were used
in a standard manpower planning model to determine resulting
future hiring requirements. This model requires flow rate
estimates for all personnel categories in the chosen population.
These rates were simultaneously estimated with a constrained
minimum absolute deviation regression where the flow rates
previously estimated for work wages availability data were fixed
at predicted values. The following model was used to estimate
the flow rates:

$$\text{minimize} \quad \sum_{i,j,t} c_1 \left| \alpha_{ijt} \right| + \sum_{i,t} c_2 \left| \beta_{it} \right| + \sum_{j,t} c_3 \left| \gamma_{jt} \right|$$

subject to: $\quad m_{ij}\, P_{it} + \alpha_{ijt} = f_{ijt}$

(9) $\qquad \sum_{j} m_{ij}\, P'_{it} + \beta_{it} = P_{it}$

(10) $\sum\limits_{i} m_{ij} P_{it} + \gamma_{it} = P_{j(t+1)} - h_{jt} + r_{jt}$

$0 < m_{ij} < 1$

Here f_{ijt}, P_{ijt}, and m_{ij} are as before, h_{jt} and r_{jt} are hires and fires in category j, time t, and α_{ijt}, β_{it}, γ_{jt} are error terms. Relative weights are given by P_1, P_2, P_3. Separate flow rates were estimated for men and women. Expressions (9), (10) represent requirements for flow balance that the predicted flows should satisfy.

HUMAN RESOURCE PLANNING MODEL

     The flow rates produced by this process were then used in the planning model to determine projected annual recruiting requirements for all categories over the 1981–1987 period under the three alternative scenarios. Constant requirements were assumed for this example. These results for mid and senior level male technicians are shown in Table 3.

     The hires for mid–level technicians increase over time by amounts that vary according to the particular scenario, but do not exceed the largest number of hires observed over the five immediately preceding years. However, the senior level technician hires show a marked increase over time in response to technological change in the economy, even for the low technology scenario. In all cases the maximum for the preceding five-year period is exceeded.

     The impact of technological changes in the economy upon the U.S. Navy's work force could be further assessed by adding hiring constraints to the manpower planning model. These constraints would place an upper bound on number of technicians that could be hired from the outside economy under alternate scenarios. Values for these bounds could be estimated by wage availability and past recruiting success rates. These estimates have not yet been developed. The effect of such an upper bound can be illustrated by choosing an arbitrary bound such as the maximum observed number of hires over the last five years for male senior technicians, 529. With this bound we would have significant shortfalls for senior technicians under the base and high technology scenarios. These shortfalls could be met in part by allowing increased flow from other categories, such as crafts journeymen, into the technician occupation. Increased hires for crafts

TABLE 3

Projected Hires (Constant Requirements)

Mid-Level Technicians

YEAR	Base Case	High Tech	Low Tech
81	331	331	331
82	332	332	332
83	341	372	336
84	351	374	341
85	360	385	347
86	370	405	352
87	381	419	359

Senior Level Technicians

81	427	427	427
82	404	435	403
83	453	670	403
84	533	743	442
85	605	813	480
86	680	976	528
87	764	1,076	570

journeymen would then be required. The manpower planning model could be modified to allow flexible flow between selected categories to evaluate this possibility.[7] In this case the projected number of crafts journeymen who would be wage available for the technician jobs would provide an upper bound for these flows. If the models show that technician requirements could not be met through external and internal recruitment then other alternatives should be considered. These would include restructuring tasks and increased reliance on military or contractor personnel.

Finally, the separate flow and wage availability rates developed for each minority and sex group could be used to evaluate the impact of alternative economic scenarios upon the U.S. Navy's equal employment opportunity program.

CONCLUSIONS

We have presented an example of how a new application of reservation wage econometric techniques can be used to project

personnel flows between specific occupations and sectors under
alternative views and economic conditions. Further, we have shown
by example how these projections, based upon supply considerations,
may be integrated with organizational demand factors within a
standard human resource planning model. This integration provides
a new way to assess the impact of projected economic and societal
conditions upon the human resource decisions of a specific
organization.

We have also indicated some ways in which our limited example might be extended to further demonstrate those capabilities.
As more occupations and internal flows are included in the econometric analysis the advantages of using the human resource plan-
ing model with its multi-objective optimization structure as a
framework will become more evident. Additional analysis will
provide hiring constraints that reflect wage availability of
external human resources. The use of flexible flow features in
the planning model will further enhance evaluation of the use of
internal resources.

FOOTNOTES

[1]See chapter 10 of R. J. Niehaus, Computer-Assisted Human
Resources Planning, (New York: Wiley Interscience, 1979) for a
discussion of the issues of integrated human resource planning
systems. The bibliography to this volume provides over 200
references to related work. Also see D. Bryant and R. J. Niehaus,
Eds. Manpower Planning and Organization Design (New York: Plenum
Publishing Co., 1978)

[2]A more comprehensive version of this report will be
prepared at a later date so that all the results will be
explicitly available.

[3]See D. M. Atwater and J. A. Sheridan, "Assessing the
Availability of Non-workers for Jobs", Human Resource Planning,
Vol. 3, No. 4, 1980, pp. 211-218 and D. M. Atwater, R. J. Niehaus,
and J. A. Sheridan "Labor Pool for Anti-bias Program Varies by
Occupation and Job Market", Monthly Labor Review, August 1981

[4]In this case a goal programming model was used. See
A. Charnes, W.W. Cooper, and R. J. Niehaus, Studies in Manpower
Planning Available from the National Technical Information Ser-
vice, Springfield, Va., Accession No. A 066952

[5]Constrained regression models were introduced in A. Charnes,
W. W. Cooper and R. Ferguson, "Optimal Estimation of Executive
Compensation by Linear Programming", Management Science, 1,
No. 2, January 1955, 423-430.

[6]There are no coefficients for flows from craft journeymen
to mid-level technician because wage availability rates were
always zero.

[7]See A. Charnes, W. W. Cooper, K. A. Lewis, and R. J. Niehaus,
Eds., Manpower Planning and Organization Design (New York: Plenum
Publishing Co., 1978) and A. Charnes, W. W. Cooper, A. Nelson
and R. J. Niehaus, "Some Prototype Studies in Goal-Arc Network
Approaches for EEO Planning", INFOR (forthcoming)

5130

PERSONNEL PLANNING: THE IMPORTANCE OF THE LABOUR MARKET

Richard Pearson

Institute of Manpower Studies

Brighton, U.K.

ABSTRACT

Employing organisations cannot isolate themselves from their external environment. Many of the techniques and approaches currently used by personnel planners, however, focus specifically on the internal management of human resources. This paper illustrates some of the main ways that the external labour market can impact on the organisation, and suggests the need for personnel planners to better understand such interactions. It goes on to identify the main information needs for personnel planners to do this and the sources of the relevant data. It concludes by recommending that personnel planners widen their perspective and planning processes to take account of the growing importance of the external labour market.

INTRODUCTION

Employers are becoming increasingly aware of the need to manage their manpower resources more effectively in a fast changing society. The approaches and techniques available to personnel planners to help them assess their organisation's manpower situation and improve their manpower management are numerous and well-documented. Rapid advances in computing facilities and parallel reductions in costs allow an increasingly wide range of employing organisations to install computerised information systems and to take advantage of manpower models that deal with such complex issues as career planning and analysing the effect of differing

wastage and recruitment patterns (1). The vast majority of these
techniques, however, centre on the management of manpower resources
within the organisation. There is now a growing realisation that
employing organisations cannot isolate themselves from their external
labour market. If successful manpower management policies are to
be adopted in the future, due consideration needs to be given to
the interaction between the organisation and its external labour
market.

     This paper illustrates the main ways in which an employing
organisation can interact with the external labour market and suggests
that personnel planners need to understand such interactions better
in order to improve their own planning processes. Such an approach
requires a detailed knowledge of both the existing manpower resources
within the organisation and the nature of the external environment
in which they operate. The paper therefore goes on to identify the
main types of information needed and illustrates some of the
key sources of information available, both within the organization,
and from external agencies in the U.K. The paper concludes by
suggesting the need for personnel planners to understand and monitor
changes in both their internal and external labour markets if they
are to improve their planning activities.

THE ORGANISATION AND THE LABOUR MARKET

     Employing organisations cannot isolate themselves from their
external environment. Their minimum involvement is that they will
be seeking to recruit some new staff in the future. Knowledge of
the external labour market not only helps them to identify whether
suitable candidates are likely to be available, and at what price
and under what conditions, but can also ensure that the recruitment
policy is cost effective in being directed to the most appropriate
parts of the labour market.

     However, the labour market is not only the source of future
recruits, it also has a major impact on such supposedly "internal"
employment patterns as wastage rates and absenteeism, on pay negotiations and on industrial relations. For example, in the same way
that the current economic climate is conducive to lower pay settle-
ments, lower wastage rates and the like, so any increased local or
national demand for manpower - e.g. due to economic upturn or a new
office complex being built locally - is likely to have the opposite
effect. It is, therefore, not only the recruitment specialists who
need to be aware of such changes, but also personnel and line
managers if they are to manage their manpower resources effectively.

Even if the organisation is in an apparent state of equilibrium,
with well established manpower policies, changes are likely to be

required in the future as a result both of internal developments -
e.g. the introduction of new technology requiring the recruitment
of new skills, or the need to retrain existing staff - and of
externally generated change. An example of the latter is the marked
reduction in the number of school leavers available for recruitment
in many areas next year and for the next decade due to the fall in
the birth rate 18 years earlier.

The essence of good planning is to anticipate, to be prepared
for change and to try and avoid or minimise future problems, not
wait for them to arise and then seek their resolution. What then
are the issues and changes that personnel planners may face in the
future, what information is available by which the implications of
such changes on the organisation can be assessed, and how can they
monitor and be prepared for such change?

Let us look first at events and issues generated from within the
organisation. In recent years the Institute of Manpower Studies has
been involved with personnel planners seeking to answer the follow-
types of questions for their own organisations:

-    Will we be able to recruit enough school leavers in the future?

-    Will our current recruitment problems over electronics engineers
     get better or worse?

-    Will we be able to rely on a continuing supply of trained tech-
     nicians leaving the armed forces in the future to meet
     recruitment needs?

-    Which of three possible locations for a new factory will offer
     the best recruitment opportunities?

-    Will we be able to staff a new hospital that is being built?

-    With impending company redundancies how easily can the local
     labour market absorb the redundant workers?

-    Wastage rates are rising; is it because of increased local
     competition, declining relative pay or what?

-    Absenteeism in one factory and location is particularly high;
     is this due to conditions in the factory, or is it a feature
     of the location, the industry, or what?

Such questions may well prompt a special enquiry or study by
the personnel planner involving analysis of the local or national
labour market. Knowledge of the information available, its applicability and reliability, will be of crucial importance.

There are, however, many changes going on outside the organisation which could also have a major impact on future manpower
policies, and especially recruitment policy, and it is equally
important to be aware of these. Some of the main ones which are
already apparent may be generated by:

-      Demographic Change: for example, the growth in the size of the
       workforce is likely to continue until the 1990's. Much of
       this increase is likely to be accounted for by married women
       entering the labour force which suggests possibly a new source
       of recruitment. At the same time, however, the number of
       school leavers will fall by 35% over the next decade, suggest-
       ing potential recruitment difficulties.

-      Educational Change: the reduction in the number of qualified
       school leavers is a result of demographic change, but this
       factor together with constraints on public expenditure may
       also mean a sharply reduced number of vocationally trained
       graduates qualifying in the latter half of the decade (2).

-      Technological Change: the continuing decline of the traditional
       industries is likely to increase the number of unskilled male
       workers becoming unemployed. If future growth is in the
       service sector and the "new technologies" such people may not
       be suitable recruits, leading to the coexistence of high
       unemployment and skill-shortages.

-      Economic Change: the current recession has seen the intake of
       apprentices fall nationally by as much as 40 per cent or more
       for some occupations. Will this mean a major shortage of
       skilled craftsmen and technicians five or ten years ahead?

-      Social Change: high levels of unemployment will increase
       pressure for some form of work sharing, which together with
       pay bargaining may result in a rapid reduction in basic
       working hours, increased holidays and restrictions on overtime
       working.

The impact of such changes on organisations will vary greatly
according to their industry, location, occupational composition and
so on. Personnel planners will need to assess the implications of
these and other changes and monitor them in the future if they are
to recruit and manage their manpower resources effectively.

THE INTERNAL LABOUR MARKET

Given the potential range of pressures developing both internally and externally, how can an organisation understand its
relationship with the external labour market, and respond to, and
manage changes, such as those illustrated above? In the same way
that the internal labour market is not a homogeneous grouping of
people but subdivided into a number of subsystems characterised by
occupations, grades, locations and departments, so is the external
labour market a series of interconnecting "sub-markets", which
develop and respond to change in different ways. The key, therefore,
to developing an understanding of this relationship is to know the
constituent parts and make up of both the internal and the relevant
external labour markets.

The first need of the personnel planner is to know the composition of his own workforce and the relative numbers and importance of
each of the sub-groups, and then those points of direct interaction
with the labour market through wastage and recruitment. Looking at
the way in which people "flow" about the organisation and in which
they can be clustered into broadly homogeneous groups is known as
analysing the manpower system. By drawing these "homogeneous"
groups of people as boxes, and the flows (recruitment, promotion,
transfers and wastage) as arrows, the manpower system of any organisation can be diagramatically represented. (See Fig. 1 for an
example). Knowledge of the relative sizes of these boxes and flows
in terms of numbers of people, the skill groups involved, together
with their relationship to the activity of the organisation, can be
used to help determine the most important features.

Perhaps the most common reason for wanting to know about the
external labour market is in relation to recruitment policy and
its likely effectiveness for one or more "occupational" groups.
This aspect of personnel planning is therefore used in the subsequent parts of this paper to illustrate the need for, availability
and usefulness of labour market information.

Having identified those groups of most interest or concern,
e.g. there may be an increased need for craftsmen, the next need
is to know which part of the external labour market is likely to be
of most relevance when considering the future recruitment strategy
and prospects. Here, there is a need for two separate pieces of
information. First, what type of person is thought most suitable
to fill those posts? Analyses of data on past recruits and their
subsequent performance will help identify the types of candidate
thought most suitable. Thus the "successful" craftsmen may come
from two alternative sources, those who are ready trained, the
others via an apprentice intake for which the most suitable
candidates are school leavers. Second, what is the catchment area
on which the recruitment strategy should be focussed? There is no

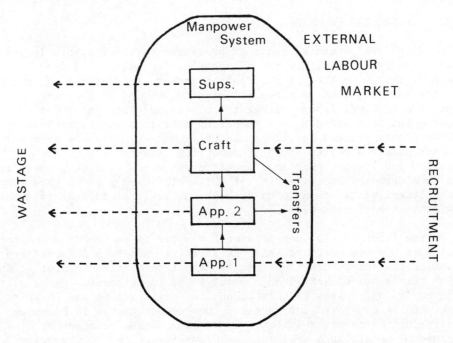

Figure 1.   The Manpower System and the Labor Market

sense in national advertising if all the recruits are likely to come
from within a ten miles radius. Analysis of the home addresses of
existing staff in these categories will show the most heavily used
labour markets but may reflect the recruitment pattern of five or
more years ago, the home addresses of recent recruits will show where
the latest recruitment campaigns have been most successful (and the
addresses of applicants will show where it has had most initial
impact).

Knowledge of the existing catchment area can, however, suggest
two potentially conflicting conclusions. First, that it is the
proven, best and only source of recruits or, alternatively, that
certain locations provide the vast majority of recruits simply
because previous recruitment activities may have missed others
altogether. The correct conclusion can only be identified by reference to the external data. If there are not enough existing staff
on which to base such an anlysis, then reference to published data
on travel to work patterns in the locality can provide an alternative
bench mark.

As the main source of such data, the Census of Population (3),
is now ten years out of date, the patterns shown will probably need
reassessing in the light of known changes in transport facilities,
roads, competing firms, housing and the like, to bring the data up
to date. Other research studies can illuminate the picture for
particular groups and localities. For example, the National Travel
Survey (4) shows that over 80 per cent of all employees lived within
a radius of eight miles of their place of work and within 35 minutes
travelling time. It also revealed major differences between the
patterns of the sexes and between occupational groups. For example,
professional managers had a travel to work distance of eleven miles,
while the average for all women was only five miles.

These are figures for existing workers. It is likely that
most new recruits will also come from within a similar catchment
area as it is only a small minority of workers - mainly in managerial
and professional occupations - who are prepared to move house in
order to change job. The craftsmen and school leavers in the example
above are both likely to be operating in a fairly small local labour
market. If there is to be "local" recruitment and in the past few
applicants or recruits have come from certain localities, can the
reason be identified, e.g. a lack of advertising in the relevant
local paper, or poor transport facilities, making the company inaccessible (only one in three semi or unskilled workers travelled to
work by car in the mid-1970s) and the problem remedied? Use of
internal data and where appropriate reference to external material
can pinpoint the answers to these two key questions - what type of
recruits are most suitable and from what catchment area are they
likely to be drawn? Assessment of their likely availability in the
future requires the use of externally generated data.

THE EXTERNAL LABOUR MARKET

The range of potentially useful information produced about the
external labour market is enormous yet the coverage of the published
data is still incomplete. Much of it is collected for the purposes
of government and its subsequent dissemination only a subsidiary
activity of the statistical services. As a result, the data of
most relevance to the personnel planner may be scattered across
many different publications and in a format that needs careful interpretation. For example, statistics relating to the supply of people
leaving the education system - schools, colleges and universities -
are spread across 15 separate volumes, each published at a different
point in the year, by many different organisations and sometimes up
to four years in arrears. In addition there are numerous sources of
unpublished information as well as a multitude of ad hoc studies
that may be of relevance. How then is the personnel planner to find
his way around all these sources and to know which is most relevant?

```
┌─────────────────────────────────────────────────────┐
│           Population and Projections                  │
│           Activity Rates                              │
│           Travel to Work Patterns                     │
│                                                       │
│           Employment Structure                        │
│           Local Employers and Competitors             │
│           Unemployment                                │
│           Vacancies                                   │
│                                                       │
│           School Leaver Supply                        │
│           College & University Graduates              │
│           Youth Opportunities Schemes                 │
│                                                       │
│           Redundancies                                │
│           Earnings and Costs                          │
│           Trade Unions                                │
│                                                       │
│           Education Facilities                        │
│           Training Facilities                         │
│           Labour Mobility                             │
│           Housing                                     │
│           Transport                                   │
└─────────────────────────────────────────────────────┘
```

Figure 2.   Key Labour Market Information. Source: (6)

Figure 2. illustrates the main items of information likely to
be of relevance in assessing a labour market. Work over the last
two years, for the IMS Co-operative Research Programme, has resulted
in a guide to all the major national and local sources of information (5), as well as a digest of the most important educational
statistics for personnel planners (6); so the key sources of
information have now been documented for the U.K. In the case of
a local labour market analysis then it is useful to proceed through
a five stage process covering

   - the size and nature of the local population

   - the structure and level of employment

   - the educational supply and facilities

   - employment conditions and the

   - infrastructure

taking note both of the current position and the trends and changes
over recent years.

First the characteristics of the local community; is it densely
or sparsely populated, is the population young or old, expanding or
contracting? What proportion of local people are in, or available
for work, and where do they normally travel to work? This information is available at quite a detailed local level but unfortunately
is only collected via the decenial Census of Population (3) and can
therefore be considerably out of date. Estimates are, however,
made of population changes in the intervening periods by the Office
of Population Census and Surveys, who also provide projections of
future trends. Likewise, more regular estimates are made of activity
rates for women by the Department of Employment while special studies
may be carried out by local authorities or other groups who can
provide further, more detailed, local information. Such an analysis
can show how the overall level of labour supply is changing in the
area in question. For example, in inner cities the working population may be falling while in the New Towns and many other areas the
rate of expansion may exceed the national average.

Having identified the overall labour supply, the next stage is
to look at the nature of the local employment and level and type of
demand from other employers. Unfortunately, it is in this respect
that the local data are weakest. Information is available about the
industrial structure of an area and changes in recent years, but
nothing, beyond the Census (3), is generally available about the
occupational breakdown. There are, however, good data on local
unemployment rates by occupation (as well as by duration and age)
and on vacancies notified to the employment service (the latter
represent about one in three of all vacancies). Local can be defined
in many ways and it should be noted that there can be remarkable
variation even within an apparently small area, e.g. in a recent
study of Stansted (7) some towns had unemployment rates two and a
half times that of other towns, all within a 20 miles radius.
Changes in unemployment and vacancies over the last few years can
show how demand for labour has varied and due account should be taken
of reports and information about labour shortages and their osten-
sible causes in the area. Other particularly useful information,
especially in assessing possible competition, can be found through
a listing of local employers and their relative sizes, types of
employment and possibly their current recruitment levels. For
example, if the locality is dominated by public service employers
their recruitment needs are likely to be very low in the short term,
but if there is a predominance of electronics firms or defence
contractors then competition is likely to get more severe.

If recruitment is likely to focus on school or college leavers
then relatively good data are available locally from careers services
or local authorities about the numbers leaving the education system
in the next few years, their broad levels of educational attainment
and in some cases the historic pattern of subsequent employment or

otherwise, e.g. the numbers going into apprenticeships, further or higher education or into different types of jobs. In times of high unemployment the fact that special Youth Opportunities Schemes may cover a major proportion of local school leavers means that this is an additional recruitment source or may indeed be a source of competition for those who have just left school.

The above are the most relevant data likely to be available locally by which recruiters can assess the availability of future recruits. Other information may be available on a less complete basis (but can nevertheless be of importance) and includes that relating to local earnings, and conditions of employment, while the pattern of redundancies may highlight the structural changes taking place locally. If trainees are being recruited, or recourse is made to taking on recruits who require further training, then knowledge of the local further education and training facilities will also be important. Similarly if an attempt is to be made to recruit workers from outside the area then knowledge of local housing conditions and its potential availability for newcomers will also be vital.

At a local level much of the above information can be collected rapidly. Local authorities, through their structure plans and planning departments, collate and publish summaries of much of this information, including an indication of the types of new employment that are or are not being encouraged e.g. on new industrial estates or office complexes, together with plans for new housing or population expansion. The Regional Manpower Intelligence Units, and Jobcentres (of the MSC) offer further help and information, especially about the current state of the labour market. There are many other local and national organisations (5) who also have and make available labour market information. Finally, much can also be done on a co-operative basis. Many informal groupings exist to discuss such things as salaries and industrial relations while the notion of "local labour market consortia", groupings of local organisations who collate and share labour market information of mutual interest, based on the California model, are to be encouraged. (8)

CONCLUSIONS

Employing organisations cannot isolate themselves from the labour market. There exists a wide range of information, both within and outside the organisation, that can be drawn together to provide a better understanding of the way organisations interact with their labour market(s). The key issue facing personnel planners is what information is it most useful to collect, what will it cost and how long will it take? Obviously the answers to these questions

depend on the specific need for the information and the circumstances
of the organisation. There is certainly no sense in collecting manpower information for its own sake. However, evidence from previous
IMS studies of local labour markets suggest that an investment of
about ten man days can provide a much enhanced perspective of an
organisation's position in its local labour market, while an
investment of a few hours a month reviewing key publications and
regular informal contact with local manpower agencies can yield
a high return in being able to anticipate and prepare for future
pressures and opportunities.

The key needs for personnel planners then are:

i)   to understand the manpower system of their organisation

ii)   to monitor its main changes and critical features

iii)   to know the external labour market(s) in which the organisation
       operates and the key points of interaction

iv)   to monitor those aspects of the external labour market likely
       to have a significant effect on the organisation (an activity
       coming to be known as environmental scanning) and finally

v)   to know the availability and usefulness of other manpower
       information so that it can be readily utilised when a new
       problem or issue needs analysis and resolution.

The labour market is an imperfect, changing place. A broad
understanding of its operation, together with an ability to
analyse and interpret key changes is an essential input to any
personnel planning process.

REFERENCES

1. Bennison, M., The IMS Approach to Manpower Planning, Institute
    of Manpower Studies, Sussex, 1978.
2. Pearson, R., Education to Employment, the Mix and the Market,
    Personnel Management, London, July 1979.
3. OPCS, Census of Population, HMSO, London 1971.
4. OPCS, National Travel Survey, HMSO, London, 1976.
5. Walsh, K., Izatt, A., Pearson, R., The U.K. Labour Market,
    the IMS Guide, Kogan Page, London, 1980.
6. Hutt, R., Parsons, D., Pearson, R., Education and Employment
    1980, New Opportunities Press, London, 1980.
7. Pearson, R., The Impact on an Airport on a Local Labour Market,
    Personnel Management, London, April 1981.
8. Grimm, K.D.G., An International Perspective of Local Labour
    Markets, Manpower Studies, Sussex, Spring 1981.

A SYSTEM FOR ASSESSING THE FEASIBILITY

OF U.S. NAVAL OFFICER MANPOWER PLANS

Murray Rowe and Joe Silverman

Management Systems R&D Program
Navy Personnel Research and Development Center
San Diego, California   92152

ABSTRACT

The Structured Accession Planning System for Officers (STRAP-O) is a set of models for determining the feasibility of proposed manpower plans or programs and indicating directions likely to achieve those plans. STRAP-O integrates accession and promotion planning functions with manpower requirements both technically and organizationally. This paper describes the system's components, its architecture, the flow of information, and outputs.

INTRODUCTION

Among corporate managers, there is a growing recognition of the need for "strategic planning". This "course plotting" involves anticipating future products, services, and markets. It is then necessary to evaluate production capacities and productivity, capital improvements and acquisitions, and labor and material availability, among other factors, to accommodate expected demand. An accurate determination of the appropriate path to follow has often been made more difficult by organizational constraints, but also by the lack of sufficient management tools, typically in the form of computer-based corporate planning models. Companies are often functionally organized to meet current operating needs (sales, advertising, operations, etc.), but have limited interaction among the functional groups when mid- to long-range plans (5-7 years out) are considered. While many of these organizational constraints remain, advances in computer technology have reduced the barriers to widespread use of corporate planning models. This has resulted from more sophisticated software, as well as reduced computer turnaround that enhances the value of information provided for decision-making.

The U.S. Navy's military personnel system is an organization with similar strategic management requirements. Annually, the Navy participates in an exercise called "programming", where evaluations of future Navy missions (3-7 years hence) are made in light of resource constraints. A critical event during the manpower portion of the programming process is the assessment of alternative officer force sizes and configurations. To properly address this issue requires the simultaneous consideration of officer <u>manpower requirements</u>, the projected <u>supply</u> of officer candidates and those in the training "pipeline", and the existing and projected officer <u>personnel inventory</u>. A parallel issue concerns the adjustment of the personnel inventory to achieve desired officer manpower levels.

Conceptually, all of these officer management functions are interrelated, but they are organizationally distinct and lack the coherent linkages necessary to adequately and rapidly respond to programming questions. Each of the three key areas of officer management are represented by numerous organizational units, by models that make conflicting assumptions and provide different degrees of detail, and by incompatible data support. This can result in costly disagreement over administrative and procedural matters, rather than consideration of substantive manpower questions. In programming, where a timely response is often the only acceptable response, the Navy suffers from a <u>linear</u> approach to manpower issues, where one organizational unit manually passes its response on an issue to the next group, and so on. To come closer to the way the system really works and improve response time, officer manpower management must function dynamically using a common set of models, data, and policy parameters.

Such a linked set of models is currently under development at the Navy Personnel Research and Development Center (NPRDC). It is referred to as the Structured Accession Planning System for Officers (STRAP-O).[1] Structurally, STRAP-O will be an integration of existing and planned officer manpower and personnel management models. These models could be used independently in the planning phase (8-15 year horizon) or for budget execution (0-2 years out), but as part of STRAP-O, the emphasis is on programming.

This paper describes the components of STRAP-O, the system architecture and flow of information between modules, and resulting outputs. The operation of the system will be described by a brief scenario.

---

[1] A similar system, for U.S. Navy enlisted personnel, the Structured Accession Planning System, Enlisted (STRAP-E), is operational and described in Silverman, et. al. (1979).

STRAP-O SYSTEM COMPONENTS

The main purpose of STRAP-O is to determine if a desired force level of officers is feasible in terms of expected attrition, the number of accessions required to achieve force levels, the available supply of officer candidates to support accessions, the promotion (between pay grades) and lateral transfer (among skills) plans required, pay grade limitations, the demands on the training establishment, and the manpower overhead (students, transients, patients, etc.) needed to sustain the force.

Since the purpose of STRAP-O is to assess the feasibility of attaining and maintaining alternative manpower levels, its central focus is on the personnel inventory and accessions necessary to achieve those manpower requirements. This aspect of the STRAP-O system consists of two models -- the Accessions Into Designators (AIDS) model and the Officer Projection Model (OPRO). (The entire STRAP-O system of models and their relationships are diagrammed in Figure 1.)

AIDS is a goal programming model that determines the optimal number of officers to access each year (for up to 10 years) from each commissioning source to achieve future force structure goals specified by total force, as well as by individual "community" (e.g., aviation, surface warfare) (Bres, et. al., 1979). Specifically, the Navy operates a number of commissioning programs to meet its officer manpower requirements. Officers produced by these programs enter a variety of occupational specialty areas. The U.S. Naval Academy (USNA) and Naval Reserve Officer Training Corps (NROTC) programs supply officers for a wide range of specialty areas, while other sources, such as the Naval Flight Officer Candidate (NFOC) program, supply a single specialty area. It has been observed that career continuation behavior differs according to an officer's commissioning source and specialty (Goudreau, 1977). Thus, the choices of which commissioning programs to use and how to distribute officers from these programs to occupational specialties have a major influence on the Navy's ability to meet future requirements for experienced officers. Commissioning programs also differ in other ways important for planning purposes -- cost, capacity, and length of training.

Prior to AIDS, accession plans were developed by each community independently. Those plans often identified a "choke point" in the career path, such as the department head tour, where requirements exceeded projected inventories for a critical assignment. Accession plans were developed to meet those "choke points". An overall accession planner then brought the individual community plans together in what proved to be a protracted negotiation process to produce an all-Navy plan.

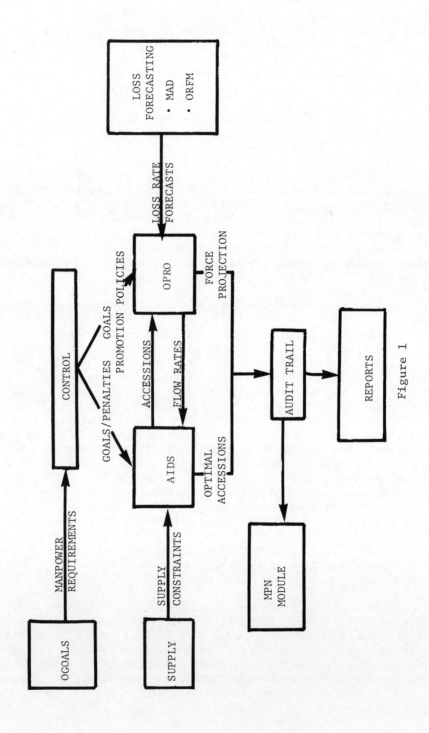

Figure 1

In contrast, AIDS simultaneously considers both community and all-Navy needs. The model identifies all feasible commissioning mixes and projects them forward (for 10 years) using continuation rates for each source and community combination. These inventories are then compared to requirements. Penalties are placed on deviations of inventories from requirements. The optimal commissioning mix is the one that minimizes the sum of penalties. Penalties are specified by the user to reflect priorities. For example, it may be more important to access to meet all submarine requirements before meeting half of aviation's needs.

The fundamental feature of STRAP-O is the simultaneous consideration of accession and promotion plans designed to achieve a particular set of manpower goals. This requires linking the accession planning model, AIDS, to the force projection model, OPRO, so that the flow rates used in AIDS do not simply reflect historical promotion policy and loss behavior, but the promotion policies under consideration and losses expected to occur in the planning horizon. In turn, an optimal accession plan will be passed from AIDS to OPRO. This process of determining accessions and flow rates will continue until the system "converges" -- the two models are flowing personnel in the same ways, hence operating as a system.

The force projection model, OPRO, is designed to forecast personnel flow behavior of officers as they are gained or lost to the system, are promoted, and "aged". OPRO is a "fractional flow model with foresight" (Grinold and Marshall (1977)). That is, it anticipates vacancies, and accesses or promotes enough replacements to fill the vacated positions. By imposing management intentions concerning accessions, expected losses, and a promotion policy, it forecasts and summarizes the personnel flows in terms of continuation rates. Those rates then become inputs to AIDS.

Loss rates, as input to OPRO, are currently forecast using an NPRDC-developed autoregressive time series, minimum absolute deviation (MAD) technique (Bres and Rowe, 1979). This will be satisfactory in an initial version of STRAP-O, but because it has no policy-testing capability (it simply projects historical data), it will give way to the Officer Retention Forecasting Model (ORFM) upon completion. ORFM is intended to estimate changes in loss behavior that are expected to occur in specific communities as a result of changes in Navy and Department of Defense programs (e.g., new or improved continuation bonuses; altered retirement benefits); exogenous events (e.g., external employment conditions, civilian wage levels); and changes in the composition of the officer force (e.g., ethnic and sexual mix, education level).

The model will attempt to capture an officer's expected life-stream earnings from the decision to remain in the military and a decision to return to civilian life. By mathematically relating the two earnings streams, as well as other economic (e.g., employment conditions) and non-economic (e.g., sea/shore rotation patterns, married/not married) variables to the current and historical continuation rates, estimates of future continuation rates can be made given the earnings streams implied by alternative compensation policies. Within STRAP-O, ORFM will permit the testing of compensation and other retention-oriented policies as means of modifying officer personnel inventory relative to requirements.

Finally, an officer supply forecasting model will be developed. It will project the expected number of qualified and available candidates to the various commissioning sources. This information will serve as constraints within the AIDS model.

The outputs of the STRAP-O system will be of two types--printed output and data stored on disk or tape. Printed output will contain information that can be controlled by the user. In addition to user input parameter data, the accession plan and a force structure matrix (paygrade by length of service) for each community for each year will be displayed.

The stored output provides input for subsequent runs and supplies several peripheral programs (MPN module, accession scheduler). It will also save data for other STRAP-O reports on a "audit trail". Examples of audit trail reports would include losses, promotions out, and laterals in, among others.

An important step in the Navy's programming process is the "roughly right" estimation of the size of the officer manpower budget associated with alternative personnel force structures. To properly cost the manpower configurations generated by OPRO, it is necessary to determine the budget consequences of promotion behavior and longevity distributions of personnel strength. To accomplish this, the MPN (Military Personnel, Navy, the manpower budget account) module produces budget line item (e.g., basic pay, flight pay, etc.) and total budget cost estimates given an officer force structure from OPRO.

The primary input to the STRAP-O system will be a set of manpower requirements. These requirements are determined as a function of workload. "Workload" is represented by the operation and support of existing fleet systems (e.g., ships) or the additions of new hardware systems at points in the planning horizon. It is these manpower requirements that describe the personnel implications of alternative Navy missions. During the programming phase, managers must modify accession and promotion policies, and expected losses to attempt to achieve these requirements.

The requirements are composed of two groups: (1) operational and support billets, known as "structured spaces" -- jobs directly associated with accomplishing service missions, and (2) manpower

overhead billets, or "unstructured spaces", such as students/
trainees, transients, and patients. These billets are not involved
in accomplishing a mission, but are necessary to keep structured
spaces filled. Officer "structured spaces" are further defined as
a combination of community warfare billets (e.g., pilot) and managerial billets that can be filled by an officer from any community.

It is the aggregation of these manpower requirements that form
the goals that the accession planning model, AIDS, and the force
projection model, OPRO, attempt to achieve. To insure that STRAP-O
is highly responsive, the process of taking a simply-stated set of
requirements from the user (at best, a set of structured spaces by
grade and community or a total strength requirement) and converting
it to a comprehensive set of goals will be completely automated in
the OGOALS program.

As a "front-end" to AIDS, OGOALS will perform a three-step
process: (1) translate the warfare-specific structured spaces from
their grade/community dimension into a community/experience level
(specific tours or length of service cells) dimension employed by
AIDS, (2) the remainder of the structured spaces, the managerial
billets, will be allocated to the specific warfare communities and
then translated as above, and (3) the manpower overhead, as a function of the size and configuration of the structured spaces, is
applied.

If the user-supplied total strength requirements, OGOALS would,
working backwards, estimate the portion of that force needed for
overhead spaces and then divide the remaining, structured spaces
into community specific goals.

Finally, the development and use of the STRAP-O system of models
implies the need for large data bases and extensive data processing
activities. Each model requires data in certain formats and employs
parameters which need to be validated and reestimated periodically.
In terms of total cost, the models themselves represent the tip of
the iceberg -- the cost below the waterline is generated by the data
processing requirements. Without systems of data support, the
STRAP-O system would soon lose its design capabilities. Currently
under construction is an automated data processing procedure, ODIN,
designed to create and maintain personnel inventory and loss files
and produce a variety of outputs. These outputs will, in turn,
become inputs to the STRAP-O models.

STRAP-O SYSTEM OPERATION

The STRAP-O system is designed to assist in personnel budget
planning over a seven (7) year period. This planning horizon is
divided into two sub-periods. The first is a two-year strength
planning period which establishes a predicted inventory for the

beginning of the second sub-period referred to as the POM or Program
Objective Memorandum years--the programming phase.[2]

The STRAP-O program will operate on an actual beginning inventory of personnel for the current fiscal year and project it (via
OPRO) two years into the future using a combination of historical
and user-supplied parameters to estimate gains and losses. The
purpose of this simulation is to establish the planned beginning
inventory for the POM years, which is then used as input to the
other STRAP-O models. The processing of the inventory through the
POM years involves interaction between OPRO and the other modules
of STRAP-O, particularly AIDS. Again, historically-derived and
user-supplied parameters are used to control the scenario for each
run of the system.

An example of a programming exercise will help illustrate how
the STRAP-O system is expected to function. Suppose two new aircraft
carriers are being considered for commissioning 5 to 7 years from
now. They will bring with them requirements for additional officers
with specific skills and experience and still more officers in the
pipeline to support those in the structured spaces. The programming
issue is, given current inventory, expected losses, promotion plans,
and planned accessions over the next 5-7 years, can the increased
demand for officers be accommodated? Then, if not, how can the personnel inventory be affected to come closer to meeting the manpower
needs?

STRAP-O would begin by estimating loss rates in either MAD or
ORFM, and then applying them to current inventories in OPRO. Some
accession level is specified (which, initially, may come from a
source other than AIDS) along with a desired promotion plan and pay
grade limits (from OGOALS),[3] and OPRO generates continuation rates
describing officer personnel flow behavior under these policy specifications. The rates are passed to AIDS and an optimal accession
plan, one that comes as close as possible to meeting the newly generated manpower requirements, is determined. The derived accession

---

[2]Programming scenarios are not carried out for the first sub-period
(current year + 1), because the force for those years is usually
"locked in" by previous programming and budgeting actions.

[3]The operations of STRAP-O assumes the intent to achieve some goal
or set of goals. Like AIDS, OPRO too has goals. It requires a
strength goal by pay grade (called "prescribed number") to use as
a target for computing vacancies and resulting promotions. This
strength target could be interpreted as authorized (paid for)
strength or desired requirements. However, it must be consistent
with the goals that AIDS is trying to satisfy through accessions.
To ensure this consistency, both AIDS and OPRO derive their goals
from OGOALS.

mix is sent back to OPRO and the force is once again projected, and the new and different continuation rates are returned to AIDS. This process of determining accessions and flow rates continues until the system "converges". Technically, it means that AIDS is producing the same accession plan and the same projected personnel force structure as OPRO in two successive iterations. Practically, it indicates that AIDS and OPRO are flowing personnel in ways which reflect identical loss and promotion behavior. In other words, AIDS and OPRO are operating as a system.

CONCLUSIONS

Within the U.S. military, where personnel supply (both new accessions and retention of existing personnel) is of real concern, there is a growing awareness of the need to assess the manpower feasibility posed by future missions (e.g., a larger fleet; more sophisticated weapons). The STRAP-O system represents a synthesis of existing personnel planning models with new techniques to meet this need. It is anticipated that a limited version of STRAP-O will be tested by the fall of 1981.

REFERENCES

Bres, E., Burns, A., Charnes, A., and Cooper, W., Optimal officer accession planning for the U.S. Navy (NPRDC TR 80-5), San Diego, CA.: Navy Personnel Research and Development Center, November 1979.

Bres, E. and Rowe, M. Development and analysis of loss rate forecasting techniques for the Navy's Unrestricted Line (URL) Officers (NPRDC TR 79-20), San Diego, CA.: Navy Personnel Research and Development Center, June 1979.

Goudreau, K. Retention of naval officers by source designation (NA76-0344.20), Arlington, VA.: Center for Naval Analyses, May 1977.

Grinold, R. and Marshall, K. Manpower Planning Models, New York: Elsevier North-Holland, Inc., 1977.

Silverman, J., et. al. Operations guide for the STRAP system: A structured accession planning system for computer-based manpower programming, San Diego, CA.: Navy Personnel Research and Development Center, October 1979.

CORPORATE TRAINING INVESTMENT DECISIONS*

Dieter Sadowski

Department of Economics and Management Science
Universität Trier
D-5500 Trier
West-Germany (FRG)

ABSTRACT

After presenting some economically surprising features of the
German apprenticeship system, the paper constructs a model of the
functions of corporate training investments, based on the hypothesis that by providing and financing training, firms may gain reputation in the labor market. This in turn induces even generally
trained employees to reduce voluntary quitting and attracts employees from outside. Besides favoring the recruitment process,
training is also assumed to improve the firm specific performance
of the trained personnel to some degree, a consequence which is
traditionally thought of as being the main rationale for providing
training. A dynamic planning model is formulated and illustrated
that formally resembles new models of budgeting the advertising
and R&D expenditures.

INTRODUCTION

In Germany apprenticeships are a major element of the *"educational"* system; in 1979 about 43% of all young people aged 15-18
were apprentices. Although education is compulsory up to the age
of 18, apprentices spend only about one fifth of their learning
time in class-rooms, because they are mainly trained and educated
in private or public firms, be it on or off a workplace. Among all
the manufacturing firms only 17% provide training as opposed to

---

*This work was supported by a grant from the German Research
Society.

25% of the artisan firms. The ratio of apprentices to the total
number of employees varies widely between firms of different size
and industry. Most of the firms with more than 2oo employees provide at least some apprenticeship training. No proportionate relationship or any other simple pattern which could explain these
variations has been recognized as yet. As the German apprenticeship system is predominantly nonunionized, no obvious institutional feature can be held responsible for these observed variations.

In 1972 a federal task force examined the cost and financing
of vocational training and found that, in some of the occupations
highly preferred by the apprentices, many training firms were able
to gain a net profit by having supplied training possibilities.In
the case of clerical workers and electrical fitters, for instance,
about 2o% managed to do so. At the same time,however, the task
force observed that the provision of training in some of the
other most preferred occupations, such as toolmakers or banking
clerks, is rarely if ever profitable up to the end of the apprenticeship but leaves the firms with a substantial net loss. (Sachverständigenkommission 1974). Given, first, the legal interdiction
of long-term contracts between the apprentices and the training
firms and, second, permanent and considerable competition for
journeymen even in recession periods, the training behavior of these
seemingly 'losing' firms needs an economic explanation. I shall
try to provide an explanation taking apprenticeship expenditures
as an example for corporate training investments in general.

THE ECONOMIC THEORY OF CORPORATE TRAINING INVESTMENTS

The microeconomic theory of training has been dominated by
Gary Becker's human capital approach for the last two decades
(Becker 1964). The human capital theory states that all *general*
training is financed by the trained employee because it is he who
benefits from the training through higher future earnings(he also
pays implicitly by relatively reduced wages). The prefinancing of
the training costs is virtually impossible for the firms in face
of the continuous fear of'poaching' firms. These firms do not in-
cur training costs but recruit personnel,who were trained elsewhere, by offering higher wages than the training firms are able
to offer. It is conceivable that employees trained in a given
firm will be more productive within this firm than outside of it.
According to Becker the value of these productivity improvements
will be shared in some way between the training firm and the
trained employee, thus basing their reciprocal commitment upon
a common interest.

Although the human capital theory thus seems to cover all forms
of corporate training it does not explain the corporate investments
in the training of apprentices, for their skills are by definition

and observation a good example for marketable skills. Furthermore,
important empirical studies reject the wage-mobility hypotheses underlying Becker's argument and stress the importance of non-monetary determinants of voluntary quit decisions (Sadowski 1980). In
this same book I have shown that, ceteris paribus, the pure fact of
financing training possibilities by the employers does reduce the
employees' propensity to quit and strengthens the firm's attractivi—
ty to potential employees outside. These effects should be stronger
the more difficult it is for employees to judge the non-monetary
qualities of different workplaces or firms. As long as the turnover
of skilled personnel causes costs to the firms, i.e.,the costs of
vacancies and replacements, corporate training investments may be
considered as *one* measure to control the turnover process by investing into the firm's reputation in the labor market.

   This concept of training expenditures is different from Becker's
argument. While he emphasizes the productivity consequences of
training with regard to the job performance, in this model the user
costs of labor not only include the wage payments but also the recruiting expenditures. This idea fits into the growing literature
on informationally imperfect labor markets and can be traced back
to Ullman's approach from 1968, which itself owes much to Stigler's
seminal paper from 1962. Ullmann developed the idea of the substitu-
tability of wage differentials, specific training, and search as
three different instruments to serve the same purpose, namely the
recruiting of qualified personnel.

   Both the empirical labor market studies mentioned earlier and
the recent developments in labor market theory at least qualify
basic assumptions of the human capital approach. These qualifications serve as the economic underpinnings of the following decision
model.

   Corporate training expenditures to foster marketable skills,
as provided by apprenticeship training, are on the one hand designated to signal the firm's concern for the non-monetary well-being
of its employees and thus are meant to increase the apprentices'
attachment to the training firm. On the other hand, the traditionally stressed effects of training, i.e.,the reduction of the operative
costs due to skill improvements, are accounted for separately.
These effects can be substantiated by the assumption of job or
team specific screening of candidates by the firm. A job rotation
procedure certainly allows for the controlled assessment of the
apprentice's skills and working habits.

   To further clarify these hypotheses in empirical terms it is
noteworthy that the reputational value of corporate vocational
training increases with its breadth and marketability. Carried to
extremes this implies that the propensity to leave the training
company decreases with the possibility to do so - as long as it is

the company which made the general training available. From an orthodox neoclassical viewpoint this may be regarded as a paradox or con-
tradiction. But the orthodox rejection of this kind of behavioral
hypotheses essentially rests in the inability to treat "trust" and
"credibility" as economic assets. Becker's theory of the allocation
of the costs and earnings of firm-specific training is a first deviation from this traditional stance, while Kuratani (1973) and Mortensen (1978) further elaborate the importance of the process of
financing for the resulting turnover behavior, and Tabbush (1977)
extends the orthodox views of vocational training by examining the
training of a firm's clients. To avoid misunderstanding I should
add that 'vocational training' in the following model corresponds
to 'search costs' in Ullman's.

A DYNAMIC MODEL OF CORPORATE TRAINING INVESTMENTS

Assume a company at the job market. The company suffers from
a continuous lack of skilled labor and cannot meet all the product
demand it faces. Therefore the company wants to keep its employees
and to fill as many vacancies as is profitable. The number of acceptable and therefore accepted job applications in period t is x(t).
Capital equipment and wage levels have already been fixed.

In this situation the company disposes of two instruments to
control its profits over time.

1. Given that the above formulated hypothesis on the rationale
of firm provided general training is valid, the demand for the
company's jobs x(t) is influenced by the height of its vocational
training expenditures incurred in the same period a(t) as well as
those expenditures incurred during the periods preceding t. They
are thought of as building up the reputation or goodwill of the
company on the job market, an effect that attenuates over the time.
For the sake of simplicitiy the value of the goodwill in period t, N(t),
is measured by the sum of the expenditures through t, and this
asset is adequately depreciated.

2. As far as the variable production costs c(t) depend on the
diligence and the job specific skills of the job holder, an improved job suitability should reduce the production costs. The expenditure intended for higher skills or a better screening in
period t is denoted by r(t). The firm specific competence level
of the company's entire personnel at time t, Q(t), depends on all
the specific training expenditures invested during the past periods,
but the obsolescence of knowledge and the turnover of the personnel
cause a steady depreciation process. It is further assumed that
c(t) depends on the size of the company, i.e., scale economies
or diseconomies may exist.

In this model, where no product market restrictions are considered, the optimal staff size is equal to the "effective" job demand. Putting aside the economies of scale issue, this implies that recruitment and hiring costs are the critical determinants of firm size. The separation of the job specific training and the general training budgets is based on the assumption that the assessment of firm specific skills is not just an unavoidable by-product of general training, but needs its own funds.

Combining these assumptions yields the following structural and evolutionary equations of the decision model.

The job market response function (embracing vocational training)

(1) $x(t) = x(a(t), N(t))$

with the *goodwill*

(2) $N(t) = \int_{0}^{t} f_N(a(s)) \, ds$

           $N(0) = N_0.$

The variable-cost function (embracing firm specific training)

(3) $c(t) = c(x(t), Q(t))$

with the *firm specific competence level* of the staff

(4) $Q(t) = \int_{0}^{t} f_Q(r(s)) \, ds$

           $Q(0) = Q_0.$

The *payoff* G for a present-value maximizing firm under an indefinite time horizon is defined by (5), with p denoting the gross earnings per staff member and period and i denoting the discount rate:

(5) $\max G(a(.), r(.)) = \int_{0}^{\infty} \exp(-it)\left\{[p-c(t)]x(t) - a(t) - r(t)\right\}dt$

           for $t \in [0, \infty).$

The control variables of the problem (1)-(5) are continuous
in time. Methods to determine optimal controls for continuous dynamic processes (the calculus of variations, control theory, and
dynamic programming) are well-known in the human capital literature,
e.g., from models that allocate the individual lifetime between
learning and working phases. Therefore it is not necessary here
to reproduce the details of the mathematical solution.

Closely following Bultez (1975) the necessary optimum conditions are found by using the calculus of variations. Assuming dif-
ferentiable functions and the existence of an inner solution the
variation of the objective function δG is calculated and transformed in such a way that the effects of the control variables
$\delta a(t)$ and $\delta r(t)$ can be separated and simply added up. In face of
the rather general specification of the equations this procedure is
direct, but as Bultez is able to reformulate the straight solutions
in terms of elasticities, the procedure yields economically sensible results.

Let denote

$\eta_a(t)$ the (short term) elasticity of the job demand with
             respect to vocational training expenditures at
             time t;

$$\eta_a^*(\tau,t) := \frac{\partial x(\tau)}{\partial N(\tau)} \cdot \frac{\partial f_N(a(t),\tau-t)}{\partial a(t)} \cdot \frac{a(t)}{x(\tau)}$$

             the (long term) elasticity of the job demand at
             time τ with respect to vocational training ex-
             penditures at time t, $\tau > t$;

$$\gamma_r^*(\tau,t) := \frac{\partial c(\tau)}{\partial Q(\tau)} \cdot \frac{\partial f_Q(r(t),\tau-t)}{\partial r(t)} \cdot \frac{r(t)}{c(\tau)}$$

             the (long term) elasticity of the variable pro-
             duction costs at time τ with respect to job speci-
             fic training expenditures at time t, $\tau > t$; and
             finally

$$CM(t) := \frac{\partial c(t)}{\partial x(t)} = c(t) + \frac{\partial c(t)}{\partial x(t)} x(t)$$

             the marginal costs of production with respect to
             staff size.

Then the necessary optimum conditions for a(.) and r(.), $t \in [0, \infty)$ read

(6a) $a(t) = \eta_a(t) \cdot x(t) \cdot [p - CM(t)]$

$$+ \int_{\tau=t}^{\tau=\infty} exp[-i(\tau-t)] \cdot \eta_a^*(\tau,t) \quad x(\tau) \cdot [p - CM(\tau)] \cdot d\tau$$

(6b) $r(t) = \int_{\tau=\tau}^{\tau=\infty} exp[-i(\tau-t)] - \gamma_r^*(\tau,t) \cdot c(\tau) \cdot x(\tau) \cdot d\tau$

(6c) $N(t=0) = N_O, \quad Q(t=0) = Q_O$

(6d) $\lim_{t \to \infty} exp(-it) \mu_a(t) = \lim_{t \to \infty} exp(-it) \cdot \mu_r(t) = 0$, with

(7a) $\mu_a(t) = \int_{\tau=t}^{\tau=\infty} exp[-i(\tau-t)] \cdot (p - CM(\tau)) \frac{\partial x(\tau)}{\partial N(\tau)} \cdot \frac{\partial f_N(a,\tau-t)}{\partial a(t)} d\tau$

(7b) $\mu_r(t) = -\int_{\tau=t}^{\tau=\infty} exp[-i(\tau-t)] x(\tau) \frac{\partial c(\tau)}{\partial Q(\tau)} \cdot \frac{\partial fa(r(t),\tau-t)}{\partial r(t)} d\tau,$

If the problem is convex, the necessary conditions are suffi-cient, and there exists a solution to the system of equations (1) -(5). Note that $\mu_a(t)$ and $\mu_r(t)$ are adjoint variables that tell how a marginal change of the optimal training expenditure changes the present value, i.e., they give shadow prices for the goodwill and the firm specific competence level respectively. As one easily can see the optimum conditions for vocational training investments depend on the short and long term elasticities of the job demand and on the cost function. The optimal job specific training ex-penditures depend on the skill elasticitiy of the production costs, on the cost function, and on the size of the firm.

A NUMERICAL ILLUSTRATION

Again following Bultez (1975) this paragraph specifies and numerically solves a discrete example with a finite horizon.

Let the job market response function take the form of a Cobb-Douglas-function and solely focus on vocational training investments. Assume an already optimal wage policy and as exogeneously given all other determinants of the job demand, such as the employment stability or the degree of monopoly on the job market. The autonomous term x_O catches these circumstances.

To keep the example manageable we only deal with the "long term" effects of training and consider the cumulated training expenditures. It is assumed that the influence of this year's training will be halved in the following year and so on, hence, at any point of time t the importance of previous training expenditures on the goodwill N(t) is determined by a weighting function w(t), in the example of a geometric distributed-lag function with a periodical bisectioning of the lag coefficients.

The job market response function thus reads:

(8) $x(t) = x_O \cdot N^{\eta_N}(t), \qquad t \in [o,T],$

the *goodwill* function:

(9) $N(t) = \sum_{\tau=o}^{t} w(t-\tau) \; a(\tau),$

the non-negativity restriction

(10) $a(t) \geq o,$ and

the *objective* function:

(11) $\max G(\, a(.)) = \sum_{t=o}^{T} \; [1/\,(1+i)^{t}] \cdot p \cdot x(N(a(.))) - a(t).$

With T = 10 years as planning horizon, the job demand elasticity η_n = .62, the autonomous term of the response function x_O = .375, a discount rate i = .07, and the weighting function described earlier, we get the following time pattern for the control variable $\{a^{+}(t)\}$ and the state variables $\{N^{+}(t)\}$, cf. tab.1.

Table 1. The Time Path of the Optimal Training Investments

year	Training Budget $a^+(t)$ [mio DM]	Goodwill $N^+(t)$ [mio DM]	Job Demand $x^+(t)$ [persons]
			a)
0	0	0	----
1	274	137	41 626
2	137	137	41 626
3	137	137	41 626
4	137	137	41 626
5	137	137	41 626
6	137	137	41 626
7	137	137	41 626
8	137	137	41 626
9	66	101	34 615
10	4	51	22 532

a) Undefined because of the multiplicative response function.

Aside from those recommendations which must be explained by the rather artificial initial conditions and the finite time horizon (an impulse investment in the beginning and a strong reduction in the last two years),the optimal training investment path demands a constant budget to permanently compensate for the goodwill losses and to guarantee an optimal job demand or, under the assumptions of the model, a stable staff size. The fact that even in period 10 there should be some training expenditure incurred, nicely reflects the hypothesis that general training expenditures immediately create a *temporary monopolistic position* for the company and aim at the actual as well as prospective personnel.

(The numerical problem, which is non-linear in the control variables $\{a(t)\}$, has been solved by a gradient technique, the SUMT algorithm.)

CONCLUDING REMARKS

This paper focusses on the interface of the external labor market and organizational *budget decisions*. The approach chosen is closer to the human capital theory and the theory of the firm than to the main body of manpower planning models and to the literature which tries to evaluate on-the-job training (cf. Carpenter-Huffman 1980). The paper stresses non-technological factors of human resource decisions. It stops short of a satisfying explication of the relationship between effective job demand and actual staff size. Lesourne (1973) as well as Chapell and Peel (1978) avoided this difficulty, but neglected explicit, non-stationary evolutions. Pencavel (1977) developed a similar model, but he restricted himself to the production cost effects of investments into the 'industrial morale'. Sadowski (1980) provides a review of the literature and further refinements and applications of the model presented above.

The method chosen here clearly resembles recent approaches to the budgeting of the advertising and R&D expenditures in companies. It is therefore unquestionable (cf.Sethi 1977) that the measurement of the type of the variables used is in principle possible. Systematic empirical comparisons will probably show that the importance of the different types of training investments varies between industries and diverse socio-economic settings. Therefore, Smith's hypothesis that employer-sponsored education mainly serves the needs of the job (Smith 198o) may be true for the United States, but perhaps is not applicable to other economies.

REFERENCES

Becker, G.S., 1964,[2] 1975, "Human capital", NBER, New York.
Bultez, A., 1975, "La firme en concurrence sur des marchés interdépendants", Unpublished doctoral dissertation, CESAM, Louvain Catholic University, Belgium
Carpenter-Huffman, P., 198o, "The cost effectiveness of on-the-job training", Rand/P-6451, Rand Corporation, Santa Monica, Ca.
Chapell, D., Peel, D.A., 1978, Optimal recruitment advertising, Management Science, 24:91o
Kuratani, M., 1973, A theory of training, earnings, and employment: An application to Japan, Unpublished Ph.D.thesis, Columbia University, New York
Lesourne, J., 1973, Modèles de croissance des entreprises, Dunot, Paris
Mortensen, D.T., 1978, Specific capital and labor turnover, Bell J. Econ., 9:572
Pencavel, J.H., 1977, Industrial morale, in: "Essays in labor market analysis," O.C. Ashenfelter, W.C. Oates, eds., Wiley, New York

Sachverständigenkommission Kosten und Finanzierung der beruflichen
    Bildung, 1974, "Kosten und Finanzierung der außerschulischen
    beruflichen Bildung (Abschlußbericht)," Bertelsmann, Bielefeld
Sadowski, D., 1980, "Berufliche Bildung und betriebliches Bildungs-
    budget,"Poeschel, Stuttgart
Sethi, S.P., 1977, Dynamic optimal control models in advertising,
    SIAM Rev., 19:685
Smith, G.B., 198o, Employer-sponsored recurrent education in the
    United States: A report on recent inquiries into its structure,
    National Institute for Work and Learning, mimeo, Washington,
    D.C.
Stigler, G., 1962, Information in the labor market, J.Pol.Ec.,
    7o:94
Tabbush, V.C., 1977, Investment in training: A broader approach,
    J.Hum.Res., 12:283
Ullman, J.C., 1968, Interfirm differences in the cost of search
    for clerical workers, J. Bus., 41:153

A MODEL OF TOTAL COMPENSATION IN

A MARKET COMPARABILITY FRAMEWORK

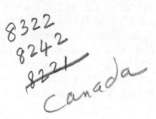

Pierre Martel
Group Chief
Analysis and Research
Personnel Policy Branch
Treasury Board Secretariat
Ottawa, Canada*

INTRODUCTION

This paper provides an overview of a model of total compensation. Analytical issues that need to be addressed to enable its implementation in an organization are discussed in a simple non-technical manner and some of the features of this management tool are illustrated by drawing from the experience of a large corporate employer.

ORGANIZATIONAL ENVIRONMENT

To fully appreciate this model, it is appropriate to first describe the environment, and the constraints under which it operates.

Collective bargaining was introduced into the Canadian Federal Public Service in 1967 with the enactment of the Public Service Staff Relations Act. This legislation recognizes Treasury Board as the corporate employer with authority to negotiate terms and conditions of employment with certified bargaining agents (i.e. unions).

Today, there are more than 80 bargaining units, structured along occupational lines, which are represented by 14 different bargaining agents. The 250,000 employees under this regime are spread across all regions of Canada and some of them serve abroad. The compensation bill of these employees reaches $6 billions ($ CAN = .84 $ US). Bargaining impasses are resolved through either conciliation followed by the right to strike or compulsory arbitration.

---

* The author is solely responsible for the contents and the views
  expressed in this text.

    For determining compensation under this regime, the government
as employer has adopted a policy of total compensation comparability
with the external market. However, in addition to the "give and
take" and other realities of collective bargaining, implementation
of this principle as conceived by the employer is further complicated
by the intricacies of compensation determination itself. Some
compensation matters, such as pensions, are provided by legislation
and are outside the scope of bargaining while others may rest within
the sole prerogative of the employer. Some other items are subject
to consultation and are universal throughout the Public Service.
Finally, those matters which are bargained on an occupational group
basis may be further complicated by the existence of zone or regional
pay systems and may have their own peculiarities or represent unique
benefits.

    Thus, in such a context, it is important that a tool be avail-
able to provide the overall perspective needed for discussions and
decisions in all these fora in order to give full expression to the
government's stated aim of total compensation comparability. The
remainder of the text focuses on such a management tool.

DEFINITION OF COMPENSATION

    Conceptually, total compensation covers all the psychological
and material rewards that parties consider relevant to the employment exchange. However, while this all-encompassing definition may
enable the inclusion of "intangibles" (such as the satisfaction of
social, ego and other higher needs, organizational climate, office
furniture, existence of quality of work life programs, etc.) as
part of compensation, it is too broad a concept to represent
something meaningful in practice in an organization. Thus, in
order to arrive at a practical definition of total compensation, a
pragmatic approach was adopted which consisted in retaining those
elements of compensation that can be translated into a monetary
value or its equivalent. Even on this basis though, extensive
literature searches failed to reveal a clear consensus on the
composition of compensation. Hence a workable definition of total
compensation for determining the appropriateness of individual items
as elements of compensation was arrived at by listing the broad
components of compensation. Specifically, compensation would
include:

    1. Basic pay (i.e. wages and salaries);
    2. premium pay;
    3. benefits and perquisites;
    4. time off with pay;
    5. payment for additional duties and responsibilities, or
        for performance of work under special circumstances and/
        or conditions;
    6. and it would take into account hours of work.

Compensation would exclude:

1. The costs of training and career development (which are
    seen as investments in human resources whose returns will
    materialize through time in various forms of greater
    effectiveness and productivity);
2. the reimbursement of expenses resulting from the perform-
    ance of work;
3. the internal administrative costs (overhead costs) to
    provide the elements of compensation to employees; and
4. universal plans required by law (e.g. unemployment insur-
    ance) as these apply to all workers in all sectors of the
    economy and would represent identical compensation for
    purposes of comparability.

The above categorization may be debatable. For instance,
premium pay may be perceived by some as a penalty imposed on the
employer to prevent abuses in arranging for extra duty hours or for
scheduling work at socially unacceptable times such as during week-
ends, at night, etc. Whatever minor adjustments are eventually
made, if any, the above serves as a useful guide to help in the
identification of the elements making up total compensation. By
actual count, it was found that there are well in excess of 150
elements of compensation in the Federal Public Service which meet
the above definition. By examination of surveys and studies, it
appears that an equally large number of compensation elements may
be provided by other employers in Canada.

However, most of these elements individually, and even collectively, count for very little relative to total compensation paid.
A dozen or so elements of compensation applicable to all Federal
Public Service occupational groups have been found to account for
over 95% of total compensation paid to virtually all these occupational groups. These same elements have been found to represent
97% of total compensation paid by Canadians employers. These
elements are:

1. wages and salaries;
2. pensions (including indexing provisions and supplementary
    retirement benefits);
3. life insurance;
4. disability insurance;
5. severance pay;
6. supplementary health insurance;
7. employer's share of provincial health insurance;
8. sick leave/sickness indemnity plans;
9. paid holidays;
10. vacation leave;
11. paid rest periods; and
12. overtime pay.

Thus, these elements, by virtue of their universality and
importance within total compensation, provide a sound basis for
analysis and comparison of compensation in Canada with reasonable
accuracy. However, given that the approach as developed for use in
Treasury Board Canada was intented for eventual implementation at
the level of each occupational group, where some elements excluded
at the overall level may become particularly relevant, care was
taken to ensure that the model possesses the necessary flexibility
to accomodate an expanded number of elements. There are currently
50 individual elements or so that are considered for establishing
compensation comparability for the 80 different bargaining units
under the responsibility of Treasury Board Canada.

METHODS OF TOTAL COMPENSATION COMPARISONS

Having defined total compensation for our comparability purposes,
next the matter of concepts and methods for measuring total compensation required attention. There are three basic approaches which
can be used. Each one is presented hereunder with the reasons for
its acceptance or rejection in the context of our model of total
compensation.

Gross Comparison Method

In this method, features and characteristics of benefits are
evaluated one by one and subjective values are assigned according
to a scoring scheme designed by the analyst. Alternatively, this
method could consist in comparing each individual benefit characteristic and assessing it as superior, inferior or equal to some standard measure or characteristic. This method, or variations thereof,
ranks benefits and distinguishes between the "best", "next to best",
etc.

Since this method can only determine that benefits are above,
below or at par when compared with others and because of the qualitative nature of the valuation procedure, this method was judged to
be inappropriate in the context of the total compensation model.
At best, this method provides only a superficial assessment of
comparability.

Cost of Benefit Method

The cost of benefit method compares the actual cost to the
employer or, stated differently, the expenditures incurred by the
employer in providing compensation in its various forms to employees.
For purposes of comparison, either total dollar cost or percent of
payroll actually expended may be utilized. The technique underlying

this method is relatively straight-forward: elements of compensation are listed and employers are surveyed to determine their expenditures on the benefits as listed. The data collected easily lends itself to processing since it is reported either as dollars expended, or as percent of payroll.

However, because this method compares only expenditures incurred by employers, there is no way of determining the amount, degree, or quality of compensation being provided to employees, i.e., an equal dollar amount expended by two different employers does not necessarily equate to similar quality and quantity of compensation to employees. For example, two organizations may provide identical retirement benefits but may have quite different expenditures either because one organization is more cost efficient or funds its plan at a much faster rate than the other; differences may exist in the composition and characteristics of the workforce (e.g., age, sex, location, type of work environment and utilization experience) which influence the level of expenditures in each organization in a quite distinct way; or a combination of the above factors differently affects the expenditures of each organization.

An additional difficulty with this method arises from the fact that many organizations combine costs of several benefits into a single item within their accounting system, making the reporting of detailed cost breakdowns difficult to survey. Furthermore, many employers consider their expenditures data confidential and are reticent to disclose their details.

All in all, the above shortcomings are serious enough to discredit the suitability of the cost of benefit method in achieving an effective measure of compensation comparability with the outside sector. In fact, this method better lends itself to financial analysis and price/cost determination than to studies of compensation comparability with the outside.

Level of Benefits Method

The level of benefits method is the valuation technique that was retained for the purposes of the total compensation model.

This method basically consists in:

(1)   calculating the cost that would be incurred by the Government if it were to provide its employees with compensation packages similar to those existing outside the Public Service, and

(2)   comparing this simulated cost with that which currently exists for public servants.

The level of benefits method, then, has the necessary flexibility to provide employees with an appreciation of the value of their compensation relative to that of others. It avoids the artificial value differences resulting from such factors as funding, utilization experience, work-force characteristics, ets. Furthermore, this method requires from outside employers only information on provision and characteristics of benefits. The main disadvantage of the method resides in the complexities of the techniques that have to be utilized to evaluate benefits.

Under the level of benefits approach, each element of compensation is first valued individually and then all the element values are integrated into a composite measure of total compensation. For some elements the valuation process is relatively straight-forward (e.g. number of paid holidays), but for others like pension and insurance-type elements, the valuation process can become complex and cumbersome. In these cases, it is necessary to utilize actuarial techniques and simulation models to calculate the present value of future benefits or to estimate the premium necessary to cover an expected risk. Employee contribution are taken into account to derive a measure of the net benefit to employees.

MODEL OF TOTAL COMPENSATION

Having established the value of the various compensation elements, the problem of determining the value of the compensation package as a whole arises, given that the values of the elements are expressed in different forms: hours, percentages, days, dollars, etc. As we have defined total compensation in the context of the employment exchange whereby employees provide their services in the form of productive time in return for rewards of different kinds, the value of total compensation takes its meaning when the monetary components are related to the net hours worked. This measure best reconciles the net average value of compensation to a group of employees under a level of benefits approach with the cost of compensation to the employer initiating the comparison. It also recognizes the pay-for-effort bargain, effort being measured in the form of net productive time.

As a formula, the value of total compensation (TC) is expressed as:

$$TC = \frac{\text{Monetary Remuneration}}{\text{Hours Actually Worked}}$$

In algebraic form, we can write:

$$u = \frac{Y \ (1 + \Sigma a_i) + \Sigma b_i}{(D - \Sigma l_i) \ (H - \Sigma h_i) + t}$$

where: u =   Total compensation in dollars per hour worked;
         Y =   Annual salary;
        Σa_i =   Sum total of salary-related benefits;

        Σb_i =   Sum total of flat dollar benefits;

         D =   Total number of scheduled days of work in a year;
        Σl_i =   Sum total of days of paid leave in a year;

         H =   Number of scheduled working hours per day;
        Σh_i =   Sum total of daily paid rest periods;

         t =   Number of extra duty hours in a year.

This expression of value provides a common basis for comparing
the composite value of compensation packages which might be quite
dissimilar in composition. Unlike other approaches that attribute
a monetary value to paid time off elements, this form of expression
does not overstate the monetary component of compensation. Finally,
it sustains the fundamental principle of an exchange and readily
reflects the impact of changes in the monetary and productive time
components inherent in this exchange process.

INFORMATION REQUIREMENTS

Ideally, the level of benefits method requires that all the
information pertaining to the Federal Public Service workforce
characteristics, benefits utilization and other work-related matters
(which we will collectively refer to as "employee characteristics"
later in the text) and all the features of the compensation packages
of each employer used in the comparison be known at the required
level of detail. This is a very stringent requirement which is not
met in practice owing to data volume and deficiencies, cost-benefit
considerations and constraints imposed by the "realities" of collective bargaining.

Employee characteristics

Data on the Public Service employee characteristics are generally available through internal Treasury Board information systems.
Age, sex, geographic location, length of service, shift hours worked,
etc. are examples of what is meant by "employee characteristics".

There are few instances, though, where certain employee data
problems or limitations surface or where cost-benefit considerations
prevent the collection of data at the ideally required level of

detail. These reside mostly in the attendance and extra duty areas.
In these cases, it was usually possible to use proxy variables or to
devise estimation methods that would overcome the deficiencies.
Take, for instance, the measurement of overtime pay where the premium
structure depends on the number of hours worked and the type of days
(i.e. holiday, day of rest, normal working day) on which overtime
is worked. The employee information systems collect data on the
number of overtime hours by premium rate (i.e. straight-time, time
and a half, double time, etc.). A stochastic method was thus devised
which analyzes the pay conditions stipulated in the collective agreement in conjunction with the hours reported in the information system
and, by using probable occurrences of different types of days during
a period of time (usually a calendar year), it rearranges the collected information to build a "typical" profile of frequency and
duration of overtime hours by type of days. This profile then
becomes the yardstick against which the various overtime pay provisions of the Public Service and of external employers are measured
for comparability purposes. It is interesting to note that, in the
case of the various occupational groups in the Public Service, the
results yielded by such an estimation procedure compare closely with
the actual overtime expenditures: a regression analysis yielded a
correlation coefficient of 0.99.

Features of Compensation Plans

     To measure compensation for comparability purposes, it is also
necessary to have information on the features of compensation plans
to be used in the comparison.

     The features of the Public Service compensation plans are
readily available from collective agreements, legislation and other
compensation documents.

     In the case of the outside employers used for comparison, the
information needed - i.e. discription of the features of the plans -
is by and large collected by an agency called the Pay Research
Bureau (PRB). In some exceptional cases, other sources of information such as data banks of private consulting firms, are used to
complement or to substitute the PRB data. The Bureau is an independent agency established to meet the information requirements of
the parties to collective bargaining in the Federal Public Service.
It regularly surveys Canadian establishments. It collects compensation information - i.e. salaries, benefits, conditions of employment, etc. - and it publishes reports to meet the requirements of
the parties to bargaining. The PRB surveys are conducted after job
matching has taken place. This process, which consists in evaluating
jobs in the outside market in terms of the Public Service job
classification system, ensures that only terms and conditions of
employment of like occupations and of similar classification worth
are compared.

While all the information required for an ideal application of the level of benefits method is collected and is available in the PRB data banks, Treasury Board Canada as employer cannot access all the detailed information on each external employer's plan because of a number of other reasons which have to do with the employee-employer relationships climate. As a result, methods which rely on aggregate data - i.e. incidence and benefits features of all firms surveyed - were developed. For instance, in the case of vacation leave, the average number of vacation days per year for each discrete year of service is calculated from the aggregate data. The resulting vector is then used with the Public Service employee profile of years of service to obtain the average value of the vacation plan to the group of employees. In the case of pension plans, aggregate data are interpreted and "typical" pension plans representing the varieties of pension plans offered by outside employers are constructed. Each plan is actuarially valued and then the results are integrated into an average market value using the plans incidences as weights.

Tests conducted on some elements like pension and vacation plans suggest that the results obtained through such a methodology compare very closely (a difference of 0.1%) with what would have been obtained through access to each employer's detailed description of benefits items.

All in all, these data problems and the constraints regarding their access will remain, but they are not considered, based on the tests conducted, to be serious enough as to affect and render inoperative compensation comparisons.

COMPUTER ASSISTANCE

Given the complexities of some compensation valuation procedures, the size and diversity of the data bases required and in keeping with the requirement to effectively and quickly meet the needs of collective bargaining where stringent time constraints have to be respected, it was necessary to develop a computer capability to support the model of total compensation. An interactive system based on APL is currently in operation.

The development of the concepts and detailed valuation methods and the assessment and tests on the data took place over a two-year period and necessitated the expertise of four to six management scientists. The development of the computerized capability took six months. In now costs in excess of $100,000 annually to operate and maintain this system; this is rather inexpensive in relation to the compensation bill and the number of employees involved.

USEFULNESS OF MODEL

This model with its output reports and its on-line compute-

TABLE 1

Example of a Compensation Comparability Simulation

Occupational Group: Fictitious Office Group
Number of Employees: 35,700 (as of June 1980)
Case no 1:   Comparability at expiry of current agreement,
             prior to negotiations.

		Public Service	External Market	Difference
A.	Wages and Salaries	$18,600	$20,300	$-1.700 (9.1%)
B.	**Salary-Related Elements**			
	B.1 Pension	$1,674 (9.0%)	$1,218 (6.0%)	$456
	B.2 Life Insurance	$93 (0.5%)	$203 (1.0%)	$-110
	B.3 Disability Insurance	$130 (0.7%)	$81 (0.4%)	$49
	B.4 Severance Pay	$595 (3.2%)	$305 (1.5%)	$290
	B.5 Overtime Pay	$744 (4.0%)	$1,259 (6.2%)	$-515
	B.6 Savings Plans	N/A	$142 (0.7%)	$-142
	B.7 Stand-by	N/A	$61 (0.3%)	$-61
	Sub-Total	$3,236	$3,269	$-33
C.	**Non-Salary-Related Elements**			
	C.1 Health Insurance	$150	$190	$-40
	C.2 Dental Care Programs	N/A	$265	$-265
	C.3 Profit Sharing	N/A	$50	$-50
	C.4 Meal Allowance	$10	$5	$5
	Sub-Total	$160	$510	$-350
D.	Monetary Loss Due to Unpaid Absences	0	$-75	$75
E.	Total Monetary Elements (A+B+C+D)	$21,996	$24,004	$-2008
F.	Scheduled Days Per Year	260 days	260 days	
G.	**Leave**			
	G.1 Vacation	15.6 days	16.0 days	-0.4 days
	G.2 Paid Holiday	11.0 days	10.3 days	0.7 days
	G.3 Sick Leave	5.5 days	5.5 days	0 days
	G.4 Other Personal Leave	1.6 days	1.6 days	0 days
	Sub-Total	33.7 days	33.4 days	0.3 days

(continued)

TABLE I (Continued)

H.	Net Days Worked per Year (F-G)	226.3 days	226.6 days	
I.	Net Daily Hours	7.0 hours	6.7 hours	0.3 hours
J.	Overtime Hrs. Worked Per Yr.	100 hours	100 hours	100 hours
K.	Total Hours Worked Annually (HxI=J)	1684 hours	1618 hours	66 hours

TABLE 2

Estimated Annual Cost (Expenditures) of Compensation Packages

Group: Fictitious Office Group
Number of Employees: 35,700 (as of June 1981)
Case no. 6: Employer's proposal after 3 days of negotiations

		Public Service Current Package*		Proposal	
		$ ('000)	%of Total	$ ('000)	% of Total
A.1	Pay for Time Worked	577,953	73.6	613,400	73.7
A.2	Pay for Time not Worked	86,067	11.0	89,090	10.7
A.	Straight-time Payroll (A1+A2)	664,020	84.6	702,490	84.4
B.	Salary-Related Elements	115,525	14.7	123,600	14.9
C.	Non-Salary-Related Elements	5,712	0.7	6,150	0.7
D.	Total Compensation (A+B+C)	785,257	100.0	832,240	100.0

Change between the two Packages:

Dollar Increase: $46,983,000

Percentage Increase: 6.0%

Note: Individual elements (e.g. vacation, holidays, etc.)
   making up each component (e.g. pay for time not worked)
   can be listed if deemed necessary.

*Reflects case no. 1, table 1.

assisted simulation capabilities enables the market relationship of
each compensation element as well as the whole package to be established and the financial implications of modifications to be determined. Tables 1 and 2 provide illustrations of some standard
reports. Also, because the valuation is done in terms of a standard
population, the method and model are eminently suited to answer
"what if" types of question.

The existence of this model in Treasury Board has made a more
precise quantification of the pay and benefits packages possible.
It has created among decision-makers and compensation officers a
more precise awareness of the magnitude of total compensation and
its components and of their values relative to other employers. It
has also contributed to a greater appreciation of the interrelationship of the elements. The model also provides a flexibility for
assessing the comparability and financial implications of compensation
options. It has reduced from days under a previous system to a
matter of hours (sometimes minutes) the time required to evaluate
a set of contemplated changes to the total compensation package of
a particular occupational group.

Its usefulness for negotiations and decision-making is evi-
denced by the three-year existence of the model in an operational
mode. Bargaining mandates are granted by the decision-making
authorities only after the full implications of the proposals are
assessed with the help of the model of total compensation. To our
knowledge, few, if any, organizations possess such a management
tool in support of compensation determination in a collective
bargaining context.

REFERENCES

Delorme F., "La rémunération globale, est-il possible
    de la mesurer?", Commerce, mars 1978, pp. 32-36,
    Montréal (1978).
Martel P., excerpts from an address at "One-Day Compensation
    Conference", Toronto (1978).
Statistics Canada, "Labour Costs in Canada 1976", Ottawa.
Thorne & Riddell Associates Ltd "Employee Benefit Costs in
    Canada 1977-78", Toronto.
United States Civil Service Commission, "Total Compensation
    Comparability", Washington, D.C., (1975).
Wolf G.D. and Cushing A.M., "Simulation: A Quantitative
    Analysis for Total Compensation Planning", Compensation
    Review, Vol. 1, no. 3 (1969).

NA

SECTION 4

INTRODUCTION OF TECHNOLOGY INTO ACTUAL WORK SETTINGS

New technology affects organization structure and performance
in many ways. Even if overt change is small, resultant changes
in job content, interpersonal relationships and even cultural
value can be large. This is the case of modernization of a
French oil refinery as reported by Miret. With the help of
their socio-technical model, Pasmore, Shani and Mietus study
similar effects observe in thier action research on a field
experiment conducted by the U.S. Army. The experiment involves
both the development and implementation of new technology and
the modification of organization and personnel conduct.

Progress in wholistic approaches, such as in case or field
research, will depend on mastering analysis of special aspects
of technological innovation and change in work content. In this
line of research, Gobel and Meers present an analytical framework and empirical results from studies of the interrelationship
between mechanization of office work and motivation of clerical
staff in the largest Belgian bank. In an even closer focus on
special relationships, Staehle analyses the impact of the visual
display unit on the worker and work processes. Similarly, Gutek
analyses the man/machine impacts of various new office equip-
ments combined in one work place.

THE MODERNIZATION OF A REFINERY : ITS INFLUENCE ON MEN, THEIR

WORK AND STRUCTURES

Pierre Miret

Eurequip

Vaucresson, France

ABSTRACT

A classical oil-refinery went through a modernization program aimed at a large increase in capacity while setting up accelerated automation. This modernization had only a slight influence on the organic structure, but deeply altered the content of some jobs and provoked a true cultural change with a de-structuring of the social group and the creation of isolated sub-groups when increased integration was required. Measures were taken to reduce these negative effects which may suggest ways of solving the problems created by any profound technological change.

THE REFINERY AND ITS MODERNIZATION

FIRST OBSERVATIONS

The oil-refinery we are talking about was a classical type built about fifteen years ago. Before its modernization it included :

- six production units (distillation, hydro-desulphuration, catalytic reforming...) ;

- product movement installations (mixing, stocking, and dispat- ching) ;

- one control-room.

Like most refineries, it was already automated, with many elementary regulations controlled automatically.

In the mid-seventies, the management decided to increase the capacity of this refinery by creating five new production units to complete and reinforce the existing units, and by doubling the dispatching capacity. At the same time, the management worked out an accelerated automation and robotization policy and introduced centralized operation - from a single control-room - of the whole refinery, which had become a lot more complex.

A few figures will give an idea of the automation policy followed :

- Number of regulation-loops from 200 to 700 ;

- Automatic handling of 3000 physical data elements ;

- Motorization and remote control of more than 200 sluice-gates ;

- Setting 6 computers into action.

This extension and automation evidently responded to economic objectives : increased production, rise in productivity, better returns. At the same time, the management set a social policy of shift-work reduction and task extension (even though automation is commonly accused of limiting the tasks).

Roughly, the general organization chart of the refinery looked as follows :

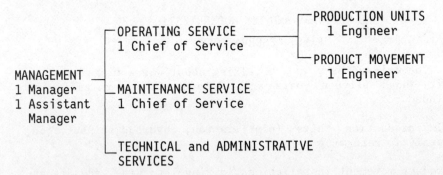

At this level, the organization chart was not touched by the alteration of the refinery.

MANPOWER EVOLUTION

D = Daytime posts, S = Shift posts (6 workers per post)

POSTS	BEFORE			AFTER		
	D	S	TOTAL	D	S	TOTAL
PRODUCTION UNITS						
Engineers	1		1	3		3
Foremen	1		1	3		3
Technician Employees	1		1	3		3
Chief Operators		1X6=6	6		3X6=18	18
Table and Desk Operators		3X6=18	18		3X6=18	18
Team Leaders					3X6=18	18
External Operators		5X6=30	30	6	8X6=48	54
TOTAL	3	54	57	15	102	117
PRODUCT MOVEMENT						
Engineers	1		1	1		1
Foremen	1		1	1		1
Technician Employees	13		13	13		13
Chief Operators		1X6=6	6		1X6=6	6
Operators		1X6=6	6		1X6=6	6
Team Leaders					3	3
External Operators	25	3X6=18	43	32	1X6=6	38
TOTAL	40	30	70	50	18	69
MAINTENANCE						
Engineers	3		3	5		5
Technician Employees	23		23	30		30
Foremen	9		9	11		11
Workers	12	1X6=6	18	12	1X6=6	18
TOTAL	47	6	53	58	6	64
GRAND TOTAL	90	90	180	123	126	249

The staff of the operating and maintenance services is indicated in the above table. It is broken down into type of post and work (day time or shift work : 3 X 8).

A glance at this table shows :

- In the production units : a doubling of the manpower, but this doubling is negligible compared to the added rise in value due to the creation of new units.

- In "product movement" : a large productivity rise ; the staff remained stable while the capacity doubled.

- In maintenance : a slight growth in manpower effectiveness due to a large increase in sub-contracting.

- Generally speaking, a relative reduction in shift workers.

- Lastly, the creation of a new post : the Team Leader for external operators (see following explanation).

POSTS	QUOTAS		
	BEFORE	AFTER	GAP
Foreman	340	340	0
Chief Operator	310	310	0
Table and Desk Operator	200-215	230-290	+50
Team Leader	0	200-230	+215
External Operator	185	200	+15

But we can see that qualification rises are far more pronounced for the control-room posts (Desk Operator) than for the external operator posts.

WORK ORGANIZATION EVOLUTION

This evolution is here investigated through four typical posts.

CONTROL-ROOM OPERATORS
(Chief and Desk Operators)

The control-room operator posts were deeply influenced by the automation policy and by the new way of displaying information.
Regulation and control take larger and more complex systems into account. More ambitious, regulation no longer deals with elementary parameters such as flows, levels and temperatures, but with more or less complex functions which are often economic : yields, recovery rates, blend compositions.
The control-room no longer includes a large number of screens showing the value of elementary parameters, but a few consoles where the operator can, at request, call for information concerning the functioning of various equipment ; this information is ordinated and allows for progressive steps, moving from general to detailed data.

The procedures are robotized and with only one move (for instance, turning a dial) the operator starts a procedure (for instance, an emergency stop) immediately carried out by the robot.
All of this implied an evolution in the post of operator which can be described as follows :

- wider complexity : the control is larger and the required regulations are more accurate ;

- larger abstraction : the reality of physical phenomena is no longer translated through elementary parameters, but through computerized models ;

- increased mental charge : the operator gets a great deal of information, from which he has to make a selection. And this information is not given automatically, he has to call for it ;

- closer attention : there are few operations to do because the robots take care of everything, but the operator must be ready to react at any time if required ; the equipment allows for incidents to be simulated in order to maintain this reactability.

The ideal profile of an operator is therefore altered and requires the following aptitudes:

- Abstraction and reasoning ;
- Synthesis ;
- Attention.

On the other hand, an accurate knowledge of the process is less important because this knowledge is included in the models used for regulation.
This enlarges the qualification level of the post, as previously indicated, and also has an influence on recruiting, which is now carried out among a highly educated population.

EXTERNAL OPERATOR

The external operator post previously consisted of following on the spot, in the units, the work of the equipment and operating machines at the request of the control-room operators (opening or closing sluice-gates, for instance), regulating and reading values (temperatures, pressures).
Robotization has suppressed these activities, which are now done by regulation devices.
The role of the external operator is then focused on the monitoring of the working of equipment and particulary the gauges. He is no longer involved in the operation of the equipment, which is now done from the control-room.
This last point is accentuated by the physical distance of the control-room ; the external operator is totally isolated and only communicates by means of radio. Therefore, it became necessary to create team-leader posts in order to manage, coordinate and assist the external operators. This, in turn, stimulated an evolution in the structure :

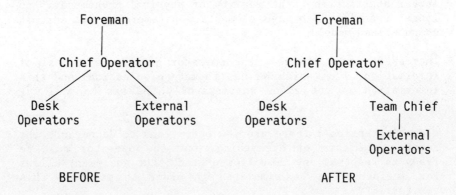

Though this structure has extended the hierarchical channels between desk operators and external operators, a direct channel of communication has been set up between them, short-circuiting the hierarchical structure.
This post evolution also has an important impact on careers : the gap between the desk operator and the external operator posts has increased. The required abilities have become different and the external operator function no longer prepares toward equipment operation. Therefore, a division of the usual career channel into two separate routes will certainly be a problem :

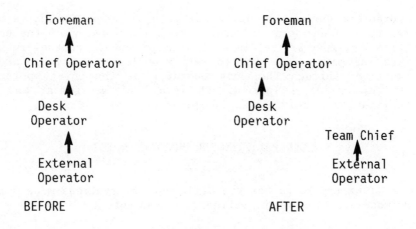

BEFORE   AFTER

MAINTENANCE PERSONNEL

The evolution of maintenance is mainly due to the mass introduction of regulation electronics (micro-processors, computers, etc.).
This introduction had two consequences :

- An increase in preventive maintenance to insure reliability (i.e. of the regulation information) ; the switch from curative to preventive maintenance calls for a whole change in state of mind, working methods and behaviour.

- An increase in sub-contracting, due to the need for special technology unknown in the refinery ; the maintenance personnel only have to be able to detect a defective electronic part and to replace it by a working part ; the repair itself is done by the constructors or by external specialists.

TOP AND MIDDLE MANAGEMENT

Two things must be pointed out in the job evolution of the top and middle management of the refinery :

. The integration of the different production units was increased considerably : the new equipment allows for an optimum functioning of the whole refinery instead of an optimum for each isolated working unit.
However, this search for an optimum leads to more complex management control, more accurate economic reasoning, and greater communication among the staff.

. Moreover, during the daily work, the engineer in charge of a group of units can no longer check the working of the units with a glimpse at the dials while he passes the control-room.
The information he used to get is no longer available and he must go through the desk operator or the chief operator, which implies a less direct method of management and the obligation to rely on his subordinates.

GENERALIZING THE PREVIOUS NOTES

If we try to gather the different observations made during the modernization of the refinery, we can note :

Productivity rises, which means relative reduction in staff numbers.

Reduction of hard and dangerous tasks (previously done by external operators, shift work).

Higher integration of the different activities leading to the product output (integration of production units, mixing units and dispatching ; search for a global optimum ; increased maintenance to secure the reliability of regulations).

But, at the same time, an increased subjection of the whole refinery to small groups of people in charge of the proper functioning of the refinery from the control-room. These people, placed in key-positions, have a large blocking power and make the system more fragile.

An increased dependence on external agents, those who possess some essential technology (sub-contracting the maintenance).

The partial loss of a certain technical culture to the benefit of the outside : the detailed knowledge of process, interfering parameters and the way they interfere has been transferred to the robot constructors and inserted into the micro-processors and computers. Finally, this knowledge is no longer really compulsory, at least when nothing happens ; it is bound to gradually disappear anyway, because the workers do not use it so often.

An increasing demand for abstraction, logic and complexity control abilities (very clearly for the operators, but also noticeable for engineers and managers) which leads to the need for a new type of manpower (better educated and prepared for such intellectual exercises).

The rejection by these men of the previous culture, with its knowledge mainly based on concrete experience, its models (i.e. the hierarchic model rejected by the desk operators), and its values (i.e. job progression becoming less useful in its traditional form).

From whence occured the split of the social group into two sub-groups : on the one hand, those who master the new tools and, on the other, those who do not and who, by the way, have fewer possibilities of career development.

Something like a cultural mutation occurs, with a de-structuring of the social group of the refinery. This group is in fact submitted to a triple tension :

- higher external dependence and decrease in autonomy ;

- integration, desired by the management and allowed by the technology ;

- differentiation, provoked by this very technology.

This tension may lead to negative social effects, mal-functionings and even complete stops. But isn't this a general phenomenon linked to the introduction of automation, robo-tization and computerization ?
For instance, let's take the case of a manufacturing and as-sembling industry like the automobile industry. There we can find productivity rises and the suppression of some hard tasks (i.e. painting now done by robots) at first.

But we also find :

- increased integration of the different workshops : the factory tends to become a single assembly-line ;

- external dependence : for the delivery of certain pieces and
parts (in modern plants a delay in delivery of some hours can
stop the output of cars at the end of the assembly-line) and
for the maintenance of automation and robots.

- the development of a group of specialists in automation,
computers and robots, calling for a new "culture" with special knowledge and new ways of reasoning ("systemics").

- Gradual disappearance of some jobs transferred to machines
(i.e. painting, welding).

- From which a trend towards segregation appears : on one side,
those who master the new technique and consequently have a
certain power and, on the other, those who submit to a technology which limits their former possibilities.

In the same way, data processing and new office equipment
have similar effects on office jobs.

In this case however, two remarks can be made :

- up to now, computer-operators have formed a separate class in
the company, holding some power due to the centralization of
their machines. But this condition is now changing with the
growth of smaller units which allow a greater decentralization ;

- just as the appearance of writing and the development of the
printing press created illiterates, so data-processing now
creates people "excluded". But one can hope that data processing systems will be within everyone's reach in the coming
years, at least for the operation of their external functions.
In the near future, however, and maybe for many years, the
previously described effects will remain. They seem to be
provoked by any technological change. Thierry Gaudin* says,
"The introduction of the knife to the Eskimos in exchange for
some seal skins, was a transaction which was seen favourably
by both sides and had the following effects :

---

* Thierry Gaudin "Listening to Silence".

- previous technique (cutting tool carved out of a seal-bone) is devalued ;

- former technique holders lose their position ;

- young people, faster in the use of the new tool, despise old people ; social relations are de-structured ;

- the old know-how is no longer accepted ; one generation is sufficient to forget it even though it had existed for thousands of years ;

- productivity grows and so does population ; but the latter now depends on an external purchase channel where it doesn't understand the terms of the exchange. It loses its autonomy, its rules and its internal balance.
  Had they been able to foresee this evolution, the Eskimos would have refused it".

POSSIBLE ROUTES OF ACTION

But it is very difficult for industry to refuse technological evolution. As it is often a question of survival, it is better to accept this evolution and try to weaken the negative effects. The procedure followed during the refinery modernization may suggest a more general route toward solutions.

When the decision was taken to increase production capacity through accelerating automation, the management first reflected on the human and social consequences of the mutation. It also sharpened its policy on two points :

- reducing the number of shift workers as much as possible ;

- without lowering job qualifications.

This second point led to focusing the external operators' knowledge on machines (with a share of the maintenance function).
One can, however, notice that this reflexion did not lead to altering the elected automation policy. The management could have examined, in particular, the influence of two components of automation on men and organization :

- automation level ;

- centralization level for managing the operations (influencing the "distance" between desk and external operators).

A second step in study and reflection carried out with the help of a consultant, led to a more detailed awareness of the risks created by the effects eventually provoked by automation. The chosen policy was maintained, however, and the corresponding cultural mutation accepted as desired.

The question was then to secure the staff's agreement on the project, to make the introduction of the new "culture" easier, without provoking any fear or blockages, and to maintain cohesion by weakening the split between external and control-room operators. The design of a general plan aimed at :

- informing the refinery staff ;

- giving them a share in the elaboration of solutions ;

- favoring internal promotions, particulary through the career development channel :
  External Operator ———> Desk Operator ———> Chief Operator

- making a large training effort, either to help in the introduction of the new "culture" among the existing personnel or to make promotions easier.

INFORMATION

Throughout the operation (about 18 months), the refinery management kept the personnel informed first about the project, then about the successive steps of realization. This information was mainly passed on through personnel representatives.

PARTICIPATION IN WORKING
OUT THE SOLUTION

Ten working groups were created to study the components of the whole project (posts, control-room, new fittings, labour organization).
In each case, the successive steps were roughly similar :

- definition of objectives and explanation of constraints by the management with, sometimes, the fixing of a budget ;

- creation of a working group made of people elected according to the chosen subject (technically competent or directly concerned people) ;

- work by the group, who can call for external experts or visit similar installations ;

- solutions proposed to the management who accepts or asks for alterations ;

- setting up the solution with, in some cases, the group's help.

PERSONNEL MOVES

Together with the elaboration of the future organization and posts, a manpower analysis was carried out (original education, professional experience, behaviour, potential). From this, a planning of moves was established with an accurate time-table which allowed any post to be filled at the right time and the realization of the highest number of promotions based on required abilities.
Apart from the posts' re-evaluation, it was made possible to carry out 28 promotions, among which only 4 were due to the move "External Operator ——> Desk Operator".
To fill all the control-room posts, 11 people (25% of the personnel) had to be recruited from outside the refinery.

TRAINING

Together with the personnel moves, detailed training planning was established, providing different training according to the posts to be filled and the characteristics of the trainees (education, experience).

The following figures give an idea of the amount of training undertaken :

- Training of a Chief Operator   : 3 to 8 months

-    "   Desk-Operator   : 3 to 5 months

-    "   Team Chief   : 2 to 5 months

-    "   External Operator   : 3 to 5 months

On the whole, 90 different training cyles were carried out, which makes a total of 650 man - months.

CONCLUSION

Thanks to all these efforts and precautions, the refinery modernization was carried out without any major difficulties on the human and social sides and the start up of the new units occurred under positive conditions.

Are we allowed, then, to conclude that the target is definitely reached ? Certainly not. As time goes by, the new "culture" strengthens its position and the novelty wears off. However, time is not always an ally, and some effects, weak at first, can gradually grow and finally become unbearable. Saying that, we particularly think of the new control-room operators with a higher education level who are no longer hired through the post of External Operator. The ratio is now quite small, but it can only grow over the years. This stimulates at least two questions :

- what type of relations are being established between the control-room Operators and the External Operators ?

- what can the career development of these young and ambitious control-room Operators be ?

These questions have no answer today. Moreover, a lot more could be brought up that remain unanswered. But wouldn't it be either naive or presumptuous to pride oneself on the total mastery of a complex human system ?

TECHNOLOGICAL CHANGE AND WORK ORGANIZATION IN

THE U.S. ARMY: A FIELD EXPERIMENT

William A. Pasmore, Abraham R. Shani and
John Mietus

Case Western Reserve University
Army Research Institute

ABSTRACT

The military, like other institutions of our day and age, is
facing a rapidly increasing rate of change on a number of fronts.
Two areas of particular concern for the next decade and beyond
will be changes associated with (1) the development of new and more
highly sophisticated technology; and (2) alterations in the skill
levels, needs and make up of the human resources who will staff
our military organizations. Although it is yet to be determined,
we believe strongly that changes on these dimensions will require
the development of new organizational forms to deal with the issues
of training, motivation, and overall preparedness that these
changes portend. The research project described herein is the U.S.
Army's first attempt to experiment with sociotechnical system
methods of work and organization design, and the results discussed
reflect the many issues involved in the use of such innovative
responses to the critical problems confronting our armed forces.

INTRODUCTION

Sociotechnical system interventions are organization development techniques that typically involve the restructuring of
work methods, rearrangement of technology, or the redesign of
organizational social structures. The objective is to optimize the relationship between the social or human systems of
the organization and the technology used by the organization
to produce output. When these systems are arranged optimally,
the organization runs more smoothly than when they are not;
output is higher, employee's needs are satisfied better, and
the organization remains adaptable to change (Pasmore and
Sherwood, 1978, p. 3).

A good deal of research has been done in the last four decades
to support the statement made by Pasmore and Sherwood above regarding the effectiveness of sociotechnical system interventions. In
reviewing the literature on organization development, for example,
Friedlander and Brown (1974) conclude that:

> Of the technostructural approaches, sociotechnical systems has
> the clearest effect on performance, while all three methods
> (sociotechnical, job design and enlargement, job enrichment)
> tend to increase satisfaction with work (p. 334).

Similarly, Srivastva, et al. (1975), in reviewing the results
of 16 sociotechnical system experiments, state that:

> The results from sociotechnical experiments are highly posi-
> tive. Seventy-five per cent of the experiments show totally
> positive results. The percentage of studies showing positive
> findings for each dependent variable were as follows: pro-
> ductivity, 93%; costs, 88%; quality, 86%; withdrawal, 73%;
> attitudes, 70% (p. 103).

More recent studies by Pasmore and King (1978), Pasmore (1979),
Tichy and Nisberg (1976), Walton (1978) and others have reaffirmed
the continuing success of sociotechnical system interventions in
this country and abroad.

Although sociotechnical system approaches to work design date
back to the 1940's, their application to the military is new; until
recently, the military had not experienced the same pressure to
deal with the fit between its human and technical resources that
competitive industrial firms had. However, the development of new
weapons and information systems has now begun to impact seriously
on the Army's need for skilled manpower; it is becoming increasingly apparent that increased skills will be required on the part of
many of our forces to man, maintain and operate the technology of
warfare in the eighties and beyond. At the same time, the average
skill level of incoming recruits has declined, in general, over the
past decade. This increasing gap between technological demands and
available skills has stirred interest in learning from industry's
earlier lessons in bridging the gap through innovative work and
organization design.

The goal of the current research effort was to develop a model
of sociotechnical system change specifically designed for the military, to develop materials to assist in the dissemination of that
model, and to investigate the issues associated with its application in an actual military organization. However, this research
should be of interest to members of a broad range of organizations
facing similar challenges stemming from social and technological
changes as well as to researchers from a variety of applied behavioral science disciplines.

The Development of the Model

The papers by Pasmore, et al. (1978) and Pasmore (1979) outline shortcomings of the traditional models of sociotechnical system analysis and interventions. Specifically, they argue that the traditional focus on the relationship between an organization's social system and its physical technology has severely limited the scope of sociotechnical system applications. Without revision, the models cannot be applied to settings in which the predominant technology is undergoing rapid change and development, as in the military. It is argued that a more fundamental examination of the processes set in motion by sociotechnical system intervention is called for. In this perspective, traditional sociotechnical system models of intervention in technical settings are viewed as a subset of a more generic type of change methology that is not constrained by contextual factors in the situation. The overall model developed by Pasmore et al. will be described next.

First, it is assumed that organizations are fundamentally cooperative systems, designed to coordinate the work of specialized individuals in the service of objectives which no individual could accomplish working alone. Specialists are necessary in a complex organization like the military because the amount of knowledge that must be available to the organization in order to allow it to function effectively is simply too great for all members to share equally. However, the presence of specialists creates the inevitable tug between integration and differentiation described by Lawrence and Lorsch (1967), which is usually managed through complex formal and informal control mechanisms.

Still, our observations lead us to believe that most organizations are not nearly as cooperative as they could or should be. Most organizations create jobs for individuals rather than paying attention to the natural interdependencies in the work among task performers; then, individuals are trained to perform these specialized tasks, evaluated on the basis of their individual performance, and rewarded for their successes or fired for their failures as if the outcomes of their work were completed independent of others. Territoriality develops among different departments; hierarchies become rigid and reduce the flow of communication throughout the organization; and conflict ensures as critical tasks fall between the cracks and scapegoats are sought out to carry the blame. Without cooperation, the organization becomes little more than a collection of individuals, each working to maximize his own gains.

Our model, in recognizing these shortcomings of traditional work design, is based on the premise that the degree of collaboration achieved by organizational members has a direct impact on organizational performance. To the extent that critical resources can be brought to bear in the solution of emergent problems, know-

ledge is shared freely which enhances the coordination of effort,
and people pitch in to help each other out, the performance of the
organization should improve. Such behaviors cannot be legislated
by traditional means (job descriptions, policies and procedures,
direct supervision) because many of the tasks which must be performed
to maintain or enhance an organization's viability cannot be
specified in advance. They are emergent in nature, stemming from
the unpredictability inherent in the pursuit of organizational goals
in a turbulent environment. Technology breaks down unexpectedly;
old members leave and new replacements must be trained; new ways of
performing tasks are discovered and must be tried out; demands
placed on the organization by its environment change, necessitating
changes in the product or processes of producing it; societal values
change, requiring the development of new roles for workers; and so
forth. No amount of prior planning, specification, or imposition
of top-down control will be effective in dealing with these issues;
they can only be solved by the commitment and discretionary efforts
of individuals throughout the organization, working in a cooperative
manner to help make the overall organization successful. Moreover,
we should note that the need for spontaneous cooperation and commitment to organizational success increases under conditions of rapidly
changing social values, technological development, and environmen-
tally induced uncertainty. The reader should recognize these as
the conditions we spoke of earlier which are facing the military
and many other of our organizations and institutions today.

What would an organization look like that fostered the commitment and spontaneous cooperation across levels and among groups
needed to respond to the challenge of adapting to change? While
this is not yet completely clear, it would seem that at a minimum,
such an organization would emphasize the fulfillment of human needs
that are the source of motivation and involvement. Research indicates that such needs include the opportunity to learn new skills
and develop new talents; to develop satisfying social relationships;
to receive feedback concerning performance; to be rewarded fairly
and equitably for efforts put forth; to have some say over how the
work is performed; to perform tasks that require some skill and
are worthy of respect; to be able to see one's contribution in the
end product or service; and to be involved in decisions which have
a major impact on the opportunity to satisfy these needs.

Furthermore, such an organization would be effective in the
process it uses to produce its goods and services to the extent
that it is staffed by individuals who possess a wide range of skills
that would allow them to respond to change, solve old and new
problems, and enhance mutual understandings of the interdependencies
that exist in the work itself. When these skills, abilities and
understandings are applied spontaneously throughout the organization, problems can be solved at their sources in less time and with
fewer ramifications throughout the rest of the organization than
would otherwise be the case.

Such an organization would design its technologies and performance procedures to allow its social and task systems to function together smoothly; would sculpt its leadership, structure and policies to create a climate in which cooperation, learning and problem solving could take place; would strive to minimize the barriers to cooperation imposed by traditional hierarchies and departmentalization; and would be capable of detecting and responding to relevant changes in its internal and external environments which threaten to make existing arrangements ineffective if not obsolete. Finally, the organization would need to be able to simultaneously provide clear goals to direct the efforts of its members toward common purposes, while maintaining the flexibility to consider those goals in light of new information.

Sociotechnical system interventions have helped organizations to create these conditions, in that these interventions pay explicit attention to the dependencies among groups of people created by the technology used in performing tasks. Quite often in sociotechnical system interventions, autonomous work groups are formed to oversee such relatively whole tasks and control the problems associated with their performance. The performance of the task becomes more dependable and effective as members of these groups are able to train one another, fill in for each other during absences, share information, and put their heads together to think of better ways of doing the job. As the spirit of teamwork increases, cooperation becomes more spontaneous and carries over into dealings outside of the group; relationships with supervisors change from being stilted, formal and ineffective to open, communicative and mutually supportive; important social needs, such as the need to belong and the need to be recognized by one's peers for one's contributions are met; and the organization as a whole becomes smarter, more flexible, and more integrated.

A model for introducing these changes in an existing organization would need to recognize the different assumptions and values upon which the new design would be based, and help all members deal with the transition from traditional ways of working to their roles in the new system. The model would need to reflect the values on which the new organization is based in the way it involves people in the change process itself; and it should involve analyses of current goals, methods and procedures in light of existing and forecast changes in the social and technical subsystems of the organization as well as in the environment. It was our intention through this research to create this model; but first, we needed to assess the feasibility of creating and maintaining a sociotechnically designed organization in the military, given the climate of the military environment and traditional views on how to design and command military units. Some of the issues we expected to encounter and indeed did are discussed below.

The Military as a Setting for Sociotechnical System Intervention

From the outside, the Army appears to some as a highly structured, inflexible, hierarchical bureaucracy. Malone and Penner (1980),
point out that there is no "union contract" for members of the Army
which specifies fair treatment of rewards; rather, individuals are
expected to follow orders and fight to win at any cost. Malone
and Penner go on to say that in the Army, class and status differences are enforced by law; that the Army is characterized by extreme
standardization and centralization of authority; and that members
of the organizations are not permitted to leave until their tour of
duty expires.

Similarly, Turney and Cohen (1978) note that the military
structure is explicit and visible; that most support systems for
housing and other services are run by the military and create a
"company town" atmosphere; that the military is plagued by the
constant turnover of personnel at all levels; that conflict often
exists between military and civilian personnel who are governed by
different rules; and that the objectives of the military shift from
maintaining readiness in peacetime to performance of specific
missions in wartime. Mulder, Ritsema and DeJong (1971) note that
in crisis situations, military leadership becomes even more directive than at other times.

To say that these stereotypical views represent the viewpoint
of an insider in any particular unit would be misleading. Just as
there is really no "average American family," there is no "typical"
army unit. In truth, each unit is a distinct combination of the
people, management system and technologies it uses in accomplishing
its mission. While many constraints do exist that limit the freedom of a commander to experiment with the design of his organization,
a good deal of latitude still exists. This latitude is especially
apparent in units that operate highly sophisticated and rapidly
changing technologies, in which soldiers have expertise that rivals
that of civilian professionals and the climate of the unit is more
flexible and business-like than stereotypically military. In these
settings, according to Umstot (1980), the general process of performing organization development, emphasizing organization improvement, client ownership and action research may be quite applicable.

Further, when one considers that the military faces manpower
shortages in several critical skill areas, increasingly sophisticated
combat technologies such as M-1 tank components, computers, and
missile systems it becomes apparent that the stereotypical Army is
outdated. The twenty-first century will demand more efficient use
of existing resources, more knowledgable personnel, more sophisticated weapon systems, and new organizational forms that allow faster response and greater ability to solve emergent problems. In
short, the Army of tomorrow will have more of the characteristics

associated with sociotechnical system design while maintaining the
benefits of centralization and standardization that it enjoys today.
Given this perspective, the present research can be viewed as a
first step in examining the obstacles that currently restrict the
transition of the military from old to new modes of operation, and
an attempt to outline a model for intervention that would make such
obstacles less formidable to overcome.

The Research Setting

The site selected for this research was a data processing center located at USAREUR headquarters in Heidelberg, West Germany.
The unit employed approximately eighty personnel, most of whom were
highly educated data processing specialists. The site employed
civilian as well as military personnel, and experienced turnover in
it human resources at all levels ranging from twenty to sixty percent per year during the study.

Several aspects of the site made it an appropriate arena for
the Army's first investigation into sociotechnical intervention.
First, the technology used in the data processing activity was
advanced and undergoing nearly constant change. Second, the site
was not so constrained by performance specifications that it could
not experiment with new ways of operating, within limits.

Third, the relatively small size of the unit allowed much more
thorough examination of the organization and closer contact with
the people in it than would have been possible in other, larger
military units. Fourth, the unit was stationary, and co-located
with the Army Research Institute field office, reducing the need
for travel and increasing available contact and observation time.
Finally, the unit was commanded by a Colonel known for his progressive orientation and prior use of organization effectiveness
(development) techniques. Future efforts in larger, more mobile,
more combat-oriented and less technologically sophisticated units
may call for re-examination of the results we obtained here or
variations in the model of intervention we created; but all in all,
the site turned out to be quite appropriate to satisfy our research
needs.

The Inquiry Process

As we noted earlier, any model of sociotechnical system intervention based on underlying values of cooperation, participation
and multi-level problem solving should itself reflect these goals
in its process. So too, we wished our research effort to incorporate these values and serve as a model for how the organization might
operate. To this end, we designed a plan for inquiry that we hoped
would maximize the involvement of unit personnel in conducting the

diagnosis of the organization; would provide opportunities for
learning and experiences with effective teamwork; that would emphasize client ownership of data and responsibility for decisions regarding the future of the organization; and generally serve as a
model of the collaborative problem solving process underlying the
sociotechnical approach. This plan called for an initial series of
meetings, interviews, and data feedback to managers in the organization at all levels in order to clarify understandings surrounding
the investigation, followed immediately by the formation of two
"parallel groups" that would collaborate in the study and formulation of changes. These parallel groups were composed of ten members each from all levels and parts of the organization, and were
composed of roughly one-third workers, one-third middle management,
and one-third top management.

     These groups were given the tasks of helping to design the
diagnosis of the organization, responding to the data collected,
formulating or reacting to recommendations for change, and constantly communicating to others in the organization regarding the
purposes and progress of the study.

     These groups were to meet with the researchers periodically
and to report the results of their activities to a steering committee
consisting of the top-level management team of the organization.
The dynamics surrounding the creation and operation of these groups
over the course of the study are richly reflective of the issues
associated with the development of a sociotechnical climate in the
military, and will be discussed in greater detail shortly.

     Data for the diagnosis of the organization were collected
through a survey and interviews conducted with all personnel; observation; archival records, particularly of unit policies and
procedures; and an analysis of the nature of tasks performed using
an analytical guide based on the theory of sociotechnical organization outlined earlier. Data from all of these sources were collected by the researchers and then summarized and shared with the two
parallel groups. Later, a series of recommendations for change
developed by the researchers based on the data were also shared with
the parallel groups for their consideration, in addition to any
recommendations they might develop on their own. Eventually, the
steering committee was to decide which of the recommendations
supported or developed by the parallel groups would be implemented.

Issues in the Utilization of Parallel Groups

     The formation and operation of the parallel groups in the unit
was perhaps the most visible sign to employees that the organization was in the process of undergoing a major transition in work
organization. For some, the groups meant an opportunity to be

heard at the top level of the organization without intermediate interference; for others, the groups represented the potential erosion of their authority, discipline or control.

Because it was anticipated that the utilization of parallel groups would be viewed by some as contrary to traditional leadership philosophy in the military, it was decided to run the two groups differently. Each group would appoint a rotating chairperson, secretary and critical observer; but one group would receive team building and problem solving input designed to help it operate in a business-like manner while the other group would operate as it naturally would within the military environment.

The results of this small study within the larger study indicated that the group receiving the additional training was more cohesive, offered more suggestions, and felt more successful after several weeks of work than the "regular military" group (Shani, 1981). This, along with the data collected through the surveys of unit personnel (Pasmore, Nogami and Shani, 1980), suggested that this unit was more similar to than different from many industrial settings in which similar projects have taken place.

Still, the issue of the conflict between the status and rank-free discussion of the parallel groups and the traditional emphasis on military hierarchy in the unit was smoldering just below the surface. Later, when the researchers were absent from the site and the groups met with the steering committee to discuss the recommendations, the conflict burst into the open. Despite long hours of discussion, the groups has been unable to come to consensual agreements on most of the recommendations; and because the groups had been warned of the pitfalls of resorting to simple majority voting, the groups were not prepared to tell the steering committee and the Colonel what they expected to hear - namely, clear cut recommendations for action.

The meeting between the groups was stormy, and filled with frustration for all concerned. The Colonel ordered the groups to go back to work and make decisions, which the groups proceeded to do in a compliant but perfunctory manner. It became evident that the members of the groups felt uncomfortable in recommending courses of action for the organization, even though members of the steering committee had served on the groups since the beginning and were part of the process. It also became evident that some people felt forced to "volunteer" to serve on the groups, and unable to extract themselves once the groups began meeting. Slowly, the parallel groups were being transformed by the dynamics of the traditional organization into powerless committees paralyzed by the combination of frustration and fear experienced by the participants who served on them. Some people began to express doubts about the entire project, saying that all the unit needed to succeed was a bit more

discipline and respect for authority. Others felt the ideas
behind the project were sensible, but that the organization simply
was incapable of enacting them. A few, including the Colonel,
were not yet prepared to give up trying. The Colonel merged the
two groups into one, allowing some group members to bow out grace-
fully in order to make the remaining group small enough to be
workable. The Colonel then joined the group personally, and gave
it the direction it had been lacking previously. He was careful
not to shortcut discussions or exercise his formal authority in the
group except when a policy decision needed to be made to allow the
group to work further. Still, his presence created a whole new
set of issues, including the isolation of some members of the
steering committee from the decisions being made in the parallel
group and suspicions on the part of some that he was slowly manipu-
lating the group to accept his ideas of what should be done.
Nevertheless, the group began to make progress and eventually took
the steps necessary to create the new organization.

The Old and New Organizations

     Prior to the intervention, the unit consisted of two distinct
customer product oriented divisions, a highly skilled group of
systems analysts, a comparatively unskilled group of computer
operators, and various staff departments. This structure had
evolved over time in response to various pulls and tugs on the
organization. Still, the design highlighted the differences among
the various divisions rather than the similarities; supported the
traditional hierarchy of command even though in many cases lower
ranking personnel were more expert in computer programming on
operations than their temporary officers; blocked the sharing of
knowledge, inhibited the application of skills learned through
classroom training; and separated the workers from the users of
their services. The survey and interview data pointed out these
problems and others, to which the new organization would need to
respond.

     The tasks analysis conducted during the data collection effort
revealed corresponding problems in the way work was being performed.
A variance analysis (shown in Figure 1) indicated that key variances
(problems) in the provision of data processing services existed in
the areas of interpretation of instructions for projects, project
planning, the level of teamwork achieved, workload surges, program
documentation, program quality, user acceptance, user knowledgabil-
ity, preparation for releasing programs as finished projects and
program maintenance.

     These variances were related to tasks that were mainly cooper-
ative and discretionary in nature; i.e., tasks that the design of
the formal organization mitigated against and were performed
mainly because of the dedication of unit personnel. Together, the

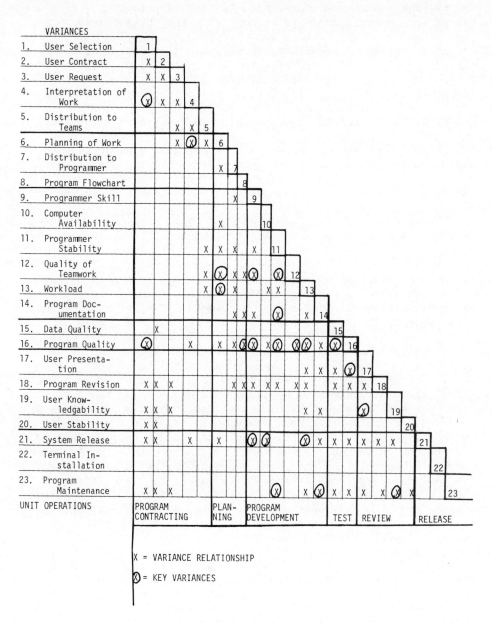

Figure 1.   Variance Matrix

analyses of the social and technical systems pointed toward the
need to reconfigure the organization in order to emphasize training
and the sharing of knowledge, make it more cooperative, more rewarding to its members, and more in tune with its environment.

The new organization, as it was eventually implemented, was a
matrix structure of sorts. The two distinct project divisions were
merged into one under the direction of a Project Manager whose
responsibility was primarily for the accomplishment of tasks in a
timely manner and in accordance with customer needs. However, the
people he would need to work on the various projects were "owned"
and assigned by a Human Resource Manager, who was responsible for
the training and development of unit personnel. The Human Resource
Manager would assign people to projects in ways that increased
their skill levels, facilitate the sharing of knowledge about programs and ideas about programming, and match together experienced
and inexperienced personnel.

The parallel groups worked out elaborate descriptions of the
roles of all persons in the organization at all levels in the new
system, paying particular attention to what workers could and should
expect from managers and vice versa. Policies regarding movement
within the organization were drafted; means of assessing performance
and discussing it in meetings of the worker, Project Manager and
Human Resource Manager were developed; the continuing role of the
parallel organization in the unit was outlined; and a performance
contracting process was instituted which clarified goals as well
as rewards for goal accomplishment desired by the worker (within
limits of course). Many agenda items relating to policies, procedures, the organization of social events and a broad range of
other targets for change remain to be worked out in the future by
the parallel groups; and many issues relating to the day to day
role of supervision, project development and customer relations
have yet to be settled. Despite the unfinished quality of the meta-
morphosis however, the unit put in a sterling performance during
its last major exercise, tripling the number of users it served
with an even greater diversity of services being offered than the
year before. Some members of the unit are still set strongly
against the changes that have been introduced, fearing that the
new system might raise expectations to an unrealistic level or
eventually impair the operation of the unit when a quick response
is called for. A roughly equal number seem to favor the changes
with a comparable positive passion; and the largest number of
people continue to view the changes with mixed emotions. Most
agree that when the Colonel leaves in a year and a half, much of
what has been accomplished will be wiped out in a series of swift
orders by the new commanding officer; and realistically, in an
organization as large as the Army and with such a long tradition
of centralized control from the top down, they may be right. It
is difficult for a large organization to accept a small sub-unit

that operates as an aberration from the norm, even if it is doing
so within regulations.

The Future of Sociotechnical System Interventions in the Military

Since this effort began, other projects have also been undertaken which demonstrate that it is indeed possible to restructure
military units using sociotechnical system principles. As in
industry, most of the informal reports have been quite positive.
Furthermore, the Army's school for training Organizational Effectiveness Staff Officers has introduced sociotechnical system theory into
its curriculum. From all signs, the Army as a whole is becoming
increasingly aware of the need to cope with the demand to do more
with less, particularly in the area of training and utilizing
human resources to operate new and highly sophisticated technologies.
Because the world is becoming more complex rather than more simple,
and because societal values continue to shift in ways that emphasize
the legitimacy of fulfilling human needs at work in organizations,
the trend toward increased usage of sociotechnical methods in the
military should continue as it has in industry.

Although space will not permit a discussion of the detailed
model for intervention developed as a part of this effort, we should
point out that it is currently being prepared for publication by
the Army Research Isntitute for dissemination and that this too
should speed the diffusion of sociotechnical methods in the military.
We are already aware of at least one large scale project being
undertaken which has the potential to effect thousands of jobs;
it and others like it should serve as the stimulus to test the
model, refine it, and confront the issues associated with creating
new organizational forms within a very traditional context. Much
work remains to be done; but with careful conservatism, research
and exploration, it would seem that the future for sociotechnical
system interventions in the military is bright despite the obstacles
that are sure to be encountered along the way.

REFERENCES

Friedlander, F. and Brown, L.D. Organization Development. Annual
        Review of Psychology, 1974.
Lawrence, P. and Lorsch, J. Organization and Environment,
        Cambridge, Massachusetts, Harvard University Press, 1967.
Malone, D. and Penner, D. You can't Run an Army Like a Corporation.
        Army, February, 1980, 39-41.
Mulder, M., Ritsema, V. and DeJong, R. An Organization in Crisis
        and Non-Crisis Situations. Human Relations, 1971, 19-41.
Pasmore, W. Roadblocks in Work Restructuring. Presented to 39th
        Annual Academy of Management, 1979.

Pasmore, W. and King, D. Understanding Organizational Change: A
    Comparative Study of Multifaceted Interventions. Journal
    of Applied Behavioral Science, 1978, 455-468.
Pasmore, W., Nogami, G., Shani, A. Sociotechnical Approaches to
    Change in USAREUR. Presented to the American Psychological
    Association, Montreal, September, 1980.
Pasmore, W. and Sherwood, J. Sociotechnical Systems: A Sourcebook.
    University Associates, 1978.
Pasmore, W., Srivastva, S. and Sherwood, J. Social Relationships
    and Organizational Performance: A Sociotask Approach. In
    W. Pasmore and J. Sherwood, Sociotechnical Systems: A
    Sourcebook, University Associates, 1978.
Shani, R. Understanding the Process of Action Research in Organiza-
    tions: A Theoretical Perspective. Doctoral Dissertation,
    Case Western Reserve University, Cleveland, Ohio, 1981.
Srivastva, S., Salipante, P., Cummings, T., Notz, W., Bigelow, J.
    and Waters, J. Job Satisfaction and Productivity. Department
    of Organizational Behavior, Case Western Reserve University,
    Cleveland, Ohio, 1975.
Tichy, N. and Nisberg, J. When Does Work Restructuring Work?
    Organizational Innovations at Volvo and G.M. Organizational
    Dynamics, 1976, 63-80.
Turney, J. and Cohen, S. Organizational Implications for Practicing
    OD in the Army. Personnel Psychology, 31, 1978, 731-738.
Walton, R. Innovative Restructuring of Work In Pasmore, W. and
    Sherwood J. Sociotechnical Systems: A Sourcebook. University
    Associates, 1978.
Umstot, D. Organization Development Technology and the Military:
    A Surprising Merger" Academy of Management Review, 2, 1980,
    189-201.

IMPACT OF TWO SUCCESSIVE MECHANIZATION PROJECTS ON MOTIVATION

AND WORK ORGANIZATION IN A BANK

Raoul Gobel and Andre Meers

Studies and Personnel Forecasts Unit
Direction of Social Affairs
Generale Bankmaatschappij
1000 Brussels, Belgium

ABSTRACT

The structure of the Bank is such that in the branches mainly commercial work is done, whereas the administrative tasks are carried
out by large, centralized operational departments. In the latter,
an interdisciplinary study on the medium-term repercussions of
routine screenwork showed that mental and physical health, as well
as work satisfaction, were more negative when the work is more mechanized. In the future the Bank plans to equip the branches with terminals connected to its data-processing network. Thereby administrative tasks would mainly be carried out in the branches, so that the
large operational departments would virtually be eliminated. As such,
one meets the principle that repetitive work should be shared out
amongst as many people as possible. However, transferring hundreds
of employees from the administrative to the commercial sector raises
enormous problems, e.g. retraining.

INTRODUCTION

Structure of the Bank

The Bank in question is the "Société Générale de Banque -
Generale Bankmaatschappij". This purely Belgian Bank is the largest
in Belgium with total assets of 30,000 million dollars. Apart from
the Central Administrative Office, which employs about 10.5 % of the
total personnel, the Bank is divided into 16 Regional Offices. These
Regional Offices, which employ between 300 and 4,800 staff each, are
in turn divided into an administrative and a commercial sector. The
commercial sector, comprising about 50 % of the personnel, is divided
into branches. The total number of the Bank's branches is 1,169 ;

their task is more or less polyvalent, although the number of their
personnel varies from 1 to 200 (average 4).

Present functioning

These branches principally carry out commercial work, which is
supported to a limited extent by information about the client available on microfiches. A large part of the administrative task consists
in the recording of incoming and outgoing amounts on paper tape in a
mechanical accounting-machine. The real administrative processing takes
place in the central operational departments of the Regional Offices.
Punching-machines were used for this purpose until 1973 and were then
replaced by video screen units. The data are recorded on a cassette
and grouped at Regional Office level on magnetic tape. The tapes are
in turn processed in one of the six computer centers. In these centers, the actual accounting operations are carried out, the client's
statement of account is prepared and the microfiches are produced for
the branches.

Personnel data

About 1/3 of the 16,192 employees are women, mainly - though not
exclusively - concentrated in the administrative sector ; 27 % of
these women work part-time. The average age of all employees is 36,
the average seniority 13.6 years. 12 % Have a university or high
school degree, 43 % have completed secondary education. Turnover is
1.7 % ; resignation is practically nil. Absenteeism is 4.5 % ; the
average cost per employee is among the highest in the world, namely
34,000 dollars.

INVESTIGATION ABOUT THE SOCIAL IMPLICATIONS OF WORKING WITH VIDEO
SCREEN UNITS IN A TRADITIONAL ADMINISTRATIVE DEPARTMENT

Description of the situation

The group investigated is that which uses video screen units for
the input of simple alpha-numerical data in a work cycle of 15 to 30
seconds according to the type of operation (e.g. check or transfer).
We are thus dealing here with purely mechanized work, carried out in
large departments which are structured in such a way that employees
either undertake this type of work only or alternate (in the course
of each day, week or fortnight) between this and traditional administrative work (pen and paper). There are slight differences in the
kinds of mechanized work carried out, according to the sort of program
(operation) and according to whether the work is input or checking.
Depending on the organization of the Regional Office, there is either
mixed or specialized allotment of mechanized tasks. About half of the
employees concerned have worked previously with punching-machines.

The Problem

About 1 year after the punching-machines were replaced by video
screen units, a short inquiry was carried out among the employees
concerned. The results were unequivocally positive : the new technology was seen as a clear improvement. Two years after the introduction
however came the first complaints, put forward either directly by
employees or indirectly by their union representatives. The continual
increase in these complaints, mostly very vague in nature, but above
all concerned with general fatigue, eye-strain, nervousness and feelings of monotony, led to this inquiry being carried out. An obvious
explanation for this increase in complaints could be that the further
removed in time from the earlier (and manifestly less good) technology, the weaker the referencial aspect becomes. Moreover, a number of
employees recruited in the intervening period had never worked with
punching-machines. Although this (formal) factor did undoubtedly play
a role, we concentrated on the time, and more precisely on the number
of years people have been doing this sort of work and endeavoured to
determine whether that might not sui generis be behind some of the
complaints. The problem was, more precisely : "What is the influence
in the medium term (3 to 4 years) of this sort of routine work with
screens on the physical and mental health and on the job satisfaction
of those involved ?"

Approach

This consisted in a comparative investigation : groups working
exclusively with the screen (N1 = 51) were compared in various aspects with groups alternating this sort of work and traditional administrative work (N2 = 36) and with groups carrying out the latter
type of work exclusively (N depending on the type of measurement).
For this purpose, groups belonging to each of these 3 categories were
examined in 3 different Regional Offices. The characteristics of the
groups were recorded and (from a qualitative point of view only)
taken into account when interpreting the results.

Although the type of work, resulting from the technology used,
was the most important independant variable, we cannot in practice
isolate this factor from the overall working situation. Certain other
aspects are indissolubly linked to this main dimension and co-deter-
mine physical and mental health and job satisfaction. Attention was
therefore also focussed on the physical, ergonomic and psycho-social
working environment (job content evaluation, working atmosphere , style
of supervision). In view of this fairly broad approach of the problem,
an interdisciplinary method was necessary. The medical, psychological
and sociological, as well as the technical and organizational aspects
were therefore taken into consideration. (*)

*Apart from the 2 authors, the following also worked on the investi-
   gation: Prof. Verhaegen, Dr. Van Mulders, Mr. Deschamps and
   Dr. Vervinckt.

The investigation was problem-oriented and this subsequently made it somewhat difficult to reduce the results to their correct proportions when they were reported to certain parties (unions). As such, the investigation did not represent a true balance of all positive and negative factors of the work situation. All those involved were given the necessary information beforehand and afterwards.

Results of the investigation

Physical working comfort

Physical working comfort is dependent on the physical working circumstances and the ergonomic characteristics of the working place. Here the following elements were assessed by means of physical measurements, questionnaires and observation : noise, lighting and visual comfort, absolute temperature, humidity, ventilation, quality of the air, the overall working environment and the ergonomic characteristics of the individual working place.

The most important findings in this respect were :

Noise. Workers with screens reported significant progress compared with the previous punching-machines. The noise level of the screens is in fact around 60 db. There were complaints about the monotonous noise of the fans in the video screen units. It was interesting to find that physical factors sometimes conceal psychological causes. Certain complaints about the disturbing noise of the fault tone or the clicking of the keys appeared to be induced by a lower performance in comparison with other workers. Certain equipment installed in the environment (printers, telex-machines, photocopiers, microfilm appliances) drew complaints about the absolute noise level, the measurements show values between 70 and 80 db. Ultimately however, it can be stated that the noise problem was not a dominant factor in workload. Most of the difficulties could be helped by applying sound-absorbing material or by re-arranging certain equipment.

Lighting and visual comfort. This is an extremely important factor in work with screens and in this respect, the situation seemed to be anything but perfect. The most common problems referred to :
- the general illumination being too strong (500 lux and over)
- high reflection on the screens producing poor contrast
- inadequate maintenance of the screens and ergonomic shortcomings of the equipment in question.
Measures taken in this respect included reduction of indoor lighting, reorientation of the screens, use of blinds and increased maintenance.

Temperature, humidity, ventilation and quality of the air. These problems are of particular importance for work with screens since the latter is exclusively sedentary, which increases sensitivity to the environment. The main comments referred to :

- the poor functioning of the air conditioning system and in particular, the problem of air currents
- the general dislike of carpeting in office buildings (causing dust
  concentrations)
- the impossibility of controlling heating in individual working
  areas.

Overall working environment. The idea of landscaped offices is
accepted. It could, however, be noted that over the years this layout
has become less functional.

Ergonomic characteristics of the working place. The equipment
(video screen units) shows the ergonomic shortcomings characteristic
for the period in which they were designed. Possibilities of improvement in this case are limited. On the other hand, it was striking
to find how very simple measures could greatly improve the physical
working conditions.

Conclusions about physical working comfort. We originally
thought that in the past, serious efforts had been made to improve
physical and ergonomic working comfort. Thus we were suprised by the
relatively high number of complaints. The most important complaints
related to the lighting and air conditioning system and the noise
overload. Certain of these complaints were clearly well-founded.
The situation could then be improved by a number of corrections
(such as re-arrangements). We had the impression, however, that there
was hypersensitivity to this external environment and that a shift
had taken place in the pattern of complaints. This impression was,
in fact, confirmed in the further course of the investigation.

Fatigue_and_health

Momentary fatigue. An attempt was made to produce a fatigue
curve for each of the 3 groups (exclusively machine work, work combined with traditional administrative work during the day, exclusively traditional administrative work). On various days, 4 times per
day, fatigue measurements were taken for this purpose, using :
- the critical flicker fusion frequency
- a graph on which the person indicated his degree of fatigue
- a form containing 30 questions about fatigue.

One of the first findings was that there was a fairly high correlation between the different measurements of fatigue. Although the
number of cases of extreme fatigue showed no significant variation
among the different groups, the fatigue score at any moment of the
day was slightly higher in the exclusively-machine-work-group than
in the other two. Moreover, the fatigue score in the first group
showed certain peaks:

- after two hours uninterrupted work with screen there was a considerable increase in fatigue
- in a real work situation we always find certain technical or organizational bottlenecks which cause a temporary increase in fatigue.

The most frequent complaints in the "mechanized" group were : tired eyes, heavy feeling in the head, pain in the back. On the basis of the results obtained we endeavoured to work out a suitable arrangement of breaks and to flatten out peak moments and days by organizational and technical measures. Generally speaking, it cannot be said that the situation in this respect was alarming.

Permanent fatigue. An extensive inquiry in this connection showed that for 19 of the 87 persons (groups 1 and 2) we may speak of more or less permanent fatigue and more than normal strain. As in other investigations, it appeared that in most cases this was due to a combination of professional and non-professional causes. It remains however, that in a number of cases, the work situation also played a role in this respect, sometimes even a dominant one. Moreover, it is significant that of these 19 persons, a relative majority belonged to the exclusively-screen-work-group. As the main causes of this fatigue and nervousness, this group stated : high workload on busy days, the monotonous rhythm of the job and the feeling of carrying out a not very enriching task.

Health. The results of the questionnaire, comprising 48 possible health complaints, showed that :
- in general the average number of health complaints was surprisingly high
- female employees had more complaints than male colleagues doing the same sort of work
- nevertheless, the work factor had a greater influence on the number of complaints than the man-woman factor : the number of complaints was in proportion to the volume of mechanized work.

The nature of the complaints may be grouped as follows : nervousness, eye troubles, physical discomfort, headache, backache. In particular, the first 2 types of complaints appear more frequently in group 1 (exclusively-screen-work). A close analysis of absences in the different groups over 3 years confirmed the findings of the health questionnaire. There was a high correlation between the 2 parameters concerned.

These data as a whole were interpreted as follows. A number of the complaints expressed (physical discomfort, backache) were clearly connected with physical and ergonomic imperfections in the working environment. This can be remedied fairly easily. The fact that there are more eye complaints in the exclusively-screen-work-group than in the others is understandable. Further investigation revealed that these complaints referred to reversible and functional eye complaints

and not to eye injuries.(*) More important was the very high number
of complaints of strain and nervousness (confirmed by a high degree
of absenteeism) in the exclusively-screen-work-group. In a limited
number of cases, particularly for middle-aged persons, a fundamental
lack of adaptation to the new technology was observed. In a large
number of cases, the mechanized nature of the work had an adverse
effect on mental and physical health. The cause did not appear to
lie primarily in an exaggerated work-rhythm. Most of the persons
involved, confirmed that the rhythm, apart from certain peaks, was
reasonable. The main cause was the monotonous and repetitive character of this sort of work, resulting in a lack of job satisfaction
for many people. This basic situation was aggravated by certain
other negative factors, which are likely to cause a decrease in
physical and mental well-being.

Psycho-social factors in the work situation

Evaluation of job content and expectations for the future.
Detailed inquiries in this respect clearly confirmed the hypothesis
already formulated : job satisfaction was more negative as the job
was more mechanized. In the exclusively-screen-work-group, the following 3 aspects were highlighted :
- the work does not require much capability ; it even involves atro-
  phy of existing capabilities, and this reduces the possibilities
  for the future
- the monotony and routine are an autonomous source of fatigue and
  nervousness
- the work is also regarded by others as inferior, and this is da-
  maging to the self-image.

In such a context, a request for variety is expressed in 2
different forms :
- some demand variety in order to increase their polyvalency and
  thus secure their own future. The alternative task must thus be
  at a certain level and the cycle of rotation must be reasonably
  long (e.g. 2 weeks) ;
- others aim more at interrupting monotony and fatigue during the
  day. Any different task (whether mechanized or not) might be considered in this connection.

Working atmosphere. Apart from some small problems, the working
atmosphere was said to be good. It was surprising to find that this

---

* In a longitudinal investigation over a period of 3 years among
  another large group of screen workers, this finding was confirmed.
  Given that the work situation is ergonomically correct, it can
  be stated that, over a relatively short period (3 years), there
  is no causal connection between screen work and eye damage.

working atmosphere was given a great deal of weight in the overall
job satisfaction.

   Style of supervision. An equal amount of weight was given to
the style of supervision. In general, this can be interpreted as an
illustration of the fact that the extrinsic elements of the task
compensate for the lack of motivating intrinsic elements. Nevertheless, attitudes towards the style of supervision are not exclusively positive. Supervision is regarded as disfunctional in cases where
the superior assumes a strict controlling attitude. In fact, it can
be objectively labelled as disfunctional. In a strongly structured
work situation with in-built technical controls, a softer style of
supervision with a mainly sustaining function is recommended.

Conclusions

   In general, it may be said that the more mechanized the work,
the less favorable the situation is in respect to both job satisfaction and mental and physical health. Neither working with a screen
nor the work-rhythm is the decisive factor, which rather resides in
the fact that the job is devoid of content and is repetitive. The
demand for variety is therefore widespread. A further inquiry among
those working exclusively with a screen revealed the existence of
3 different groups :
- a limited number of persons ($\pm$ 1/4) looked on full-time screen
  work from a neutral to positive stand-point, because of limited
  capabilities and/or ambition
- another group ($\pm$ 1/4), the "extrinsically motivated", were not
  satisfied with the job content, but saw that it was sufficiently
  compensated for by other elements in the work situation (working
  atmosphere, style of supervision etc.)
- a third group ($\pm$ 1/2), the "job-oriented," no longer accepted this
  situation and wanted a more enriching task.

Possibilities of improvement

   Improvements in the physical, ergonomic and psychological work
environment may not change the overall work situation fundamentally
but they do make it more acceptable for a number of persons. The
other measures all tend to offer more variety. These include : de-
specialization within the mechanized sections, decentralization of
a number of subsidiary activities (control activities, microfilming),
delegation of tasks. A more fundamental improvement consists in
merging mechanized and traditional administrative sections. The
problem here however, is that the overall volume of non-mechanized
work is continually decreasing. On condition that transfer-pos-
sibilities still exist, the newly recruited, after a period of 2 or
3 years of mechanized work, can be allowed to pass on to more
enriching tasks. A further possibility would be to take on unem-
ployed young people under a one-year contract under the Ministry of

Labour scheme. In practice, the solutions proposed here are only
partially feasible. Our hopes therefore lie in the introduction of
terminals in branches, which will eliminate many of the purely mechanized tasks.

INVESTIGATIONS ABOUT THE SOCIAL IMPLICATIONS OF EQUIPPING BRANCHES
WITH INTELLIGENT TERMINALS

Within a period of 4 years, all branches are to be equipped
with a mini-computer ; each counter is to have a video screen terminal and 1 printer will be installed for every 2 counters. All branches will be interconnected through a network and all internal and
external banking functions, such as home-terminals, are to be integrated. This branch terminal is to serve a threefold purpose :
- the system should permit the greater part of the administrative
  data to be processed at source, in the branch
- since information will be constantly updated, the branch will have
  a greater degree of self-management than before
- commercial capacity should increase thanks to the information
  (client profiles) available to the counter-clerks and branch
  manager.

It should also be noted that the project is being implemented
in phases: after a two-year study period, 2 branches were equipped
with test equipment. The latter was afterwards extended to 20 branches. After this test phase, which will last for a total of 3 years,
all 1,169 branches will be equipped with terminals over a period of
3 years, beginning next year. This phasing is important because it
allows for gradual evaluation and adjustment. We shall briefly deal
with the social implications of the project.

Ergonomic aspects

Ergonomics and equipment. In recent years the Bank has attached
increasing importance to ergonomic aspects. Ergonomic requirements
for equipment are considered to be as important as technical performance and price. The manufacturers too, are becoming increasingly aware of the ergonomic dimension.

Ergonomics in implantation. The infrastructure of the branches is
not ideally suited to this new technology. The lack of flexibility
involves additional difficulties.

Ergonomics and dynamic working situation. More and more, attention is shifting to the dynamic aspects of the working situation.
Commercial application of the terminal system requires an optimum
combination of the following elements :
- the equipment and its accompanying data processing logic (hardware and software)

- the logic of the internal user
- the administrative and legal regulations to be respected in banking operations
- the environment, including the external user and his logic.
Thus we are concerned with the ergonomic and psychological aspects
of the sofware. Little progress had been made in this area. The motivation, however, is powerful, since not only the comfort of the
internal user but also the commercial performance of the system is
at stake.

Effects on traditional personnel policy

The branch terminals will mean the loss of about 500 jobs in
the large operational departments. Only a fraction of these persons
can be absorbed into the additional administrative work in the branches ; in fact, the new technology should permit the greater part
of this work surplus to be done in the same time. The solution is
therefore to use the extra staff to increase the commercial capacity of the branches. The underlying assumption is that there are
still possibilities in the market and a greater refinement of the
equipment should permet further penetration of this market. This
hypotheses is not completely borne out in the present situation ;
part of the excess personnel should therefore be absorbed by natural release. But several hundreds of employees will still have to
be moved from the administrative to the commercial sector. Others
will have to be given additional training. This will be attempted
in the ways mentioned below.

Recruitment. For the last 3 years, recruitment has been almost
exclusively limited to highly educated personnel with an aptitude
for commercial tasks. They must temporarily carry out simple administrative tasks, which poses some problems, but this will make the
eventual switch-over to branches easier. Those who easily make
human contact, tend to be most quickly frustrated in the monotonous
working situation described in the first part.

Recycling and retrainging. A number of persons who do not have
the positive characteristics described above would also have to be
reassigned, although a direct transfer to the commercial sector does
not always take place. It is, in fact, possible to transfer someone
to another administrative department in order to replace someone
else who has an interest in and potential for a commercial post.
All this neccesitates extensive theoretical and practical retraining
in order to gradually familiarize administrative staff with the
commercial sector. An inquiry among a number of persons who had
been transferred, showed that the transition had, up to now, been
successful. However, it must be said that we were dealing with a
selective group; it remains to be seen whether this transaction
would have been successful in the case of persons who were already
older and had spent more years in the administrative post and had

less potential for a commercial post. In this respect, special attention must be given to staff in higher grades.

   Training. In addition to retraining for new jobs, training is given to enable staff to carry out modified jobs. The latter process is steadily gaining in importance, for example :
- those who remain in the administrative sector must be trained to cope with the growing proportion of monitoring work as automation is extended
- counter staff must be taught how to use the terminal as an administrative processing instrument, but particularly as a commercial aid
- branch managers must also be trained to exploit the available information commercially.

Effects on labour relations

   Employment. This is the most important source of concern to the personnel and their representatives. Certain actual developments, as well as sensational reports in the media, have given rise to a very emotional and critical attitude towards any form of mechanization. This general attitude naturally makes it more difficult to discuss more "neutral" aspects of mechanization. As regards the situation in the Bank, mechanization in itself is not the cause of any loss of jobs. Employment is almost exclusively dependent on the Bank's general business activity. For the moment, in any case, there is no threat to individual jobs : naturel release of personnel is exceeding the loss of jobs. Personnel representatives will understandably ask questions about the profitability of projects of this kind.

   Centralization versus decentralization. Due to the fact that the system of branch terminals is to be integrated into the whole Bank network, a very basic question arises about the degree of centralization or decentralization of responsibilities and supervising authority. The unions are actively interested in these problems since their impact is lower in autonomous and decentralized units. However, decentralization is also felt by the unions to have a positive effect on the quality of the work.

   Classification. One question which arises here is whether the counter employee, who can carry out more complex transactions using the terminal, is entitled to a higher classification. This question is not unimportant from the point of view of the cost-effectiveness of the whole project and taking account of the already very high personnel costs.

Effects on consultation procedures

The importance of a project of this kind and the demands made
by a number of parties imply that unilateral introduction is out
of the question. Therefore an extensive system of information must
be worked out, among other things by setting up working groups
(according to specialized knowledge, geographical distibution and
hierarchical level).

5131
~~5130~~
6210
8331
W. Germany

TECHNOLOGICAL AND ORGANIZATIONAL CHANGE IN OFFICE WORK

THE CASE OF THE VISUAL DISPLAY UNITS

Wolfgang H. Staehle

Freie Universität Berlin
Institut für Unternehmungsführung
Garystr. 21, 1000 Berlin 33

ABSTRACT

The aim of this paper is to create a situational and organizational frame of reference in order to specify the effects of VDUs on the individually perceived work situation. This specification will be effected by way of hypotheses, which will be tested by means of empirical data. The starting point will be the proposition that office technology itself is not necessarily inhuman, but that the reorganization of work initiated with its aid may lead to intolerable working conditions. It is also assumed that the work situation is not determined by technology but that the manager or organizer has a certain amount of freedom in the structuring of work organization, which can be used to prevent inhuman working conditions.

1. Introduction

The proportion of costs accounted for by adminstration has more than doubled in the last 30 years. The trend of productivity in the field of administration of both private firms and public institutions has lagged far behind that in the field of production for decades. The main reason for this is considered to be the different rates of technological development. The fundamental prerequisites for increasing the efficiency of office work only came into being with the introduction of electronic data processing (EDP). In each organization which has installed an electronic data processing system, the data and information to be processed must be 'translated' from normal to machine language. A wide automation gap which has to be filled by people still yawns here.
The first mechanization measures were introduced by Hollerith

(1889), who invented the punched card as a data carrier, allowing
mechanical data evaluation (the first application was for censuses
in the USA). Owing to the difficulty of locating and eliminating
input errors, the relatively slow input rate and the extremely high
noise level of the punching machine, card punching increasingly
became a problem job. Since about 1970, a new technology has been
available which not only provides a significantly better solution
to the problem of data acquisition than the perforator but also has
other adventages. The full potential of EDP can only be fully
utilized when visual display units (VDUs) are employed. Data banks
can be quickly built up, modified, checked and edited by means of
decentralized VDUs. The VDU comprises:

(a) A keyboard (as with the perforator)
(b) A display screen (cathode ray tube)
(c) A control unit.

     VDUs can be connected direct to a central processor (e.g., to
provide a continuous updating service), or they may be decentralized,
being installed independently of the computer and having facilities
for autonomous data recording on magnetic tape cassette or floppy
disk (a small magnetic disk).

     The control unit constitutes the actual advance in the field
of automation; it transfers the data entered to the magnetic data
carrier or to the CPU. It first checks the plausibility of the
input (number of positions, digits or figures, and in some cases
the order of magnitude of the figures) and displays the status of
the input to the operator on the screen at all times. A positioning
symbol (the cursor) displays the status of the input on a "mask"
corresponding to the document to be entered on the screen. Operating
mistakes are indicated to the data typist by an audible bleep. In
units connected to the CPU, the nature of the input error is
displayed by a code word on the screen.

     In the Federal Republic of Germany about 300 000 persons are
working on these units and the use of VDUs at office work places
is likely to increase at a very fast pace. This trend justifies
the degree of public attention now being focused on the spread of
EDP in general and VDU technology in particular, as well as on
their consequences. However, the interest of manufacturers, users,
trade unions and scientists has been largely confined to two kinds
of consequential effects of VDUs: At macroeconomic level, attention
is drawn to the ultimately adverse employment effects of EDP, in
particular using VDUs. At micro level, the ergonomic problems of
VDUs have aroused widespread public interest, which has been further
increased, for example, by the results of various empirical
investigations of psychological troubles in employees working
with VDUs. According to our estimates, about a quarter of all
forcasts, investigations and reports concentrate on the macroeconomic

aspects, while well over half are devoted primarily to ergonomic
facets of VDU work; however, for a long time, the aspect of <u>work</u>
<u>organization</u> was disregarded. It is only recently that the psycho-
logical troubles of VDU users have ceased to be attributed exclusively to ergonomic deficiencies (or fear of losing their jobs).

2. The Concept of Dual Work Situation Analysis

2.1. Sociotechnical Analysis

Since the ergonomic problems associated with VDUs have been
more or less fully researched now the implications for work content
and organization are becoming the new focus of interest. It is now
being conceded that changes in work organization due to the introduction of VDU work places are at least partially responsible for
ailments or negative attitudes to VDUs.

If the causes of ailments or negative attitudes to VDUs are
to be analysed in detail from the viewpoint of organizational
theory, an initial requirement is a situational description of
the sociotechnical system:

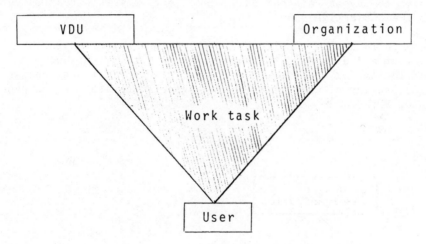

Fig. 1: Components for analysis of the sociotechnical system

As fig. 1 shows, the VDU with its technical and ergonomic components
acts together with the organizational structure as a determinant
for executing the work task which the user is required to perform.
In addition to these effects, the perception and action competence
of the user is relevant to the formulation of ailments and negative
attitudes. This results, for example, from work experiences,
vocational qualification, motivation, and attitudes.

From the viewpoint of a behaviorally oriented science of
business administration, attituded, action and behaviour of
organizational members are central aspects for analysis. However,
these variables are not influenced direct by the objective,
technological and organizational circumstances but are coloured by
the work situation as subjectively redefined by the individual
(HACKMAN 1969, STAEHLE 1977, 1980). Both the direct (e.g. objective
task, VDU technology) and the indirect determinants (e.g. work
environment, organizational climate) of the work situation must be
analysed for description and evaluation of the work organization
under consideration. Work organization here means the relations
between people and machines; its central function is the allocation
of tasks to people and machines (GAITANIDES 1976).

Depending on the form assumed by the variables, there will be
different forms of work organization, constituting the objective
work situation. Subjective work situation then means the elements
of the objective situation as perceived by the individual. This
subjective work situation is the central empirical parameter for
explanation of the actions and behaviour of those responsible for
the performance of tasks. Therefore, in dual work situation analysis
we have to investigate not only the objective work situation but
also the subjective side of it.

A third complex decisively influencing the perception,
behaviour and action of the user in a new work situation is,
specifically for VDU work, the process of organizational change
(restructuring of work organization associated with the introduction of VDUs).

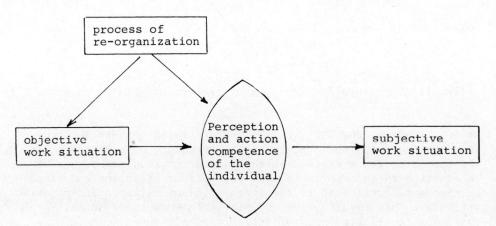

Fig. 2: Conception of empirical survey

To sum up, the objective work situation is changed in the
restructuring process when VDUs are introduced. In addition, the
process has a direct impact on the perception and action competence
of the user (see fig. 2). The three dimensions of fig. 2 are each
surveyed by one questionnaire:
- objective work situation
- subjective work situation
- process of re-organization.

2.2. Objective Work Situation: Typology of VDU Work Places

A VDU work place is one whose determining component is a VDU.
Publications on VDUs are generally based on three types of VDU
work places (see, for example, CAKIR et al. 1978, WIESNER 1978):

(a) Data acquisition position: data input with visual check by
    data typists
(b) Dialogue position: direct two-way communication with central
    processor, mainly by persons responsible for the specific
    matter
(c) Data output position: data recall by one or more users.

In our opinion, the objective work situation is not
sufficiently described by a typologization based solely on the
fundamental type of activity (input, dialogue, output). Other
important criteria for a situational analysis of the effects of
VDUs are the concrete work content, work organization within the
department or group, the relative proportions of working time
accounted for by work with and without the VDU, quality of work
without VDU, etc. These objective features of the situation affect
the attitudes and behaviour of organization members at work
places incorporating a VDU. Important situational features can
be summarized as the nature of the activity, the manner of
organizational restructuring at the time of VDU introduction and
the intensity of interaction of the worker with the technical
system represented by data processing. Among the indicators of
this intensity is the working time spent daily at the VDU. Thus,
there is a strong possibility of correlation between the frequency
of psychophysical troubles and working time at the VDU (DAG 1979).

In addition, we consider it appropriate to subdivide the type
"dialogue work place". In accordance with the empirical differences
in the qualities of dialogue work places, we therefore distinguish
the type "dialogue 1", characterized by data input and recall with
checking facilities, principally in the field of commerce, from
the type "dialogue 2", which can be described as "interactive
working" (e.g., interactive programming).

The nature of the activity and VDU working time are the
principal parameters of the work situation and as such must be

understood as continua. However, a provisional evaluation of the
available investigation results showed that six types of VDU work
places were empirically significant (see fig. 3), so that the
following types of work situations are assumed:

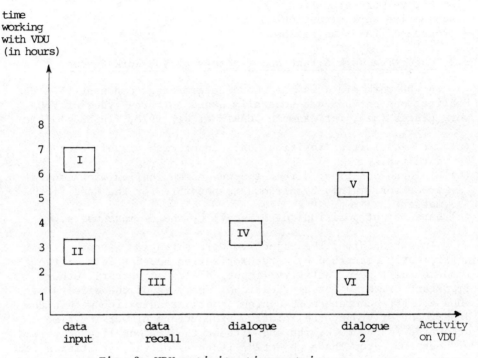

Fig. 3: VDU activity-time matrix

2.3. Subjectively Perceived Work Situation

Our starting point will be that time and the nature of the
activity at the VDU, of all parameters of the work situation, are
the ones which predominate in perception of the work situation by
the VDU user, the result of the individual processes of perception
being influenced primarily by the following factors (see HACKMAN
1969):

    Comprehension of the task to be performed with the VDU
    Experiences with past reorganizations
    Individual needs and values of the VDU user
    Exchanges of experience with other organization members
    Acquired attitudes to VDUs (e.g., public opinion)
    Position in the reorganization process (e.g., pilot
    group, group with late VDU introduction).

Account is taken of the subjective perception of the (new) work situation, for example, with regard to the perceived importance of the VDU for the work of an organization member, his contribution to task performance, work requirements imposed on the VDU user, the character or the activity, personal status, communication with colleagues, supervisability, freedom of action, efficiency, etc.

3. Hypotheses concerning VDU use in offices

The following central hyptheses on office use of VDUs are assumed:

The degree of work satisfaction depends on the work content assigned to the VDU at the work place.

Dissatisfaction arises in the introductory phase as a consequence of late and insufficient information, inadequate training and excessively rapid introduction (e.g., overloading and stress due to initial duplication of work − manual plus DP) and failure to take account of wishes concerning work place configuration, work content and the sequence of work operations.

Inadequate programming and scale of computer facilities (e.g., complicated VDU operation, excessively long computer answer times, high level of standardization and routine) reduce the benefit of the new technology and hence its acceptance.

Fear of downgrading and salary reduction and of changes in the structure of the working group reduces acceptability.

4. Empirical Investigation

Collection of data for the primary survey was completed on 31 August 1980. The methodology used and the sample are outlined below.

4.1. Methodology

The hypotheses formulated are tested by means of data from a cross-sectional examination. In view of the fundamental distinction made between the objective and the subjective work situation, the survey covered both data on actual features of the work situation and data on the subjectively perceived work situation. Three questionnaires were compiled for this purpose.

The questionnaire on "Description of the objective work

situation" was submitted to group leaders and department heads in
whose organizational units VDUs are used. VDU workers were asked
to complete a questionnaire on "Description of the work situation
by those concerned". The process of organizational change was also
recorded by means of a questionnaire submitted to the competent
organization control departments or planning and decision personnel
or offices.

Owing to the current political, social and economic relevance
of the subject and its potential for conflict, it was feared that
firms would be reluctant to participate. This fear was confirmed
by the large number of refusals to provide information (about
40 %). Nevertheless, 14 organizations and 247 organization members
declared themselves to be willing to participate in our survey.

The sample of this empirical investigation covers organizations
in the following sectors:

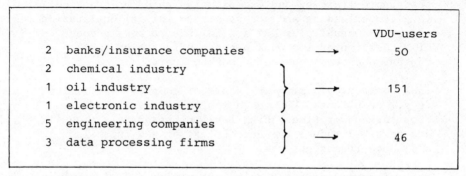

Fig. 4: Participating organizations and their members by sectors

The results presented here are based on 247 questionnaires
completed by VDU users on "Description of the work situation by
those concerned." Results from the questionnaire on "Description
of the objective work situation" and that on the process of
reorganization are used only where quantifiable.

4.2. Results

Evaluation of the general results allows the construction of
a VDU user typical of our sample; although such a user is unlikely
to exist in this form in reality, the notion may be useful as a
summary of the principal findings.

The following is true of the typical VDU user in our survey:

He is male and 33 years old.

He has a vocational qualification and earns over DM 2800 per month.

He has been employed by his firm for 8 years, 5 years of which have been spent in his present job, the last 3 of them with a VDU.

He works for 3 1/2 hours per day in a Dialogue I work place.

He has an unit of his own, used by him alone, at his work place.

He was not informed about the introduction of the new technology and not trained before its introduction.

He needed a familiarization time of 2 weeks.

He had expressed wishes on the configuration of the work place, the work content and the working process, some of which were satisfied.

He states that a specific project group was formed for VDU implementation in his organization.

He does not know whether a seperate works agreement was concluded.

Other findings

(1) User attitudes:

On the whole we found that positive statements about VDUs tended to be agreed with, while negative statements on VDU working were more likely to be rejected by the workers questioned.

However, the respondents disagreed with many positive statements about VDU working: they do not believe that the VDU has enriched their work, contributes to work satisfaction, permits a freer organization of work, clarifies the context of the work, or increases the significance of their work or recognition by colleagues. Even stronger disagreement is registered with statements to the effect that the VDU promotes recognition by friends, allows the boss to assess one's work more fairly, improves the working climate, increases scope for individual organization of the work place, and affords better chances of promotion.
On the other hand, no negative statement about VDU work

is agreed with. The average VDU user assumes an attitude
between agreement and disagreement only with respect to
the following two statements: Work on the VDU tires him
quickly, and the VDU demands unusually strong concentration.

(2) Other changes due to introduction of the VDU:

Change in the composition of the working group in the
case of one third of the respondents.

Occurence of serious or slight eye pain in 41 % of the
VDU users questioned; eye examinations were carried out
in less than half the respondents.

Improvement of salary grade in 28 % of those affected.

The following general hypotheses on VDU use in the office
were confirmed more or less definitely:

The degree of work satisfaction depends on the work content
assigned to the VDU work place; contrary to expectations,
those engaged in text processing proved to be relatively
satisfied.

A long period of training and the quality of the information
given to those concerned about the introduction of the VDU
has a positive effect on work satisfaction.

The degree of interest presented by the work, the effect
of the VDU on the importance of the work and enlargement
of their field of work have a positive effect on the
satisfaction of the users.

The following results were obtained in connection with the
hypotheses concerning situation class I, which included workers
solely engaged in data acquisition and text processing (4-8 h/day):

Persons in this work situation were trained more than
others for rapid operation of the VDU.

They agreed most strongly with the statement that the
VDU makes their work more interesting, although they
consider that the VDU does not provide them with more
information than previously.

They tend to agree that the VDU leads to a reduction
in social communication.

The following results were obtained in situation class II,
with the same activities but only for a maximum of 3 h/day:

The VDU is central to the work and does not convey to its
users in this work situation the impression of increased
work standardization.

The majority of those concerned do not feel constrained by
the VDU as regards individual organization of work.

As regards situation class III (data recall, 1-4 h/day), the
following hypothesis may be regarded as empirically confirmed:

The VDU is considered by its users to be a useful tool
at their work places.

In situation class IV, in which the respondents work for
2-4 h/day on commercial tasks with the VDU in the dialogue mode,
the following results were obtained:

Disagreement with the statement that the VDU relieves too
little of the users' work load.

VDU users in this situation class do not feel that their
responsibility is increased.

The statement that the VDU stifles communication with
colleagues is unequivocally rejected, particularly as the
unit is often used by more than one person in this
situation class.

In the case of VDU users working in the technical dialogue
mode and for programming (situation class V), it was assumed that
their active involvement in the introduction of the VDUs was
particularly important for their subsequent work satisfaction:

In this situation class, wishes were expressed when the
VDU was introduced more frequently then in other work
situations, if not by the majority. Where wishes are
expressed, failure by management to conform to them
leads to dissatisfaction in those concerned whereas if
their VDU needs are met, this increases work satisfaction.

The same results as for the hypotheses in situation classes
II and V were obtained for the mixed work places of situation
class VI (dialogue operation in the technical field for 1-3 h/day).
All results, whether confirming or refuting the hypotheses, were,
however, more unequivocal than in these classes.

REFERENCES

Cakir, A., Reuter, H.-J., Schmude, L.v., and, Armbruster, A.,
      1978, Anpassung von Bildschirmarbeitsplätzen an die physische
      und psychische Funktionsweise des Menschen, Forschungsbericht
      des Bundesministers für Arbeit und Sozialordnung, Bonn.
DAG, 1979, Arbeitsbedingungen an Datensichtgeräten, Auswertung
      der DAG-Umfrageaktion, Deutsche Angestellten Gewerkschaft,
      Hamburg.
Gaitanides, M., 1976, Industrielle Arbeitsorganisation und tech-
      nische Entwicklung, Berlin und New York.
Hackman, J.R., 1969, Towards understanding the role of tasks in
      behavioral research, in: Data Psychologica, 31:97-128.
Staehle, W.H., 1977, Die Arbeitssituation als Ausgangspunkt von
      Arbeitsgestaltungsempfehlungen, in: Personal- und Sozial-
      orientierung in der Betriebswirtschaftslehre, Bd. 1, 223-249,
      Reber, G. ed., Stuttgart.
Staehle, W.H., 1980, Management, Eine verhaltenswissenschaftliche
      Einführung, München.
Staehle, W.H., Hattke, W. and, Sydow, J., 1980, Die Arbeit an Da-
      tensichtgeräten aus der Sicht der Betroffenen, Ergebnisse
      einer empirischen Erhebung unterschiedlicher Arbeitssitua-
      tionen, Berlin.
Wiesner, H., 1978, Arbeit und Arbeitsgestaltung an Bildschirm-
      Arbeitsplätzen, in: Rationalisierung, 29(8):179-181.

EFFECTS OF "OFFICE OF THE FUTURE" TECHNOLOGY

ON USERS: RESULTS OF A LONGITUDINAL FIELD STUDY

B.A. Gutek

Xerox Corporation (PARC/ADL)
701 South Aviation Boulevard, M.S. C3-50
El Segundo, California 90245   USA

Advanced office technology and The Office of the Future
have developed with minimal regard to the attitudinal and behavioral
effects of new office systems on the users. Will the new office
technology raise job satisfaction and improve the quality of working life? Will it even increase productivity?

The rationale for introducing the Office of the Future is
increased efficiency. Coopers and Lybrand's (1979) evaluation is
typical:

The target of office automation and the office of the future
can be simply put: to boost white-collar productivity and
effectiveness by systematically applying the appropriate
technology to the restructuring of the office environment.
(Coopers & Lybrand, 1979, p. 4)

The fact that users of new technology will be affected is
readily acknowledged. Coopers and Lybrand wrote:

Probably the most critical aspect of office reorganization
is its impact on human resources.   ...Longstanding work
patterns and reporting relationships may have to be modified.
(Coopers & Lybrand, 1979, p. 4)

How people will be affected by office technology has not been
systematically addressed by research although conjecture about how
people will have to adjust abounds. (cf. Coopers & Lybrands 1979,
a and b; Dickson & Simmons, 1970; Connell, 1980)

This paper reports the results of a two year research project
on the effects of the Senior Manager's Work Station (SMWS) on the
principle users: a manager and secretary (Gutek, 1981). The SMWS
was developed at the Advanced Development Lab, Palo Alto Research
Center, Xerox Corporation.

The SMWS is a software package to be used on a Xerox ALTO
computer. It was designed to help a manager and secretary operate
a two-person office. The manager and secretary could each create
forms for use in the office, could share forms or designate forms
as private, could send messages to oneself or the other SMWS user
and the like. Thus, the SMWS was designed to help organize and
maintain the two-person office and was not designed to help the
manager or secretary handle problems external to the office.

METHOD

Research Design

Although the project had many constraints (e.g., on-going
office site, only one research dyad necessitated by experimental
equipment), the intent of the research design was to retain as
many principles of experimentation as possible. The research
consisted of two methods of data collection: a series of questionnaires filled out over a 20 month period that were analyzed
statistically (a quantitative method),as well as observation and
interviewing that were interpreted in an intuitive, non-statistical
(a qualitative method) manner. The questionnaire data were
collected from the two research subjects, a manager and a secretary,
across four periods of time. The initial or pre-installation phase
consisted of the four month period before the SMWS was installed.
Data were collected during 4 one-week intervals. The second data
collection period was the learning phase, a 6 month period of
learning and adjusting to a system that was undergoing modifications and improvements. Data were collected during four one week
intervals. The third or stable phase of the project consisted of
6 months of stable use of the system. Data were collected for
shorter time intervals, 2 or 3 days rather than a week,during that
period. More data were collected in Phase III than in the other
phases so that those results could be correlated with use of
specific system features as described below. The fourth phase,
a post-SMWS phase, consisted of the 3 month period after the
system was removed from the manager's and secretary's offices.
Data were thus collected across all days of the week and throughout each month, thereby eliminating effects of weekly and monthly
work cycles in responses.

The transitions between phases I and II and between III and IV were clear. But the transitions between the learning phase and the stable phase were less clear. Four criteria were used to mark the transition between phase II and phase III: both subjects felt they knew the system, all features had been added to the system, the number of system crashes reached an acceptably low level, and the number of forms created by the two subjects approached an asymptote. During Phase III, use of specific features of the system was automatically collected on the system. Thus, user reactions could be correlated with use of specific system features.

Besides filling out questionnaires, subjects were periodically interviewed and/or observed. A summary and synthesis of the notes generated from observation and interview constituted a second, qualitative, method of studying the impact of the SMWS. The present report focuses on quantitative results of the research.

The research subjects were a senior level manager and his secretary. While the manager had extensive experience with computer systems and their functions, the secretary had none. The manager and secretary had a good, mutually supportive working relationship with a comfortable division of tasks that had evolved over more than 15 years of working together.

The secretary's office is entered from the main hallway. A door from the secretary's office to the manager's office provides the only access to and from the manager's office. Thus, anyone who wishes to see the manager must pass through the secretary's office. The layout of each office was modified slightly to accommodate the ALTO, CRT and keyboard.

Instruments

In the pre-installation and post SMWS phases, the two subjects each filled out five one-page forms; in the learning and stable phases, they filled out a sixth one-page form on their attitudes toward the SMWS. The manager and secretary filled out identical forms except for one form predominantly concerned with theirr contact with each other.

Several considerations guided the development of specific questionnaires for the project. First, an effort was made to use or devise questions that would not require a substantial amount of time to complete. Second, an effort was made to use existing validated measures. Third, an effort was made to ask questions that apply to both the manager and secretary in order to see whether the SMWS affected them differentially. Fourth, since there were only two subjects, an effort was made to avoid questions that could cause embarrassment or invasion of privacy.

Four of the questionnaires were based on two well-known validated instruments, the Job Description Index (Smith et. al., 1965)
(JDI) and the Sherwood Self Concept Inventory (Sherwood, 1962).
Both of the instruments were altered for this project. Only two
of the JDI subscales, work and people, were used in this project.
The other three subscales, supervision, pay and promotions, were
not relevant to the present study. The format of the JDI was
altered from a simple yes-no response to a Likert-type response
option in order to increase variance in responses.

The Sherwood scale is a measure of self-esteem. It contains
16 items and has a semantic differential format. In the present
project, two minor modifications were made. First, three items
were added. They are productive/unproductive, efficient/
inefficient, and organized/disorganized, aspects of self-concepts
that might be related to use of the SMWS. The second minor
modification was to have only 5 points between the anchors instead
of 11 points as Sherwood used, since there is evidence that
people have difficulty making distinctions among more than 7 or 9
categories when making a judgment and that five categories yields
virtually the same results as 7 or 9 categories (cf. Moser & Kalton,
1972, p. 351).

The measure assessing attitude toward the SMWS is the JDI work
subscale but subjects evaluated the SMWS rather than work in
general. The other two forms assessed the subjects' behavior
rather than attitudes. One asks subjects to evaluate the percentage of time spent on various activities that might be altered
with the introduction of the SMWS. The other form assesses the
number of times and the reasons why the manager and secretary go
into each other's offices or contact each other. It also measures
their feelings of being interrupted and being pressured.

Subjects filled out the forms at the end of the working day,
or occasionally the following morning.

RESEARCH RESULTS

Overall, the manager filled out questionnaire forms for 59
days over the entire study period; the secretary filled out forms
on 65 days. None of the data generated during the phases was
analyzed until all data were collected. This was done in order
to eliminate the possibility that knowledge of early results by
either the subjects or the researchers might influence subjects'
subsequent behavior (cf. Rosenthal, 1976).

Measures of Attitudes and Behaviors

The means and standard deviations for all items on the four
attitude scales across the four phases of the study were calculated.
A perusal of the mean scores suggested that the subjects reacted
positively toward their work, the people with whom they work, the
SMWS and they have positive self concepts. Unfortunately, the
usefulness of mean scores on all attitude variables was limited
by the fact that the two subjects had significantly different
scores on most items. A discussion of differences between subjects
on all items except evaluation of the SMWS scores would violate
their confidentiality. An examination of the mean SMWS scores
across Phases II and III showed that the secretary evaluated the
SMWS significantly more favorably than the manager on all 19 items.

The two subjects in this study interacted with each other
frequently. They went into each other's offices up to 13 times
in one day, in addition to contacting each other by phone, and
using the SMWS for communicating when it was available. The
secretary went into the manager's office most frequently to empty
the out-basket. The number of times per day ranged from 0 to 6,
with a median of 2. She also went into the manager's office
frequently for specific work information (range of 0 to 7 times
per day; median of 1 time per day). She went into his office less
frequently to socialize (at least once on 27% of the days studied),
to take dictation (at least once on 26% of the days studied),
mark the calendar (at least once on 3% of the days studied), or to
get a signature (at least once on 18% of the days studied). She
averaged one trip a day into the manager's office for some other
reason.

The manager also contacted the secretary frequently by going
into her office. Perhaps reflecting the status differences in
their positions, the manager went into the secretary's office
to socialize significantly more than she went into his office
to socialize. The manager also went into the secretary's office
to get mail (range = 0.4 times a day; median = 1 time a day).
Less frequently he went into the secretary's office to give
dictation (on 7% of days studied) and to get a phone number
(on 27% of the days studied).

Besides physically going into each other's office, the manager
and secretary contacted each other frequently by phone. The
manager rang the secretary 2-3 times a day whereas she contacted
him 9-10 times a day on the average.

Not only did the manager and secretary contact each other often during the day, but they both contacted other people often as well. For example, the manager went out of his office about 8 times a day and the secretary went out of hers about 5 times a day. The range for both of them was 0 to 12 times a day. In addition, the secretary reported that her door to the main hallway opened repeatedly; the range was 24 to 61 times a day. People would only come into her office to see her or the manager.

Given the amount of interaction both subjects have with other people, it is not too surprising that both occasionally felt interrupted an average of once a day with a range of 0 to 3 times. The secretary reported that she felt interrupted an average of twice daily with a range of 0 to 9 times daily. Both subjects also reported feeling pressured an average of once a day, with a range of 0 to 5 for the manager and 0 to 4 for the secretary.

The two subjects were asked to estimate how much of their day they spent on various activities that might be affected by the SMWS. The activities rated were not mutually exclusive; the amount of time could sum to more than 100%. There were statistically significant differences between the two subjects in time allocated to each activity. The manager spent more time in general than the secretary on wage and salary, the budget, on the phone, talking inside and outside his office, writing memos and notes. On the other hand, the secretary spent more time than the manager on preparing hard copy for internal use, calendar activities, filing, logs, travel forms, and working with or using machines. ("Machines" is used broadly and included typewriter, telecopier,, SMWS, telephone, etc.)

In summary, the behavior reported by the subjects involves a lot of interaction with people, including each other. That interaction is direct as well as indirect; e.g., by phone or memo. The daily tasks of the manager and secretary differ significantly; the manager's activities are centered around communication with other people, whereas the secretary's activities are more concerned with maintaining the office. They both report feeling pressured and interrupted at least once a day.

Attitudinal Dependent Variables

Four measures of attitudes were assessed: attitudes toward work, attitudes toward co-workers, attitudes toward the SMWS and self-esteem. Except for the SMWS scale which was an application of a work scale to a specific aspect of work (the SMWS), the scales had been previously used as were validated as undimensional scales. Each of the four scales was factor analyzed using the SPSS factor analysis program to verify that the items were measuring one dimension in this study. Unfortunately, factor analyses of these data did not yield a clear one-factor solution for any of the four variables, nor did clear two-factor variables emerge. Therefore, for analyses of phase effects, the items in the scales were analyzed individually. However, mean scores on the variables were used for the analyses relating attitude to use of specific features of the SMWS. This was necessary because of the instability of simple item responses and the low frequency of use of specific SMWS features.

Analysis Strategy

     Two sets of analyses will be reported. First are the analyses of variance testing effects of the SMWS on users reactions. Second are the analyses that relate user attitudes to use of specific features of the SMWS.

     The purpose of the first set of analyses was to see if the subjects responded differently about aspects of their work, their co-workers and themselves in different phases of the project. In addition, their responses to the SMWS were measured in two phases and were assessed for differences. In order to examine differences in effects across phases, the initial strategy was to perform analyses of variance with the four-category phase variable as the independent variable and the individual scale items as dependent variables. Two subjects had significantly different mean scores on most of the items on all four attitude scales, which were verified by two-way Anovas with subject and phase as the independent variables. The results reported here are from one way Anovas, run separately for each subject. Because degrees of freedom are much less in the separate one-way Anovas, they provide a more conservative test of effects.

     Although the dependent variables were not related strongly enough to form one factor, they were not independent of each other. Running multiple Anovas on correlated items increases the probability of having a chance finding appear to be statistically significant, of course. Nevertheless, Anovas, rather than canonical correlations were used. In these analyses, statistical significance level was used as a guide to understanding subjects' responses rather than as a definitive statement about cause and effect. There were too many sources of invalidity present in this study to have a great deal of confidence in any single, isolated statistically significant result.

The same analysis strategy was used to examine changes in behavior. Analyses of variance were run, controlling for subject differences in response, to show the effect of phase of the study on behaviors such as using the phone or going into the other person's office.

The second set of analyses is a set of correlations between user attitudes toward work, co-workers, self and the SMWS and use of specific features of the SMWS.

Changes in Manager's and Secretary's Attitudes Across Four Phases

The following summarizes the results of a series of analyses of variance using the 19 attitudes toward work:

1. Routine. Work was not uniformly routine across the four phases for the secretary ($F(3,59)=2.37, p<.10$) or for the manager ($F(3,53)=2.55, p<.10$). A priori contrasts showed that the training and removal phases were significantly different from the other two.

2. Creative. Work was not uniformly creative across the phases for the secretary ($F(3,53=3.23, p<.05$). Work was the least creative during the pre-installation phase for the secretary.

3. Healthful. Work was viewed as significantly less healthful by the secretary ($F(3,59)=4.24, p<.01$) and somewhat less healthful by the manager ($F(3,53)=2.39, p<.10$) during some phases than others. The lowest mean scores for both were in the stable use phase, with the second lowest mean scores in the training phase.

4. Challenging. Challenge was also only significant for the secretary ($F(3,59)=8.77, p<.001$). A priori contrasts showed that work was significantly more challenging for the secretary after the SMWS. The least challenging phase was the pre-installation phase.

5. On My Feet. This variable also was significant only for the secretary ($F(3,59=51.5, p.<.001$). A priori contrasts showed that the secretary reported that she was on her feet significantly more during the pre-installation phase than she was during the three subsequent phases.

6. Simple. The work of both the manager ($F(3,53)=2.88, p<.05$) and the secretary ($F(3,59)=8.44, p<.001$) varied in simplicity. Work was simplest before the SMWS was installed and least simple after it was removed.

7. Endless. The secretary's work varied on this dimension
($F(3,59)=12.1,p<.001$) Mean scores increased through the four
phases of the project (Xs=2.3,2.8,3.0,3.1), showing that her work
became increasingly endless as the project progressed.

8. Accomplishment. The secretary reported that her work varied
in providing a sense of accomplishment over the project
($F(3,59)=3.4,p<.05$). She reported the lowest sense of accomplishment before the SMWS was installed.

9. Complicated. The secretary reported varying levels of
complicatedness in her work ($F(3,58)=9.8p<.001$). Work was least
complicated before the SMWS was installed and, surprisingly, in
the training phase.

In summary, the secretary's attitudes toward aspects of work
varied more with phase of the project than did the manager's For
the secretary, work was least creative, least challenging, required
being on her feet more, and provided the least sense of accomplishment, before the SMWS was installed. For both subjects, change
made work less routine; the presence of the SMWS was associated
with viewing work as less healthful, and the post-SMWS phase was
the least simple.

The following summarizes the results of a series of analyses
of variance using the 18 items measuring attitudes toward other
people at work. Six of the 18 items showed significant effects:

1. Ambitious. This variable was significant for the secretary
only ($F(3,60)=2.9,p<.05$). She perceived other people to be more
ambitious during Phases II and IV.

2. Fast. The manager's evaluation of other people varied on this
dimension across the four phases ($F(3.53)=7.8,p<.001$). Other
workers were rated the "least fast" during the two SMWS·present
stages, Phases II and III, and fastest in Phase IV.

3. Intelligent. The secretary's evaluation of other workers
varied on this dimension ($F(3,59)=3.2,p<.05$). Other people were
rated as increasingly intelligent through the four phases
(Xs=3.2,3.5,3.6,3.8). The manager also rated people differentially
on this dimension ($F(3,53)=3.05,p<.05$) with people rated most
intelligent before the SMWS was installed, and least intelligent
in phase III.

4. <u>Smart</u>. The secretary showed a similar pattern of responses to "smart" as she did to "intelligent" $(F(3,60)=4.6,p<.01,Xs=3.1,3.5, 3.6,3.8)$.

5. <u>Lazy</u>. This variable was significant for the secretary only $(F(3,60)=3.06,p<.05)$ with people rated the least lazy in Phases II and III respectively, although other people in all four phases were evaluated positively.

6. <u>Unpleasant</u>. The manager's evaluation of other people at work varied across phases $(F(3,53)=4.7,p<.01)$ with people rated least unpleasant before the SMWS was installed $(X=1.6)$ and most unpleasant after the SMWS was removed $(X=2.3)$.

In summary, evaluation of other people at work showed little variation during the phase of the study. The results that were significant were not consistent across the two subjects.

   The following summarizes the results of a series of analyses of variance using the 19 <u>self-esteem</u> items. Sixteen items were significant.

1. <u>Self-confidence</u>. The secretary's self reports of self-confidence varied across the four phases of the project $(F(3,60)=11.2,p<.001)$. She increased her self-confidence through the first three phases of the project. Self-confidence remained at the Phase III high level in Phase IV.

2. <u>Tolerant</u>. The secretary's reports of tolerance also varied $(F(3,60)=10.5,p<.001)$ such that tolerance increased monotonically over the four phases of the project.

3. <u>Productive</u>. The secretary's reports of productivity changed over the project $(F(3,60)=8.7,p<.001)$, again with feelings of productivity increasing monotonically through the project. The greatest difference in mean scores was between Phase I and Phase II, with smaller increases in perceived productivity in Phases III and especially Phase IV.

4. <u>Efficient</u>. The secretary's reports of efficiency varied across the phases $(F(3,60)=10.1,p<.001)$, with perceived efficiency increasing in the first three phases and leveling off in Phase IV.

5. <u>Able</u>. The secretary's ratings of being able to do most things well was significantly related to the phase of the project $(F(3,60)= 14.3,p<.001)$. The relationship between phase and ratings was positive, with the highest increase between Phase II and Phase III.

6. Enthusiastic. Enthusiasm of the secretary correlated positively with project phase ($F(3,60)=7.4, p<.001$) with enthusiasm increasing the most between Phase I and Phase II, an increase in mean score of almost .5.

7. Likable. The secretary reported changes in this variable across phases ($F(3,60)=12.9, p<.001$). Again, the relationship between phase and the rating was positive with the highest change in mean scores occurring between Phase I and Phase II.

8. Cooperative. The secretary varied in her reports of cooperation across the four phases ($F(3,60)=26.2, p<.001$). Her scores on cooperation increased through the four phases. The biggest change occurred between Phase II and Phase III.

9. Satisfied. The secretary varied in her reports of frustration/ satisfaction ($F3,60)=12.0, p<.001$). She reported that she was more satisfied in Phases III and IV, than in the earlier phases with the highest score in Phase III. The difference between scores in Phase III and Phase IV was not significant.

10. Intelligent. This variable was also significant for the secretary ($F(3,59)=6.5, p<.001$). Her reported feelings of intelligence increased monotonically throughout the project, with the largest increase occurring between Phase I and Phase II.

11. Friendly. Self reports of friendliness were significantly different across the four phases for the secretary ($F(3,60)=8.6$, $p<.001$). She rated herself increasing more friendly throughout the project.

12. Calm. Both the secretary ($F(3,60)=8.5, p<.001$) and the manager ($F(3,53)=3.6, p<.05$) varied in their reports of being calm/anxious in the project. The secretary's scores moved in the direction of being increasingly calm during the project. The manager reported more anxiety in Phases II and IV, the two phases in which change occurred (introduction and removal of the SMWS).

13. Useful. The secretary also showed a significant difference in a rating of being useful during the project ($F(3,60)=16.0$, $p<.001$). Her rating of usefulness increased monotonically through-out the project with the bigger changes occurring between Phases I and II and between Phases III and IV.

14. Know myself well. This variable was significant for both the secretary ($F(3,60)=14.9, p<.001$) and the manager ($F(3,53)=7.7, p<.001$). Scores for both subjects changed monotonically through out the project in the direction of increased self-understanding.

15. Organized. Both the secretary ($F(3,60)=14.5, p<.001$) and the manager ($F(3,53)=3.4, p<.05$) varied in their reports of being organized. The secretary's scores increased montonically in the direction of being more organized with the greatest changes occurring between Phases I and II and Phases III and IV. The manager reported being more organized when he had the SMWS (Phases II and III) than he did in the other two phases.

16. High self-esteem. Significant effects for phase were found
in self-esteem for both the secretary $(F(3,60)=5.8, p<.01)$ and the
manager $(F(3,53)=4.4, p<.01)$. Both had monotonically increasing
self-esteem scores throughout the project.

    In summary, the measures of self-esteem varied with phase
more for the secretary than the manager. There were also more
signifcant effects on the self-esteem measures than were for the
work or people scales. In general, the secretary reported increased
levels of self-esteem and its various components throughout the
project. In general self-esteem did not decrease in Phase IV and
only occasionally did it even level off. The effects on the
manager were neither as strong nor as consistent. His general self-
esteem and self-understanding increased throughout the project.
He felt more organized with the SMWS than without it and somewhat
more anxious during the training and post-SMWS phases than during
the other two phases.

    The same 19 items that are in the work scale were used to
evaluate the SMWS in Phases II and III. The following summarizes
the results of a series of analyses of variance using those items.
Ten of the 19 were significant

1. Routine. The secretary reported that using the SMWS was less
routine in Phase II than Phase III $(F(1,34)=22.8, p<.001)$. The
manager's ratings were similar, but do not quite reach statistical
significance $(F(1,30)=3.7, p=.06)$.
2. Boring. The secretary reported that using the SMWS was less
boring during Phase II than Phase III $(F(1,34)=5.4, p<.05)$.
3. Respect from others. The manager reported more respect from
others during Phase III than Phase II $(F(1,26)=22.8, p<.001)$. The
secretary's scores show a similar though non-significant pattern
$(F(1,34)=3.0, p=.09)$.
4. Useful. The manager reported that using the SMWS was more
useful during Phase III than Phase II $(F(1,30)=6.0, p<.05)$.
5. Tiresome. The manager's scores showed that using the SMWS
during Phase II was more tiresome than using it in Phase III
$(F(1,30)=7.3, p=<.05)$.
6. Challenging. The manager reported that using the SMWS was
slightly more challenging in Phase II than Phase III $(F(1,30)=3.8,$
$p= .06)$.
7. Frustrating. Using the SMWS was viewed as significantly more
frustrating during Phase III than Phase II by the secretary
$(F(1,34)=5.0, p<.05)$ but significantly more frustrating during
Phase II than Phase III by the manager $(F(1,30)=6.9, p<.01)$.

8. Simple. Both subjects felt that using the SMWS was simpler in
Phase III than Phase II, although the relationship was stronger for
the manager (F(1,30)=6.1,p<.01) than the secretary (F(1,34)=3.4,
p= .07).
9. Sense of accomplishment. The manager felt a greater sense of
accomplishment from using the SMWS during Phase III than Phase II
(F(1,29)=5.9,p<.05).
10. Alienating. The manager reported that using the SMWS was more
alienating in Phase II than Phase III (F(1,30)=4.9,p<.05).

        In summary, both the manager and secretary reported differences
between Phase II and III in attitudes toward using the SMWS, although
there were more significant effects for the manager than secretary.
The manager reported more positive effects in Phase III. In the
stable use phase, he felt the SMWS was more useful, less alienating,
less frustrating, and he achieved a greater sense of accomplishment and respect from others. It was also viewed as simpler to
use and less challenging in Phase III. The secretary seemed to
view the SMWS somewhat less favorably in Phase III than Phase II:
the SMWS was reated more routine, more boring, more frustrating
and simpler in Phase III. It should be remembered that overall,
the secretary rated the SMWS characteristics more favorably than
the manager, although her scores were not affected by phase of the
project as much as the manager's scores.

Changes in Manager's and Secretary's Behavior Across Four Phases

        The analysis strategy used in the previous sections was also
used for the two pages of self-reported behaviors. The independent
variables were phase and experimental subject and the dependent
variables were the items listed on the two single page questionnaires.

        In general, there were few significant differences across
phases of the study on manager/secretary interaction with each
other or with general movement around the work environment. The
manager reported differences across phases in going to the
secretary's office to get a phone number (F(3,55)=3.2,p<.05). He
went into the secretary's office to get a phone numbers less in
Phases II and III. The two subjects also reported that they went
into each others' offices for "other" reasons more during Phases II
and III than in Phases I and IV. The differences among the four
phases were signficant (F(3,116)=5.46,p<.01). Presumably, these
"other" reasons--other than to take dictation, empty out-basket,
mark calendar, and the like--were predominantly concerned with
SMWS, for example, asking a question, seeing if the other system
was turned on, etc. The final significant effect relating to this

interaction involved number of times they contacted each other
through the intercom (F(3,116)=64.4,p<.001). They tended to contact
each other through the intercom the least in Phase I, and contacts
via the intercom system increased across the four phases. One
other item on that page of questions varied with phase of the study.
Subjects reported feeling pressured the most in Phase IV, although
the results were not quite statistically significant (F(3,116=2.4,
p= .07).

In summary, the results of analyses on the subjects' interactions and reasons for interactions showed that the SMWS had very
little impact. The manager did not go into the secretary's office
as often to get phone numbers when he could look them up on SMWS
phone directory form, and they also went into each others' offices
for "other" reasons more during Phases II and III than in Phases I
and IV. The differences among the four phases were significant
(F(3,116)=5.45,p<.01). Presumably, these "other" reasons--other
than to take dictation, empty out-basket, mark calendar, and the
like--were predominantly concerned with the SMWS, for example,
asking a question, seeing if the other system was turned on, etc.
The final significant effect relating to this interaction involved
number of times they contacted each other through the intercom
(F(3,116)=64.4,p<.001). They tended to contact each other through
the intercom the least in Phase I, and contacts via the intercom
system increased across the four phases. One other item on that
page of questions varied with phase of the study. Subjects reported
feeling pressured the most in Phase IV, although the results were
not quite statistically signifcant (F(3,116)=2.4, p = .07).

In summary, the results of analyses on the subjects' interactions and reasons for interactions showed that the SMWS had very
little impact. The manager did not go into the secretary's office
as often to get phone numbers when he could look them up on SMWS
phone directory form, and they also went into each others' offices
more during Phases II and III, presumably on "SMWS business".
But overall, the effect was minimal.

The assessment of time allocation among tasks showed greater
responsiveness to the presence of the SMWS. The results were
generated from two-way Anovas.

1. Calendar Activities. The secretary spent more time than the
manager on calendar activities in general (F(1,116)=15.51,p<.001)
and both subjects spent the most time on calendar activities in
Phases III and the least in Phase IV.

2. Filing. Filing is a task that occupied more of the secretary's
time than the manager's (F(1,116)=127.9,p<.001). Time allocated to
filing also varied across phases of the project in interaction with
subject differences (F(3,116)=4.9,p<.01). The secretary spent the
most time filing in Phase III and the least in IV. The manager
did very little filing but spent the most time filing in Phase I.
3. Logs. Handling logs is also a secretarial task. She spent
significantly more time on logs than the manager (F(1,116)=238.0,
p<.001). There was also a signifcant main effect for phase of the
study (F(3,116)=9.1,p<.001) such that the secretary spent considerably more time on logs in Phase IV than in the other three phases,
whereas the manager spent no time on this task in Phase IV. The
least amount of time allocated to logs was in Phase II.

4. Wage and Salary. The manager spent more time than the
secretary on wage and salary forms (F(1,116)=21.1,p<.001). There
was also a significant phase effect (F(3,116)=2.8,p<.05) and a
significant interation between phase and subject (F(3,116)=2.8,
p<.05). The manager spent the least time on wage and salary forms
during Phase II and the most time during Phase IV.
5. Budget. The manager spent more time than the secretary on the
budget (F(1,116)=7.8,p<.01). There was also a significant main
effect for phase (F(3,116)=3.5,p<.05). More time was spent on the
budget in Phase III than in the other three phases.
6. Travel. The secretary reported significantly more time spent
handling travel forms than the manager (F(1,116)=71.9,p<.001).
Phase was also significant (F(3,116)=15.4,p<.001). The most time
allocated to handling travel forms was in Phase I and the least
time was in Phases III and IV.
7. Working with Machines. A significant effect was found for
phase (F(3,116)=3.0,p<.05) but the two subjects did not differ
in their reports of amount of time spent working on machines.
More time was spent working on machines in Phases II and III than
in Phase IV and especially Phase I.
8. Talking on the Phone. The manager spent more time than the
secretary on the phone (F(1,116)=15.2,p<.001). There was also a
significant main effect for phase (F(3,116)=2.7,p<.05)
significant interaction between subject and phase (F(3,116)=4.7,
p<.01). Time on the phone was the greatest for the manager in
Phase IV and lowest for both in Phase II.
9. Talking to People in Office. The manager spent more time than
the secretary talking to people in his office (F(1,116)=81.2,
p<.001) but amount of time did not vary across phases of the project.
10. Talking to People Outside Office. The manager spent more time
than the secretary talking to people outside his office (F(1,116)=
81.2,p<.001). There was a significant interaction between subject
and phase of the project (F(3,116)=3.1,p<.05). Together, they
spent the least time talking to people outside their offices in
Phase II and the manager also spent less time in Phase III.

11. Working with Hardcopy. The secretary spent more time than the
manager working with hardcopy ($F(1,116)=76.4, p<.001$). There was
also a significant phase effect ($F(3,116)=3.7, p<.01$) and a
significant phase by subject interaction ($F(3,116)=9.6, p<.001$).
Overall the amount of time working with hardcopy increased
monotonically across the four phases of the research for the
secretary.

     In general, phase of the project had an effect on the way the
secretary and manager allocated time although the results are not
as consistent as one might expect. The manager and secretary
generally kept the same division of tasks throughout the project.
In no case did a task switch from being a secretarial task to a
managerial task or vice versa with the introduction of the SMWS.
The data suggest that the manager spent somewhat less time learning to use the SMWS. Both subjects spent the most time working
with machines in Phases II and III, a result that reflects their
use of the SMWS. The data also show that time spent on many of
these tasks was the lowest in Phase II, suggesting that the time
spent learning how to use the SMWS, creating forms and the like,
left less time for other tasks.

     In summary, it appears that learning the SMWS did take time
that was usually allocated to other tasks. Learning the SMWS was
associated with decreased secretarial time spent on the calendar,
filing and logs and with decreased managerial time spent talking
to other people.

Relationship of Specific System Use to Attitudes

     For approximately a six-month period in Phase III, statistics
on use of specific system features were generated on the Altos.
In order to see if specific system usage was related to attitudes
about work, co-workers, self and the SMWS, Pearson product moment
correlations were run between the mean of each scale and specific
uses of the system. These analyses were especially interesting
because, unlike the Anovas, these statistics are generated from
independent sources of data.

     In general, there was little relationship between three of the
attitude scales--work, people and self-esteem--and use of specific
features of the system. The few items that were significantly
related at the .01 level or beyond may be chance findings.

On the other hand, ratings of the SMWS were related to use of specific features. Table 1 shows the correlations between composite score on the SMWS and some statistics generated on the ALTO. In general, ratings of the SMWS were positively related to global use measures. For example, a positive rating of the SMWS was associated with high numbers of startups, endings, marks, keystrokes, and use of user keys. (These items are themselves correlated, of course.) Rating of the SMWS was not related to number of crashes, or amount of idle time. Ratings of the SMWS were more strongly related to global use measures than to specific features of the system.

TABLE 1

Correlation Coefficients Between General Rating of the SMWS and Use of Specific Functions of the SMWS

	r
Number of Startups	.32**
Number of Endings	.31**
Number of Crashes	.12
Number of Marks	.30**
Number of Keystrokes	.34**
Number of User Keys Used	.33**
Number of Alarms	.08
Idle Main Menu	.01
Total Idle Time	.03
User Keys Deleted	.15
User Keys Created	.15

*p<.01   **p<.001

In summary, attitudes toward work, other people at work, and measures of self-esteem were not related to use of specific features of the SMWS or even frequency of use. However, attitudes toward the SMWS were associated with system use. A positive attitude toward the SMWS was related to increased frequency of use of the SMWS and to use of some specific features such as amount of line scrolling and frequency of "notify" commands. Evaluation of the SMWS was not related to number of crashes.

CONCLUSIONS AND DISCUSSION

The conclusions of the project will be discussed under four headings: Subjective reactions to the SMWS; Problems of Training Users; Development of a Research Method; and The Office of the Future.

Subjective Evaluation of the SMWS

One of the strongest effects in this research project was the favorable impact of the SMWS on the self-esteem of the secretary. The evaluation of work also changed somewhat, especially for the secretary. Work was viewed as more creative, less routine, more challenging, etc., with the SMWS than without it. These results cannot be generalized from the SMWS to other office technology without further research.

The SMWS did not alter the design of work. It increased, or was intended to increase, the amount of flexibility or control the manager and secretary had over their work. A system that would alter one's job or that would control the pace of work is likely to elicit a very different response from users.

Problems of Training Users

The subjects responded differently to the SMWS during the training phase than the stable use phase. The system designer and several other people who were knowledgeable about the system were available to the research subjects. And yet learning the system appeared to create anxiety and decreased time spent on other tasks. Subjects did not use either of the two manuals that were prepared. In a setting where users do not have a choice about adopting new technology and where many people are being trained at once, the decrease in productivity and possible decrease in morale and satisfaction can be great.

The results of this study add evidence showing that training is problematic. Further research addressing training specifically would be necessary to provide some definitive answers on an optimal training program. Training might be addressed most profitably within the larger issue of how to introduce new technology into a work setting.

Development of a Research Method

In order to understand any given phenomenon, a research usually aims to control as many extraneous variables as possible and manipulate the variables that are hypothesized to be important. Many interesting phenomena, however, cannot be controlled and manipulated in the way a researcher would like. One alternative in such a situation is to abandon experimentation and instead rely on intuition, insight, clinical judgment, common sense, logic and the like. Another alternative is to retain as much experimentation as possible even though the research design departs from ideal in significant ways.

Studying the effects of experimental computer systems on users is an interesting phenomenon with direct implication for marketing systems, introducing them into settings, and training users. The experimental nature of such systems usually means that only a few are available for use in research. Furthermore, in order to understand how they are used in an on-going work environment, it is often desirable to place them in on-going office settings to conduct research. These two considerations, a limited number of available systems, and research in a functioning work environment, severely decrease the amount of control a researcher has in a research design.

In this project, a research design was adopted for a two person, one-system project that did allow for some control of relevant variables. The questionnaires filled out by subjects, and the introduction and removal of the SMWS were controlled. The quantitative statistical data uncovered some results that would not have been discovered using interviews or observation. The research design developed is applicable to other research on technological innovation. And the steady accumulation of empirical data from modest research efforts such as this, in the long run, provide data as valid as any large-scale effort, although the sources of invalidity no doubt differ.

The Office of the Future

Management consultants and other "experts" are predicting that the biggest change in the workplace in the 1980's will be the introduction of the "Office of the Future". A diverse array of products is now on the market and many more will be introduced in the near future. The prime justification for introducing such systems is to increase office productivity which is viewed as being substantially lower than it should be. Comparisons between expenditures for capital equipment in production units and office units are used to suggest that an increase in capital equipment in offices will increase productivity. This argument, like others that appear in the literature, is specious. Corporations considering the adoption of office-of-the-future technology should be concerned about the general lack of hard data on productivity increases resulting from new equipment.

Along with corporate concerns about the effectiveness of new technology is a corporate and worker concern for the quality of working life in offices. Will the full scale introduction of office technology result in few jobs overall, relatively more jobs in which work is paced by a machine, isolation from other workers, increased worker alienation and anomie? Will jobs be deskilled and fractionated?

Most people in the field agree that the technology will be adopted and work will change. How new office sociotechnical systems (the machines, the physical environment, the job descriptions, the flow of work, and worker reactions) are designed and implemented is not set in stone. But very little research has been done so far that would help an organization redesign office work using the new technology. The National Science Foundation, recognizing the importance of research in this area, has recently set up a program to fund research on this topic. Results from NSF funded projects are not likely to be available for several years. In the meantime, companies will either have to do their own action research or rely on intuition and accumulated wisdom.

REFERENCES

Bikson, T.K., Personal Communication, November, 1980.

Coopers & Lybrand, "Office of the Future Amalgamates People, Equipment, Systems; Coopers & Lybrand Studies Affirm that Automation is Worth it," Coopers & Lybrand Newsletter, 21 #10 (1979) pp. 4-5.

Coopers & Lybrand, "White-collar workers, Contrary to Current Belief, think they would work better under improved conditions," Coopers & Lybrand Newsletter, 21 #9 (1979) pp. 5-6.

Connell, J.J., "The 'people factor' in the Office of the Future," Administrative Management, 41 #1 (1980) p. 36f.

Dickson, G.W. & Simmons, J.K., "The behavioral side of MIS," Business Horizons, 13 (1970) pp. 59-71.

Gutek, B.A., Senior Manager's Work Station: An Evaluation of Attitidinal and Behavioral Effects, Xerox PARC/ADL, Internal Report, 1981.

Mintzberg, H., The Nature of Managerial Work, New York: Harper & Row, 1973.

Moser, C.A., and Kalton G., Survey Methods in Social Investigation, (2nd ed.), N.Y.: Basic Books (1972).

Rosenthal, R., Experimenter Effects in Behavioral Research. N.Y. Irvington Pub. Co., 1976.

Sherwood, J.J., Self-identity and Self-actualization: A Theory and Research. Unpublished doctoral dissertation, The University of Michigan, 1962.

Smith, P.C. et. al., Cornell Studies of Job Satisfaction: I,II,III, IV,V,VI. Mimeo, Cornell University, (circa 1965).

SECTION 5

APPROACHES FOR ORGANIZATIONAL EFFICIENCY

Exploratory research on the cause-effect patterns in the co-evolution of technology and work organizations is a necessary background for the development of the model-assisted analysis for the professional human resource planner. Systematic consideration of these causal relationships will be an important component of the overall planning process. Mackenzie, Martel and Price provide a survey of the human resource decision support methodologies available today for organizational design in the context of environmental change. A set of phenomenological approaches assists the manager in situations where only nominal and ordinal information is available for a number of relevant dimensions. These methods assist the manager to a higher or lower extent dependent on the degree to which the situations are programmable.

F. von Hayek, the Nobel laureate in economics, once characterized the rate of progress in the economic sciences over the last 200 years by the extend to which economists became able to integrate subjective elements into their theories. By the same token, advancement of the best practice frontier in human resource planning should be determined largely by success in either widening the realm of measurable subjective phenomena or deepening the information content comprised in certain measures. Two contributions are cases in point. Crawley and Spurgeon describe a task analysis methodology specially designed for the investigation of motivations and job satisfaction within the British air traffic control system. Charnes, Cooper, Niehaus and Schinnar develop and apply an algorithm for numerically evaluating the efficiency with which various units of the U.S. Navy attain their equal employment opportunity program objectives. The advantage of working on the cardinal scale manifests itself in the large scope of this human resource planning tool. The numerical analysis allows for identification of representative efficient units, which in turn provides standards for further evaluation of the inefficient programs. Hence, Charnes, Cooper, Niehaus, and Schinnar accomplish the transition from the descriptive to the prescriptive in this very important and conflict-ridden area of human resource planning.

211

HUMAN RESOURCE PLANNING AND

ORGANIZATIONAL DESIGN

Kenneth D. Mackenzie[+], Alain Martel[++],
Wilson L. Price [+++]

[+]President, Organizational Systems, Inc.
700 Massachusetts Street and Edmund P. Learned
Distinguished Professor, School of Business
University of Kansas, Lawrence, Kansas 66044

[++]Professeur, Agrégé, Faculté des Sciences
de L' administration, Université Laval
Québec, P.Q., Canada G1K 7P4.

[+++]Professeur, Faculté des Sciences de
L' administration, Université Laval
Québec, P.Q., Canada G1K 7P4

ABSTRACT

All managers are involved with human resource planning and
organizational design. This involvement varies in the extent to
which a decision support methodology is employed. Managers will
employ a decision support methodology when it is suitable.
Because human resource planning and organizational design are
supplementary technologies, managers must decide on which is the
more suitable. The suitability depends upon the organizational
situation. Organizational design is more suitable for non-
programmable situations and human resource modelling is more
suitable for programmable situations. In particular, at the
strategic apex of the organization, human resource management
problems become organizational design issues.

1. INTRODUCTION

Manpower planning and organizational design share a concern
for the use of human resources. As Bryant and Niehaus (1978) have
pointed out, the two disciplines are developing in parallel and
use different technologies. The proceedings of a NATO conference
on Manpower Planning and Organization Design (1978) helped to
identify the gaps between the two approaches and provide a basis
for future integration and for selecting the human resource
decision support methodologies.

Manpower planning and organizational design both seek to
provide a methodology to support decisions affecting personnel.
Their domains overlap on some problems and differ on others.
Human Resource Planning strives to apply operations research
methods to the management of human resources. Organizational
Design is a younger field and is less precise and quantitative.
The various approaches to Organizational Design share a common
concern for helping an organization adapt to change with the help
of behavioral theories and intervention technologies. Manpower
planning is defined by Vetter (1967) as "the process by which
management determines how the organization should move from its
current manpower position to its desired manpower position.
Through planning, management strives to have the right number and
the right kind of people, at the right place, at the right time,
doing things which result in both the organization and the
individual receiving long term benefit." Organizational design
was defined by Kilmann, Pondy, and Slevin (1976,p.1) as ". . . the
arrangement, and the process of arranging, the organization's
structural characteristics to attain or improve the efficiency,
effectiveness, and adaptability of the organization." The close
linkage between the structures of an organization and direct
experience suggests that the older definition be modified.
Organizational Design is defined here as the continuing cycle of
adapting goals and strategies, arranging and maintaining the
organizational technology to implement them, and to produce
desired results in the face of changing environments while the
organization continues to function. Organizations are dynamic and
adaptive to their environments and the newer definition gives
prominence to the need for continuous maintenance and the
processes of design.

There is no way for senior management to avoid either human
resource management (HRM) problems or organizational design
problems. Managers may lack timely, accurate, and complete
information and they are hampered by limitations in information
processing capabilities, time, resources, energy, and conflicting
pressures. Managers are simultaneously faced with more possible
decision support methodologies (DSM) than they can handle and a
lack of specific and usable ones. They must cope with this

dilemma and are very concerned with the suitability of a DSM for
the problems they face. The <u>suitability</u> of DSM for an actual
problem situation is the degree to which (a) its assumptions and
premises correspond to the actual problem characteristics and (b)
its usage requirements match available time, capacity, resource,
and executive energy.

We assume that managers will adapt a DSM to the extent to
which it is suitable and adaptable to compensate for its
imperfections with acceptable supplementary considerations. It is
therefore crucial that the underlying assumptions and premises be
understood before presenting a DSM. We assume that both Human
Resource Planning and Organizational Design represent DSMs. The
purpose of this paper, then, is to highlight the assumptions and
premises of each in order to provide guidelines for selection by a
manager.

2. ORGANIZATIONS

An organization is a type of open system and consists of a
related set of suborganizations. It has units, personnel,
structures relating the personnel, task and non-task processes, as
well as significant relationships with its environment. Each unit
has its own boundaries and there are many possible ways of
categorizing and classifying the parts of an organization.
Mintzberg (1980) provided a scheme which describes five major
parts of an organization. These are:

(a) the <u>operating core</u> which includes all the operational
    jobs;

(b) the <u>strategic apex</u> which includes all the top general
    managers jobs;

(c) the <u>middle line</u> which comprises those jobs who are in
    a direct line of formal authority between the jobs in
    the strategic apex and the operating core;

(d) the <u>technostructure</u> which consist of the analyst jobs
    required to manage the organization (e.g. accountants,
    work schedulers, etc.);

(e) the <u>support jobs</u> which provide indirect support to the
    rest of the organization (e.g. legal counsellor, public
    relations, etc.)

Mintzberg stresses his belief that the relative importance of each
of these five basic parts in an actual organization depends on
various contingency factors such as age, size, technology,
environment and power needs.

Organizations are enmeshed in their environments. Management is the art of arranging the organization's internal processes to fit the environment and of selecting and influencing the environment to fit the interests of the organization. The mutual dependence of the organization and its environment is continuing and dynamic. In practice, however, the adaptation of the organization to the selected environment is made in discrete steps in its organizational architecture. The organizational architecture comprises the formal organization authority structures, the ordering of all identified organizational processes and the specific responsibilities attached to every position. Even if the internal and external environments change continually, the organizational architecture must remain unchanged sufficiently long to permit personel to learn of the roles of its positions to detect eventual maladaptations and to search for a new architecture to remove these maladaptations. The processes on which an organizational architecture is based are defined by its organizational logic. The organizational logic describes how the many activities are to be performed, allocated, coordinated, controlled and adapted. This applies to all activities, including the HRM activities, and hence, for a given architecture, the manpower planning and control methods used by an organization are embedded in its organizational logic. The architecture is the blueprint for how the organization is supposed to work. The reality, however, rarely matches the design. As the external and internal environments evolve, the actual organization gradually drifts away from its architecture through ad hoc adjustments. These ad hoc adjustments cumulate to create significant gaps between the actual organization and the architecture (maladaptations) that have increasingly serious consequences to productivity. Hence, there is a need to continually re-design the architecture.

The frequency of revisions of the organizational architecture and the use of of human resource planning methods depend very much on the actual organizational or sub-organizational situation. In practice, these organizational situations fall between two boundary cases which have been described by Simon (1960) as "programmable" and "non-programmable". In our context, the various facets of these two boundary situations are summarized in Table I.

Organizations that are in programmable situations are goal seeking in the sense that human resource planning and control decisions are made to maintain or to attain goals which are known and permanent. The environment in which the organization evolves is taken for granted and its future states are determined by a known process. The organizational logic is known, invariant, and encompasses established administrative procedures. The

TABLE 1

Multidimensional Characterization
of Organizational Situations

	ORGANIZATIONAL SITUATIONS	
Facets	Progammable	Non-programmable
Goals and Strategies	Goal seeking	−	Purposeful
Environment	Taken as a constraint	−	Selected and influenced
Organizational logic	Known and invariant	−	Contingent upon the internal and external environments
	Accepted as a constraint	−	Adapted as necessary
Organizational architecture	Well defined, stable and hierarchical	−	Ill defined, fluid and non-hierarchical
	Good fit to actual organization	−	Poor fit to actual organization
Information	Intensive	−	Impoverished
	Easy access	−	Difficult access
Jobs/positions	Individuals inter changeable and matched to job	−	Individual not interchangeable and job matched to individual
	Job description faithfully represents the work	−	Actual work is not represented by job description
Complexity	Managed by reducing variety	−	Managed by retaining variety

organizational architecture is well defined, stable and it usually takes the form of a formal hierarchy. The information required to make decisions is clearly identified and accessible. Positions can be filled by any individuals with the qualifications required in the job description. Large groups of individuals are involved and the complexity of the situation is managed by reducing variety, i.e. by managing clusters of positions at an aggregate level.

Non-programmable situations are those in which goals are not taken for granted but are always being reconsidered and reevaluated. They can be associated to the purposeful systems described by Ackoff and Emery (1972). The environment in which to operate is selected and can be influenced. The organizational logic is contingent upon the internal and external environments and is adapted as necessary. The organizational architecture is ill defined, fluid, non-hierarchical, and inconsistent with the actual organization. The information required to make decisions is impoverished and difficult to access. Jobs are largely determined by incumbents. The individuals are unique and are usually selected for the impact they may have on the organization rather than on the basis of a job description. A small number of individuals is involved and the complexity of the situation is managed by retaining variety, i.e. by taking all the attributes of the individuals involved into account.

3. ORGANIZATIONAL DESIGN PROBLEMS

Organizational design is a top management, leadership task. The design involves all aspects of the organization including its environment, goals, strategies, vulnerabilities, opportunities, work, government regulations, its current structures, the strengths and weaknesses of its personnel, its resources, and its constraints. There are six major congruency tests:

(1) Are the goals and strategies consistent with its environment?

(2) Is the organizational technology consistent with its environment?

(3) Is the organizational technology consistent with the goals and strategies?

(4) Is the organizational technology consistent with itself?

(5) Are the results consistent with the goals?

(6) Are senior management consistent and cohesive?

Any organizational design must consider both the technical merits of the design and the internal political acceptability of the design. An excellent technical solution which is not acceptable will generally backfire. Structures represent need satisfying interaction patterns (Mackenzie, 1976a, 1978) and they will be maintained as designed if they serve the interests of the members of the organization. Most organizational design problems are created by inconsistencies between how the organization should be working and how it is actually working. One of the main issues in organizational design is defining the real problems amidst a plethora of conflicting symptoms. The six congruency tests are valuable clues about the nature of the real problems for which the new design represents a solution.

In principle, the determination of the organizational design preceeds the elaboration in the form or position of job descriptions, wage and salary systems, HRM methods, management succession, and many more. And, in principle, because change is only partially controlled by the organization, there is a necessity to view organizational design as a continual process of organizational maintenance (Mackenzie, 1978). Consequently, a major problem in organizational design is to create a mechanism for maintenance and updating.

4.   ORGANIZATIONAL DESIGN PROCESSES

A number of organizational design processes have been proposed in the literature and many have been field tested. Mackenzie (1981) reviews the principal methodologies available and shows that they can be cast in the general means and linkages methodology described by the ABC Model given in Figure 1.

A detailed description of the seven elements of the ABC model can be found in Mackenzie (1981). Each has a technical meaning and a specific technology behind it. The elements in the ABC Model, especially the B processes, are currently supported by models and an interactive computer technology. Research into the ABC processes is continual with the development being on-line with actual organizational design engagements in a wide variety of organizations. This research is part of an on-going commitment to constructing a theory of group structures.

Rather than repeat here what has been published elsewhere, a quick summary will help explain the ABC model. The A part has two elements: goals and strategies. Normally, goals and strategies are ill defined, co-determined, and are often the results of attempts to rationalize what is occurring. The selection of goals and strategies serves to select the environment and to define the direction through it for the organization.   The B processes

constitute the Organizational Technology. The ideas of the
organizational logic, organizational architecture, and the actual
organization are discussed in Section 2. The element labelled
design premises refers to the assumptions made about
vulnerabilities facing the organization. These design premises
provide the macro-structure of the organizational logic.
Generally, the chief executive officer will have direct access to
organizational units whose responsibilities involve controlling
major non-postponable recurring vulnerabilities. Well designed
organizations show a consistency between the organizational logic
and the organizational architecture. Most organizations exhibit a
marked discrepancy between the organizational architecture and the
actual organization. The highest consistency observed in the
field is 63% with most falling below 50%. This inconsistency is
usually assumed not to exist in HRM. These inconsistencies
usually represent de facto reorganizations and, in some cases,
have managed to save the organization from itself. However, the
accumulation of discrepancies when accompanied by turnover builds
up stresses, impediments, and conflicts, can become serious. One
clear signal of such problems is the prevalence of many
committees. (Lippitt and Mackenzie, 1976).

     The ABC Model for Organizational Design is a continuing cycle
that must be repeated in order to orient and maintain the
organization. Organizational situations that are programmable
will require less frequent cycling through the entire process.
Organizational situations that are non-programmable will require a
more frequent cycling through the complete ABC process.

5. HUMAN RESOURCE MANAGEMENT PROBLEMS

     Human resource management problems include those of
recruiting, selection, hiring, training, assignment, development,
promotions, transfers, separations, benefits, and maintenance and
well being of the people in the system. They are so pervasive
that all managers must deal with them at some point or other.

     There is no well defined line between problems of human
resource management and those of organizational design. The
difference resides mainly in the context in which they occur.
Problems treated by the processes of human resource planning tend
to be those concerned with maintaining the actual organization in
a state prescribed by the organizational architecture or of moving
the actual organization towards such a desired state. The human
resource planning processes themselves are an integral part of the
organizational logic.

     Key questions that a human resources planner must answer are:

     - What "stocks" of personnel should be maintained. . .

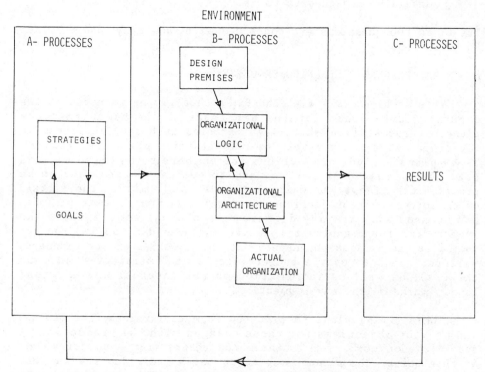

Figure 1.   The ABC Model of the Means-Ends Linkages
            in Organizational Design

how many people should the organization employ in
various categories?

- What "flows" (hiring, promotions, transfers, separations)
  should occur in the organization?

- Which individuals should be in which stock or take part
  in which flow?

- Where a given job can be done by various people, who
  should be assigned to do it?

The first two questions are concerned with "how many" and the last
two with "who".

6. HUMAN RESOURCES PLANNING PROCESSES

This section deals with the DSM's that may be imbedded in the
organizational logic for the HRM tasks. The human resource
planning process is a multi-level process such as illustrated in
Figure 2. At the top level, overall staffing plans are produced.
These plans include projections for numbers to be hired or laid
off, to be trained or re-trained, to be transferred, and to be
promoted (the flows) as well as the desired numbers (the stocks)
of employees. The projections can cover a number of time periods,
and be made with varying degrees of technical complexity. The
"memory" of the organization, which includes data bases, records
and files, as well as the experience and memories of the planners,
will be tapped to obtain budgets, goals, and strategies (manning
targets), as well as information on the internal and external
labour markets (the environment).

The staffing plans so produced form constraints (the stocks)
and provide objectives for those charged with the production of
schedules for work, for training and re-training, and for leave.
At this point, the planner begins to match individuals with what
were previously only anonymous quotas. The memory of the
organization, here, includes information on the goals and
strategies, the production or service requirements, the collective
agreements, and on individuals.

These schedules in turn constrain the solutions available to
those charged with the making of daily work assignments. These
assignments are made at the level of the first line supervisor and
draw information on collective agreements, personnel availability
and short term load forecasting from the memory. The result is,
in a sense, the "sharp end" of the human resource planning process
"spear" since it results in final decision as to who will do what
and where. Note that at both of the lower levels, information is
passed back to at least the next higher level so that performance

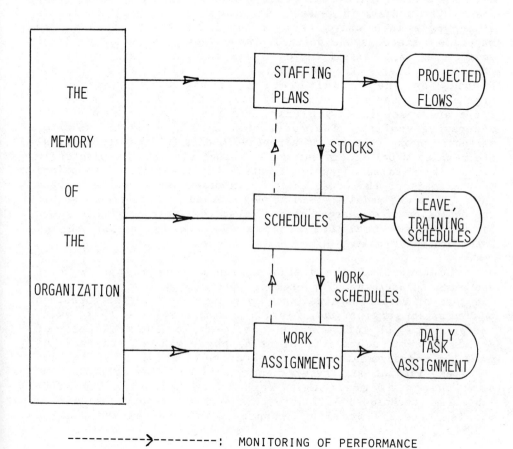

--------->--------: MONITORING OF PERFORMANCE

Figure 1.   A Human Resource Planning Process

can be monitored and changes made at either the monitoring or the
monitored level, as required.

At each of these levels of the planning process, there exists
a class of modeling methods, each the result of a set of
assumptions concerning the goals and the dynamics of the
organization. Price, Martel, and Lewis (1980) have reviewed the
extensive literature that deals with the production of staffing
plans. The models, in general, decompose the organization into a
state-space into which the personnel (or the positions) are
mapped. A state may be defined, for example, by reporting level
and occupational classification in one case, and simply by formal
rank in another. The definition of the state space depends
heavily on the actual problem facing the planner. An
organization may use a number of models for different purposes.
The state space is an arbitrary but useful description of the
underlying reality. The model must also specify the flow of
personnel among states. In normative models (of the mathematical
programming type), decisions will be suggested to the planner in
conformity with an objective function that reflects assumptions
on the goals of the organization. In descriptive models (Markov
chains, renewal models, simulations) the model attempts to exhibit
the evolution of the organization given the environment and a set
of strategies that are elaborated outside the model and
incorporated into it.

The output of the staffing plan may be in the form of
decisions (concerning, for example, promotions or hirings) or in
the form of desired states (such as desired stocks) that are then
passed as constraints to a lower planning level. At this level,
the principal objective may be, for example, the production of
schedules that ensure sufficient manpower will be available to
meet the requirements of the productive or service system, and
that take into account requirements for leave, rest periods, and
such things as training. Work force balancing and shift
scheduling techniques fall into this second category, as do more
problem specific modeling techniques (cf. Martel and Ouellet
[1981]).

At a lower planning level these schedules form constraints on
the actual work assignments. The task here is to assign
individual workers to specific tasks over a daily, weekly or
monthly period. The problems are combinatorial and can be very
difficult to solve when many of the practical constraints
(concerning, for example, collective agreements) are considered
(cf. Martel and Ouellet [1981]). It should be pointed out that in
a given organizational situation, the multi-level planning process
may contain more or less than three levels - the work assignments
might be produced at the same level as the schedules, or the
staffing plans produced in several steps at different levels of

disaggregation, each level adding successively more detail over a shorter planning horizon.

In his efforts to assist the manager in the human resource planning process, the modeler will, of course, change, correct, and refine his assumptions and the models that are built upon them in order to represent, in the best possible manner, the organizational architecture. However, the fact remains that in certain organizational situations, which tend to resemble those that we have termed non-programmable, the assumptions necessary for modeling cannot be identified, or are fluid to the point that the modeling process cannot correctly capture the essence of the problem. The models seem to provide good results where the organizational situation is programmable.

7. CONCLUSIONS

The four previous sections have outlined problems and processes of organizational design and human resource planning, discussed in the context of a characterization of organizational situations where these processes would seem the most applicable. The authors feel that the more the organizational situation is non programmable, the more frequently it will be necessary to go through the entire A-B-C cycle in establishing and maintaining the organizational design, because of changes in the environment, in the goals and strategies, in the individuals in the organization, and perhaps in the work to be carried out.

On the other hand, the more the organizational situation resembles that described as "programmable", the greater will be the opportunities for profitable insertion of human resource planning and control processes into the organizational logic. In fact, these processes then provide a regulatory mechanism which enables the organization to adapt to the fluctuations of the internal and external environments, thus ensuring that the results obtained are compatible with the given goals and strategies. These processes are sufficient as long as the goals, strategies and environment remain unchanged.

Consider the difference in the problems of choosing a chief executive officer and in choosing a regional sales manager. The choice of a chief executive officer can result in major changes in the goals and strategies of the organization and in major changes in the structures. Indeed the selection may be made in order to bring about such changes. The regional sales manager will, in all likelihood be chosen within the context of existing goals and strategies and the existing organizational architecture.

The relative importance of human resource planning and organizational design methodologies is a function of the type of

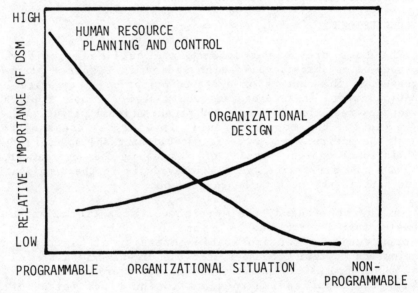

Figure 3.   Relative Importance of Decision Support Methodologies

situation as illustrated in Figure 3. Frequent cycling through
the A-B-C processes of organizational design are therefore likely
to be required for the <u>strategic apex</u> of most organizations and
for what have been termed <u>organic organizations</u>, i.e. small
entrepreneurial firms and adhocracies. In these organizational
situations, the human resource planning processes tend to be more
informal. They predominate in situations where it is necessary to
maintain the same architecture for long period of time in order to
reduce complexity. The middle management (technostructure, middle
line, and support staff) and operating cores of large
organizations are situations where the conditions are met.

<h2 style="text-align:center">REFERENCES</h2>

ACKOFF, R.L., & EMERY, F.E., <u>On Purposeful Systems</u>, Tavistock
        Publications, London, 1972.
BRYANT, D.T., and R.J. NIEHAUS (eds.), <u>Manpower Planning and
        Organization Design</u>, Plenum, 1978.
KILMANN, H.R., PONDY, L.R., and SLEVEN, D.P. (eds.), <u>The
        Management of Organization Design</u>, North Holland, New
        York, Vol. I & II, 1976.
LIPPITT, M.E. and MACKENZIE, K.D., "Authority Task Problem,"
        <u>Administrative Science Quarterly</u>, 1976, Vol. 21,
        pp 649 - 660.
MACKENZIE, K.D., <u>A Theory of Group Structures, Vol. I: Basic
        Theory</u>, Gordon and Breach, N.Y., 1976.
MACKENZIE, K.D., "A Process Based Measure for the Degree of
        Hierarchy in a Group. III: Applications to Organizational
        Design", <u>J. Enterprise Management</u>, Vol. 1, 1978,
        pp. 175 - 184.
MACKENZIE, K.D., "Concepts and Measurement in Organizational
        Development" in J.Hogan (ed.), <u>Dimensions of
        Productivity Research, Vol. 1,</u> American Productivity
        Center, Houston, 1981, pp 233 - 301.
MARTEL, A., et J. OUELLET, "La gestion Prévisionnelle des
        resources humaines", <u>Document de travail no. 81-04</u>,
        Faculté des sciences de l'administration, Université
        Laval, Quebec, 1981.
MINTZBERG, H., "Structure in 5's: A Synthesis of the Research
        on Organization Design" <u>Management Science</u>, Vol. 26,
        no. 2, 1980, pp. 322 - 341.
PRICE, W.L., A. MARTEL & K.A. LEWIS, "A Review of Mathematical
        Models in Human Resource Planning", <u>OMEGA</u>, Vol. 8,
        no. 6, 1980, pp. 639 - 645.
SIMON, H.A., <u>The New Science of Management Decision</u>, Harper and
        Row, New York, 1960.
VETTER, E.W., <u>Manpower Planning for High Talent Personnel</u>,
        Michigan Univ., Ann Arbor, 1967.

INVESTIGATING EMPLOYEES' MOTIVATIONS AND SATISFACTIONS IN AN

EXISTING COMPLEX SYSTEM AS A BASIS FOR FUTURE SYSTEMS DESIGN

R.C. Crawley and P. Spurgeon

Applied Psychology Department
University of Aston in Birmingham
Gosta Green
Birmingham B4 7ET

ABSTRACT

Maximising both social and technical aspects in the design of
complex systems presents several problems, notably the lack of
clear criteria and techniques for allocation of function between
humans and machine, and decisions about the point of input to the
design process. The argument is developed that previous approaches
fail to analyse jobs at a level appropriate to system design. A
methodology is described, developed within the context of air
traffic control, which overcomes to some extent these problems.

INTRODUCTION

In attempting to maximise both social and technical aspects
in the design of complex systems, such as air traffic control,
the social scientist is faced with several problems. Notably the
lack of clear criteria for allocation of function between man and
machine. In addition there is the crucial matter of the point of
input to the design process. Singleton & Crawley (1980) note
that "the problem with high-technology systems is that there is
usually a long period of systems research before the more concrete stage of system building and implementation is reached."
(p.163). Thus an inappropriate decision to examine and develop
a particular concept to a simulation stage may prove very costly,
since several man-years' work may have been involved. There is
the added incentive to get it right at the beginning in such
systems since they tend to be unique to one client and therefore
costs of research can rarely be shared. Whilst the justicication

for adopting a sociotechnical perspective towards system design
has been shown elsewhere (eg. Davis & Canter, 1955; Davis &
Trist, 1964; Hedberg & Mumford, 1975) the point has not been
emphasised sufficiently that such a perspective must be adopted
from the earliest stages of system design, from conception through
to implementation. If the technological imperative goes unchecked
in the initial development of a concept, the sponsoring organisation
may be faced with very heavy research and development costs that
prohibit fundamental redesign at a later stage. On the other
hand, if it remains unchecked throughout the design stage, the
organisation may well be faced with heavy costs on and after
implementation, in the form of dissatisfied employees and
associated effects.

A second problem concerns the form of human factors input,
specifically that part relating to the socio-psychological aspects
of systems design. Typically, descriptions of the system design
process (eg. McCormick 1976) fail to clarify the distinction
between, on the one hand, the allocation of function between
machine and people, and on the other hand the allocation of tasks
between the several people in the system to form a satisfactory
set of jobs. The former process is characteristic of system
design, and it requires decisions to be made from the outset about
the sort of functions or tasks that are suitable for automation.
The latter process is the substance of job design, although it
can be subsumed within system design as a later stage.

It is clear from this description that tasks are a relevant
unit of analysis central both to system and job design. Yet a
serious practical difficulty is that neither suitable techniques
nor criteria have emerged for allocating tasks between computer
and person in the context of sociotechnical system design,
especially in the type of system under discussion in this paper.
The participative approaches gaining acceptance in the sociotechnical system design field (Hawgood et al., 1978; Mumford et
al., 1972) present particular difficulties in the context of, for
example, air traffic control. Experience in the UK air traffic
control system is that operational air traffic controllers (ATCOs)
have difficulty comprehending systems research which may be based
on assumptions that are completely unrealistic in the current
environment. Thus a considerable training effort would be required
before they could provide a useful input. Of greater importance
than this is that the participative approach is based on the
assumption that the employees who join the system design team will
operate the new system upon implementation. No such assumption
can be made in air traffic control where several years may pass
before implementation is reached.

Criteria for allocating tasks in terms of adequate performance have been provided by, for example, Fitts (1951).

However, no similar set of criteria is available to help us to
allocate tasks on the basis of people's needs from work. The
participative approach of Mumford and others, where it can be
applied, leaves such criteria implicit in the decisions made by
employees about their job content in the new system. The related
research into job design (Hackman & Oldham, 1975; 1976; Hackman
et al., 1975) is directed at too general a level for it to be of
significant use in system design. Whilst the relevance of task-
specific criteria is acknowledged by Hackman in the sense that the
job characteristics model specifies that a job ought to enable the
employee to use skills that he values, the practical recommendations
for job design (Hackman et al., 1975) simply refer to the combination
of tasks to accomplish high skill variety, in the hope that high
variety will lead to the use of valued skills. In practice, of
course, this need not necessarily follow for it would be possible
for a job to contain a high variety of mundane tasks requiring
little skill.

     The research described in this paper offers an approach to the
problem that in the main provides data specific to the organisation
or system under examination. The approach itself, however, may be
applied to many different situations, although it may be more
appropriate to systems that require a lengthy development period.

INVESTIGATION OF ATCOS' AFFECTIVE REACTIONS TO EXISTING SYSTEM

     The purpose of the research carried out between 1976-80 was
to provide general guidelines for systems designers in air traffic
control with respect to the allocation of functions between human
controllers and computer with specific reference to the effects of
technological change on the job content and therefore the intrinsic
satisfactions of ATCOs. The stimulus for the research was the
potential of computer assisted systems to become involved in executive decision-making functions. There were no systems of the
type envisaged in a sufficiently advanced stage of development for
simulation trials to be conducted; neither were there any in
operation, either in the UK or elsewhere. It was inevitable,
therefore, that the research had to make its focus the affective
reactions of ATCOs towards the existing system, so that those tasks
in the system associated with strong positive or negative reactions might be highlighted and inferences drawn with respect to
future system design.

     The adopted method required analysis of job content using two
techniques, task analysis and the Job Diagnostic Survey (JDS).
Analysis was carried out on several discrete operational work
positions at the London Air Traffic Control Centre and Heathrow
Airport. The ATCOs were qualified to work on all positions
analysed for the study. Thus, having identified differences in
job content and job characteristics by use of the two techniques

mentioned above, it was possible to relate these differences to
affective reactions towards the positions which were measured at
three levels of detail. At the most general level, rank order
preferences were obtained for working on the various positions.
Secondly, reactions to job characteristics were measured by use of
the later sections of the JDS. Finally, and most important of all
for the purposes of this investigation, reactions towards and constructions of tasks from the task analysis were obtained by means
of Repertory Grid technique. Cluster analysis of the Repertory
Grid data revealed clusters of tasks which were either positively
or negatively perceived by the ATCOs.

Task Analysis

     In accord with the approach adopted here it was necessary to
overcome the imprecision of general job descriptions or indeed any
reliance upon job title differentiation at all. In contrast it was
necessary to establish the tasks required to ensure the attainment
of the overall system goals. Inevitably this level of analysis
focusses upon the process or system as a whole. The approach
advocated has been labelled Overview Task Analysis and is a
modified version of the technique initiated by Annett et al.
(1971).

     The original form of Hierarchical Task Analysis was primarily
oriented towards developing training material. The subsequent
overview was developed by Spurgeon & Patrick (1979) in an attempt
to utilise the assumptions of the technique to tackle wider issues.
Essentially the modification involves a change in the rule for
stopping the analysis procedure. The original rule involved an
estimate of the cost of inadequate performance of a task coupled
with the probability of the task being adequately performed by the
trainee group. In the modified version the analysis focusses upon
the total process rather than a particular job. In this way each
of the tasks required to undertake the process are identified and
their relationship specified.

     The procedures involved in the analysis involve a combination
of various data collection activities. Obviously any existing job
descriptions or training material is an important initial source.
This is then supplemented and elaborated through in-depth interviews
or verbal protocols (Bainbridge, 1975) from job incumbents. The
material gathered in this way is checked and validated with numerous
levels of staff involved in the job. Tasks thus identified are
then built into an hierarchical tree, such that the activities
required to ensure that a super-ordinate task is achieved, are
identified. These are built into a complete mapping of the system
tasks. (A brief first level set of descriptors for air traffic
control is presented overleaf).

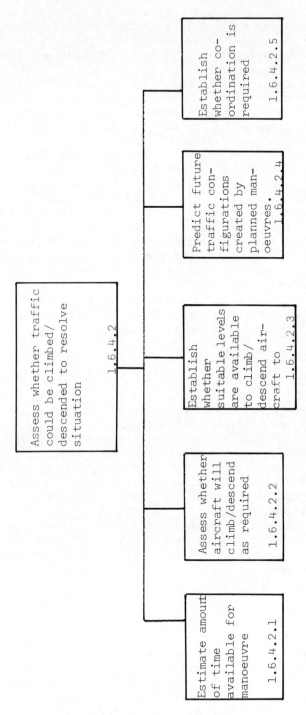

Figure 1. Extract from Task Analysis for Sector Controller at LATCC

The extended version of such a description has been found to have many industrial applications (Patrick et al., 1980). Its prime function here was one of job design. It is clear that in the light of the exhaustive set of material obtained any job title (or ATC position) may be superimposed upon the full set so that the specific tasks associated with a job can be identified. The impact of new technologies or organisation can be monitored via the changes to these task groupings. However, this is still a neutral set of descriptors and what is needed is the affective component or how people feel about these tasks. Hence the task analysis serves as a vital preliminary to identifying sources of satisfaction and reward at a task level.

It was mentioned earlier that affective reactions towards job content were obtained at three levels of detail using a different technique for each. One reason for choosing this approach was to provide an internal check on the validity of the Repertory Grid data, since the use of Rep Grid was novel in this context and to date no body of knowledge is available at the task level. By comparing task ratings with ratings of the positions on which they occur it would be possible to check the consistency of results. For example, one would not expect a highly rated position to consist only of dissatisfying tasks.

Rank Preference for Positions

Two techniques were used to establish preferences for positions, 'Simple Ranking' and 'Forced Choice Pair Comparisons'. The use of two techniques was adopted again as an internal check on the reliability of the findings. 'Simple Ranking' required each ATCO to rank a complete set of positions in order of preference in response to the question: "All things being equal, which positions do you prefer usually to work on?" The Forced Choice Pair Comparisons questionnaire presented, separately, all possible pairs of positions from the set. Each ATCO was required to indicate for each pair "Which position would you look forward more to working on generally?" An overall rank order of preference was derived from each ATCO by counting the frequency with which positions were preferred. 87 per cent of the ATCOs at Heathrow completed one or other of these questionnaires, enabling a high level of confidence to be attached to the results.

Job Diagnostic Survey

The JDS was developed in the USA by Hackman & Oldham (1975) as a diagnostic tool to aid job design and as a means of testing their Job Characteristics Model. The origins of the JDS may be found in earlier work by Hackman & Lawler (1971) and Turner & Lawrence (1965). The JDS is administered as a self-completion questionnaire to job incumbents, and it provides measures of three groups of variables

plus an individual difference measure. Firstly, there are measures
of a number of job characteristics which are hypothesised to contribute to the motivating potential of a job: skill variety, task
identity, task significance, autonomy and feedback. It should be
emphasised that it is the employee's perception of how much of
each characteristic is present in the job that is measured, since
the model predicts that employee reactions are more closely
associated with subjective perceptions of a job than objective
appraisal by outsiders. Secondly, the JDS obtains measures of
three psychological states which are posited to be elicited in an
employee by the presence of the job characteristics: experienced
meaningfulness of the job, experienced responsibility for the outcomes of the work, and knowledge of results. Finally, there are
measures of a number of personal outcomes which are elicited by the
psychological states: general satisfaction with the work, internal
motivation, and satisfaction with opportunities for psychological
growth in the job.

As well as obtaining ratings of their whole job, the JDS was
administered to ATCOs to obtain ratings of the individual work
positions so that comparisons could be made between the positions
and also with the rank preference data. It was of interest to
compare the extent to which overall preference for working on a
position was associated with its motivating potential as derived
from the JDS by summing the scores of each of the five core job
dimensions.

Repertory Grid Analysis of Tasks

The Rep Grid technique was chosen for the most detailed level
of investigation so that we might overcome a problem common to
standardised questionnaires and attitude scales: namely, that they
often assume the relevance of particular terms or ideas to the
subject, and reflect the attitudes or experiences of the invest-
igator or test developer (Shaw & Thomas, 1978). Since our research
was to investigate a novel focus of attitudes, namely tasks within
jobs, it was not possible a priori to predict either the extent to
which ATCOs discriminated between tasks or the way in which they
make those discriminations. The Rep Grid, derived from Kelly's
Construct Theory (Kelly, 1955; Fransella & Bannister, 1977),
offered an approach which not only allowed for but actually made its
focus the individual person's particular and unique experience of
his phenomenological world.

The essence of construct theory is that people look to the
future and anticipate events, rather than merely reacting to stimuli.
Each person does this via his personal construct system, a unique
network of related 'goggles' through which he views the world. A
construct is a bipolar concept which provides a means of cate-
gorising similarities and differences between objects or events.

An example of a construct is 'Interesting-Boring'. It is not an
inherent characteristic of things in nature so much as a construction of reality by the person doing the construing. By construing a job as being 'interesting', the person is formulating a
set of hypotheses about it; for example, that it may be challenging,
require him to think hard, to pay attention, and provide satisfaction. It is in this way that constructions of reality anticipate
events.

We were interested not in the entire personal construct system
of the ATCOs, but only in the typical constructs used by ATCOs to
construe the tasks in their job. We were also interested in the
relationships of other constructs with the superordinate construct
'Satisfying-Dissatisfying'. Finally, it was hoped to be able to
obtain a consensus view of the differential levels of satisfaction
derived from the different tasks in the job by asking ATCOs to rate
each task on a number of constructs. All of these areas of interest
could be dealt with satisfactorily by the use of Rep Grid technique
and appropriate analysis of the data thus obtained.

Full Rep Grid interviews were conducted with a small sample
of ATCOs to elicit constructs that would form the basis of subsequent ratings of tasks by a larger and more reliable sample. The
initial interviewees were asked to select 9-12 tasks from a full
set of over 30, which provided the focus for the interview. The
total set of constructs elicited during these interviews were
content analysed for the purpose of selecting a sample of 'common
constructs', ie. constructs that seemed to be relevant or important
to most ATCOs. Analysis suggested that there were six major
clusters of construct which subsumed most of the individually
elicited constructs. These were labelled as follows:
(a)   job satisfaction; (b) decision-making; (c) skill and
experience; (d) concentration; (e) significance/relevance; and
(f) autonomy.

From these six common construction groupings, a standard Grid
was developed incorporating eleven unipolar constructs representing
different facets of each common construct. These are shown in Table 1.
The constructs are unipolar only in the sense that one pole of each
is explicit. It was not clear whether bipolar constructs would be
more appropriate than unipolar on a priori grounds, and indeed
there were arguments in favour of each. Bipolar constructs would
have provided clearer results; on the other hand, the omission of
the implicit pole meant that the subject could create his own,
thereby making the construct as a whole more personally meaningful.
Reference to the initial Rep Grid interviews made it possible to
ascertain whether the explicit pole of each construct was preferred
or not, since this information was obtained during the interviews.
The explicit poles were phrased so that they represented an extreme:

Table 1. List of Constructs included in Standard Grid derived
          from Rep Grid Interviews.

---

1.   What I'm paid to do
2.   Gives me a lot of job satisfaction
3.   Demands a lot of concentration
4.   Requires all my skill and experience
5.   Mainly up to me whether it gets done well
6.   I have a lot of autonomy in doing it
7.   Unnecessarily time-consuming
8.   Requires a high level of decision making on my part
9.   Involves a lot of co-ordination and consultation with others
10. Very boring
11. Very annoying

---

that is, a rating that indicated a task was not accurately described
by the preferred explicit pole could be taken fairly reliably to
mean that it was less, not more, preferred.

The standard Grids were administered to 40 ATCOs, who were
asked to rate all tasks on each construct according to how accurately
the explicit pole described their perception of the task. Seven
point scales were used.

The grids were analysed by a computer program called FOCUS (Shaw,
1980) which was developed in response to the need for a method of
Rep Grid analysis that was less bewildering for the subject and the
investigators, and which allowed for further conversations between
the two based on the summary data from the initial interview. FOCUS
derives a matrix of distance measures, or matching scores, using
Minkowski's City Block metric. The program was a two-way cluster
analysis which is similar to the nearest neighbour or single
linkage hierarchical method, although as Shaw (1980) notes it is
not a hierarchical method: "The major criterion for forming
clusters is that linear reorderings of the constructs and elements
respectively will result in the final grid displaying a minimum
total difference between all adjacent pairs of rows and columns."
(p.34).

In addition to the analysis of individual ATCO's grids, a group analysis was carried out as follows. The FOCUS program derives similarity measures between variables, and it can do this equally well with a group Grid. Individual analysis revealed that the construct 'Gives me a lot of job satisfaction' was most frequently matched highest with the construct 'Requires all my skill and experience', this occurring on about 50 per cent of the Grids. Thus the group Grid consisted of the combined ratings of tasks by all ATCOs on these two constructs.

DISCUSSION

The results obtained in the research are presented elsewhere (Crawley, 1981; Crawley et al., 1980), and will not be referred to in this paper since they are specific to the air traffic control context.

The investigation has shown that it is possible to elicit employee reactions to job content at a lower, more detailed level of analysis than hitherto; namely, at the task level. This ability is a crucial prerequisite if we are to provide suitable criteria for the design of complex systems. The bringing together of Task Analysis and Repertory Grid technique provides the method that enables this level of affective data to be obtained.

The characteristics of tasks that are associated with intrinsic satisfaction for the sample of ATCOs in this investigation show strong similarities to the job characteristics specified by Hackman & Oldham (1976) in their model. However, there appears to be a much greater importance attached to the "use of skill and experience" than to other characteristics such as autonomy and task significance. This is as one might predict if affective reactions towards tasks rather than whole jobs are obtained. Previous research (Hackman & Oldham, 1976; Mumford et al., 1972; Turner & Lawrence, 1965) has pointed to the need to match employee's skills and abilities to the task demands within jobs. Thus it is predictable that employees react most positively towards tasks that require the use of valued skills and hard-earned experience. In addition, there is evidence with the sample of ATCOs that the use of their skills is more important to their intrinsic job satisfaction than any other factor even at the overall job level.

The importance of matching task demands to skill and ability reinforces the value of the methodology specified in this paper. However, the method has wider application than just this one aspect of system and job design. The recommended approach to socio-technical system design in complex systems requires movement between different levels of analysis according to the specific questions to be answered. At the outset, when a system design team or a sponsoring organisation have to decide on a suitable concept for

development, they must consider amongst other things the sorts of
tasks that are central to current job satisfaction for the
employees whose jobs are to be affected by change. The Rep Grid
data related to the more superordinate levels of the Task Analysis
will provide information relevant to this problem.

Once this decision has been made, subsequent decisions about
allocation of tasks or functions between computer and people must
be made too. For this stage of the design process, the Rep Grid
data relating to both superordinate and subordinate levels of the
Task Analysis will be of use, enabling the system designers to
specify which sub-tasks provide most satisfaction and to achieve
an acceptable balance within the range of tasks to be performed
by humans. Finally, decisions about the design of people's jobs
may require the use of data at the job level (the JDS data) and at
the task level (Rep Grid and Task Analysis). Throughout the
design process, management and researchers may need to relate the
balance of tasks at one level of analysis to the possibilities for
job rotation at a higher level. Thus loss of satisfaction within
one area of the job may be compensated for by new satisfaction from
elsewhere.

REFERENCES

Annett J., Duncan K. D., Stammers R. B. & Gray M. J., 1974, "Task
        Analysis", Department of Employment Training Information,
        Paper 6, London, HMSO.
Bainbridge L., 1975, The representation of working storage and its
        use in the organisation of behaviour, in: "Measurement of
        human resources", W. T. Singleton and P. Spurgeon, eds.,
        Taylor and Francis, London.
Crawley R. C., 1981, "Air Traffic Controller Reactions to Computer
        Assistance", PhD. Thesis, University of Aston in Birmingham.
Crawley R. C., Spurgeon P. and Whitfield D., 1980, "Air Traffic
        Controller Reactions to Computer Assistance", AP Report
        94, University of Aston in Birmingham
Davis L. E. and Canter R. R., 1955, Job Design, Journal of
        Industrial Engineering, 6:6.
Davis L. E. and Trist E. L., 1974, Improving the quality of work
        life: Sociotechnical case studies, in: "Work and the quality
        of life", J. O'Toole, ed., MIT Press, Cambridge, Mass.
Fitts P. M., 1951, ed. "Human engineering for an effective Air
        Navigation and Traffic Control System", National
        Research Council.
Fransella F. and Bannister D., 1977, "A manual for Repertory Grid
        Technique", Academic Press, London.
Hackman J. R. and Lawler E. E., 1971, Employee Reactions to Job
        Characteristics, Journal of Applied Psychology Monograph,
        55(3):259-286.

Hackman J. R. and Oldham G. R., 1975, Development of the Job
        Diagnostic Survey, Journal of Applied Psychology, 60(2):
        159-170.
Hackman J. R. and Oldham G. R., 1976, Motivation through the design
        of work: Test of a theory. Organisational Behaviour and
        Human Performance, 16:250-279.
Hackman J. R., Oldham, G. R., Janson R. and Purdy K., 1975, A new
        strategy for job enrichment, California Management Review,
        Summer:57-71.
Hawgood J., Land F. and Mumford E., 1978, A participative approach
        to forward planning and system change, from: Lecture
        Notes in Computer Science 65: Information Systems Method-
        ology, Proceedings Venice: 39-61. Springer-Verlag, Berlin.
Hedberg B. and Mumford E., 1975, The Design of Computer Systems:
        Man's vision of man as an integral part of the system
        design process, in: "Human choice and computers", Mumford,
        E. & Sackman, H., eds., North Holland.
Kelly, G. A., 1955, "The psychology of personal constructs", Norton.
McCormick E. J., 1976, Job and task analysis, in: "Handbook of
        Industrial and Organisational Psychology", Dunnette, M. D.,
        ed., Rand McNally.
Mumford E., Mercer D., Mills S. and Weir M., 1972, The human
        problems of computer introduction, Management Decision, 10:
        6-17.
Patrick J., Spurgeon P., Barwell F. and Sparrow J., 1980. Re-
        deployment by upgrading to technician, Vol. 1. Report
        submitted to the Training Services Division, Manpower
        Services Commission, Applied Psychology Department,
        University of Aston in Birmingham.
Shaw M. L. G., 1980, "On becoming a personal scientist", Academic
        Press, London.
Shaw M. L. G. and Thomas L. F., 1978, FOCUS on education - an
        interactive computer system for the development and analysis
        of repertory grids. International Journal of Man-Machine
        Studies, 10:139-173.
Singleton W. T. and Crawley R. C, 1980, Changing job and job demands
        in Europe, in: "Changes in working life", K. D. Duncan,
        M. M. Gruneberg and D. Wallis, eds., Wiley.
Spurgeon P. & Patrick J., 1979, "Grouping of Skills: Re-Deployment
        by upgrading to Technician", Report submitted to Manpower
        Services Commission.
Turner A. N. and Lawrence P. R., 1965, "Industrial jobs and the
        worker", Harvard University Graduate School of Business
        Administration, Boston.

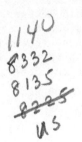

MEASURING EFFICIENCY AND TRADEOFFS

IN ATTAINMENT OF EEO GOALS[1]

A. Charnes,* W.W. Cooper*
R.J. Niehaus,** A.P. Schinnar***

    * Graduate School of Business
      The University of Texas at Austin
      Austin, Texas 78712, USA

   ** Office of the Assistant Secretary of the Navy
      Manpower and Reserve Affairs - ODASN (CPP/EEO)
      Washington, D.C. 20350, USA

 *** School of Public and Urban Policy
      University of Pennsylvania
      Philadelphia, Pennsylvania 19104, USA

1. INTRODUCTION

This paper sketches a framework for evaluating the efficiency
with which various units of the U.S. Navy attain their EEO goals.
We focus on the conversion of "effort" or "input resources" into
"outcomes" or "program outputs" reflecting the improvement in the
representation of women and minorities in the various Navy units.
To determine the efficiency of this conversion process we assume
that, where needed, an EEO program will set out to provide the
maximum amount of improvement in representation for a given level

---

[1]This report was developed as part of the EEO Policy Analysis
advanced development project sponsored by the Navy Personnel
Research and Development Center under NPRDC Work Request
N6822180W40053. It was prepared under ONR project NR 047-222.
Reproduction in whole or in part is permitted for any purpose of
the U.S. Government. The authors gratefully acknowledge the
assistance of E. Bres and A. Desai in the preparation of the
example discussed in section 3.

of resources; any such program is designated as being technically
efficient. However, in the absence of known formulae of program
operation, it is difficult to ascertain the level of outcome that
should be expected and then determine efficiency by comparing
attained outcomes with expected outcomes. In this paper, we shall
propose to compare resource utilization and outcomes achieved by
EEO programs among all units, and then evaluate the outcome levels
that any program has achieved relative to what has been shown by
other programs to be practically attainable with, at most, as
many resources.

There are two useful by-products to the proposed exercise in
measurement of EEO program efficiency. For each program being
evaluated, a representative subset of efficient units is selected
for comparison. By identifying this set of representative
efficient units, we are able to provide guidance for further
evaluation of the inefficient programs in order to help deduce
what organizational action might help improve the efficiency of
these programs.

A second useful result of the analysis is a set of tradeoff
indicators reflecting the rate of substitution among program outcomes and input resources. Tradeoffs among outputs reflect
possible substitution between the representation of different
minority population groups because of limited opportunities within
the organization, or limited supply of personnel in the relevant
labor market. Tradeoffs among resources provide for the possibility of effecting substitutions between more expensive resources
and less expensive resources, thereby achieving cost reduction.

The proposed approach incorporates a multiplicity of resource
inputs and outcome indicators that need not be dimensionally
commensurate. For example, inputs may be measured by the number
of EEO staff, cost of program operation, number of opportunities
created through internal or external recruitment, time of training
required, etc. As indicators of program outcomes we shall usually
choose, but not be restricted to, changes in the representation of
women and minorities. Considerations of parity with the relevant
external labor markets, the attainment of overall manpower staffing
objectives, and the quality of personnel recruited will also enter
into the analysis.

The proposed framework is intended to serve as part of the
Navy's presently operating EEO monitoring system (see Niehaus and
Nitterhouse [12]). Since 1975, comprehensive research into
setting and monitoring the attainment of EEO goals has been underway

at the U.S. Navy.[2] The current Department of the Navy EEO Goals
Accountability System (DONEAS) develops EEO goals based on
estimated relevant labor market supply ratios and organizational
data (see Atwater, Niehaus, and Sheridan [1] and [2]). These
EEO goals are developed separately for each organizational unit.
By way of constrast, the present paper is intended to provide
a means to effect comparisons among organizations engaged in
evaluation in order to measure the efficiency of their performance
in goal attainment.

     The organization of the paper is as follows. In section 2
we review the formulation of the relative efficiency problem,
drawing primarily on the work by Charnes, Cooper and co-workers
[5,7]. In section 3 we provide a numerical illustration of the
approach. In section 4 we introduce a criteria for partitioning
the set of EEO programs into comparison groups for further evaluation and case studies, and conclude in section 5 by suggesting
application of the method for linking analyses in internal and
external labor markets.

     While some portions of this paper are methodological, it
should be noted that our purpose here is not to provide the general
formulation of the relative efficiency problem,[3] rather, the
objective is to outline the basic framework of analysis as a basis
of experimental application within the Navy.

2. MEASUREMENT OF RELATIVE EFFICIENCY

     For any EEO program (unit) being evaluated, a corresponding
practically attainable point on the frontier of EEO program outcomes consists of the maximum output (set of outcomes, or combinations thereof) that has been shown by all units to be practically
attainable with, at most, as many EEO input resources (set of
inputs, or combinations thereof). Consider a partitioning of the
Navy into n administrative units with associated EEO programs,
each with m input resources and s program outcome indicators.
Let x_{ij} be the amount of input i to EEO program j, and y_{rj}
a measure of program outcome r (output) of EEO program j.

---

     [2]For a summary of the initial phases of this research
see Charnes, Cooper, Lewis, and Niehaus [6]. Also see
Chapters III and IV of Niehaus [11].
     [3]See, for example, Banker, Charnes, Cooper and Schinnar [3]
for a nonlinear formulation and access to measuring returns to
scale, and Schinnar [15] for a constrained formulation of
the problem.

Following the Charnes, Cooper et al [5,7] non-Archimedean characterization of the efficiency problem,[4] we can describe the EEO program frontier by a piecewise linear envelope constructed from solutions to the following linear programming problem

$$\text{maximize} \qquad z_o + \varepsilon \sum_{r=1}^{s} \delta_r^- + \varepsilon \sum_{i=1}^{m} \delta_i^+$$

$$\text{subject to} \quad -\sum_{j=1}^{n} y_{rj}\lambda_j + y_{ro}z_o + \delta_r^- = 0, \quad \text{all } r \qquad (1)$$

$$\sum_{j=1}^{n} x_{ij}\lambda_j + \delta_i^+ = x_{io}, \quad \text{all } i \qquad (2)$$

$$\lambda_j, \ \delta_r^-, \ \delta_i^+ \geq 0, \ z_o \text{ unrestricted}$$

where $r = 1,\ldots,s$; $i = 1,\ldots,m$; the symbol $\varepsilon > 0$ and less than every positive number in the base field, is the infinitesimal used to generate the non-Archimedean ordered extension field (see [4] pp. 756-757) and "o" indicates the subscripts of one of the $j=1,\ldots,n$ units that are being evaluated. The constraint set (1) envelops the outputs from above, while the constraint set (2) envelops the inputs from below. The scalar z_o provides the proportionate factor increase in outputs $\{y_{io}\}$ that has been shown by other EEO programs to be practically attainable with at most $\{x_{io}\}$ input resources,[5] while efficiency is, in turn, measured by the reciprocal of z_o^*. By the non-Archimedean efficiency theorem [5], an EEO program is efficient if and only if

$$z_o^* + \varepsilon \sum_{r=1}^{s} \delta^{*-}_r + \varepsilon \sum_{i=1}^{m} \delta^{*+}_i = 1, \text{ which implies, in turn, that}$$

$\delta^{*-}_r = \delta^{*+}_i = 0$, all r and i, and therefore that $z_o^* = 1$.

The set of weights $\{\lambda_j\}$ defines the piecewise linear frontier constructed from facets of a polytope whose extreme points are efficient decision-making units. For each unit, the corresponding basic $\{\lambda_j^*\}$ identify the representative group of efficient units on the frontier. Several different units may all be in the same cone generated from the same set of $\{\lambda_j^*\}$. In section 4 we shall

---

[4]This way of proceeding also clears up an ambiguity noticed by Färe and Lowell [9] in the pioneering work of Farrell [10].

[5]Assuming complete managerial discretionary control of input resources. See [5] and [14] for a discussion of non-discretionary resources.

refer to such sets of units as "comparison groups" of EEO programs, and describe some of their properties.

The non-Archimedean ratio form[6] of the efficiency problem is especially amenable to a "cost/benefit" interpretation of the efficiency score and is obtainable from the dual program of (1)-(2).

$$\text{minimize} \qquad g_o = \sum_{i=1}^{m} w_i x_{io} \qquad\qquad (3)$$

$$\text{subject to: } -\sum_{r=1}^{s} u_r y_{rj} + \sum_{i=1}^{m} w_i x_{ij} \geq 0 \qquad j = 1,\ldots,n \qquad (4)$$

$$\sum_{r=1}^{s} u_r y_{ro} = 1 \qquad\qquad (5)$$

$$u_r, w_i \geq \epsilon,$$

which may be transformed to the ratio problem via a change of variables $u_r = t\mu_r$, $w_i = t\omega_r$, $t > 0$

$$\text{minimize} \qquad f_o = \left(\sum_{i=1}^{m} \omega_i x_{io} \right) / \left(\sum_{r=1}^{s} \mu_r y_{ro} \right) \qquad (6)$$

$$\text{subject to:} \qquad \left(\sum_{i=1}^{m} \omega_i x_{ij} \right) / \left(\sum_{i=1}^{s} \mu_r y_{rj} \right) \geq 1, \text{ all } j \quad (7)$$

$$\mu_r, \omega_i \geq \epsilon \sum_{r=1}^{s} \mu_r y_{ro} \qquad\qquad \text{all } r, i$$

The measure of efficiency in (6) reflects the minimum of a ratio of "weighted" inputs to "weighted" outputs subject to the condition that similar ratios for every unit be at least one. These weights here are not preassigned, however; instead, they are obtained as solutions to the above optimization problem in view of the data on observed EEO program operations. Units incurring least amount of input, a so-called virtual input, per unit of (virtual) output are relatively efficient. All other units are inefficient.

---

[6]In [5] and [7] the ratio form (6)-(7) constitutes the initial formulation from which the envelopment form (1)-(2) is derived. We have chosen here to motivate the problem by the envelopment procedure and show (6)-(7) and its associated interpretations as a derivation instead.

The dual variables u_r and w_i provide access to tradeoff
interpretations. For every efficient unit found in the basis of
an optimal solution of (1)-(2), the corresponding constraint set
in (4) is tight.

$$- \sum_{r=1}^{s} u_r^* y_{rj} + \sum_{i=1}^{m} w_i^* x_{ij} = 0 \qquad\qquad (8)$$

Thus, for a given level of efficiency, defined by $1/z_0^*$, the ratio
$-w_i^*/w_k^*$ gives the trade-off rate between inputs k and i on the
corresponding facet of the frontier (holding all other inputs and
outputs constant), while the ratio $-u_r^*/u_q^*$ gives the tradeoff rate
between outputs q and r (holding all other inputs and outputs
constant).[7] Tradeoffs in the input space (2) could reflect a
substitution between, e.g., EEO program staffing levels and the
length of a training period for minority recruits. Tradeoffs in
the output space (1) could reflect a substitution between, e.g.,
the attainment of representation goals for Black and Hispanic
populations, respectively.

3. EXAMPLE

A data base on 18 of the Navy's organizational units that meet
a minimum size requirement of 1000 civilian employees is used to
illustrate the method of analysis. We focus in this example on the
representation of two minority groups, Blacks and Hispanics, in
the job category of managers and administrators, grade 9-12, as
indicators of EEO program outputs, and on the level of EEO staffing
and the recruitment opportunity rates as input data.

Table 1 contains the input and output measures corresponding
to each unit.

Input Data: The input data are designed to reflect, on the one hand,
the level of EEO program resources available and, on the other hand,
the availability of opportunities for minority personnel. The two
inputs are: EEO STAFF = The number of EEO specialists (or full-
                time equivalent) per 1000 employees for each organi-
                zational unit.
             OPP RATE = The general[8] rate of recruitment (promotion
                and hiring) into the managers and administrators grades
                (GS 9 through 12).

---

[7]For a related discussion of tradeoffs see [8]. Bear in mind
that this does not apply to efficient corner units which may be
part of several facets.
[8]Recruitment of minority and non-minority personnel.

Table 1. Input and Output Data for Eighteen Organizational
          Units in the U.S. Navy

I.D. Number	Outputs		Inputs	
	Blacks R	Hispanics R	EEO STAFF*	OPP RATE
1	1.098	0.976	0.985	0.104
2	0.997	0.884	1.897	0.110
3	1.032	1.138	0.672	0.125
4	1.133	1.070	1.589	0.115
5	1.146	0.862	0.827	0.109
6	1.651	0.978	1.875	0.108
7	1.120	1.035	1.905	0.108
8	0.993	1.227	1.769	0.108
9	1.104	0.968	0.676	0.121
10	1.114	0.634	0.000	0.140
11	1.186	1.025	1.718	0.158
12	1.062	1.000	0.472	0.135
13	0.868	1.093	0.656	0.126
14	0.971	1.040	2.386	0.128
15	0.667	0.967	0.531	0.101
16	1.262	0.443	1.292	0.175
17	0.914	1.064	1.261	0.136
18	1.362	1.000	0.882	0.145

* All quantities are stated in units of 1,000 employees

The figures reflect an average annual rate based on FY72–FY78
data. These are shown in the last two columns of table 1.
Output Data: The output data are intended to reflect relative
progress (or the lack of progress) in attainment of minority
representation goals between FY78 and FY79. In table 2 we show
the percentage of EEO[9] goals attained for Black and Hispanic
personnel by each unit in FY78 and FY79, i.e.,

---

[9]These goals are based on undifferentiated Civilian Labor
Force (CLF) ratios required for reporting by the Equal Employment
Opportunity Commission (EEOC) rather than the more appropriate
Relevant Labor Force ratios that consider occupational and wage
availability. Both RLF and CLF standards are incorporated in the
Navy's DONEAS described in [12]. The first operational version of
Navy RLF data is provided in [2].

Table 2.   Percentage of EEO Goals Attained by
           Navy Units for FY78-FY79

I.D. Number	Blacks		Hispanics	
	FY78	FY79	FY78	FY79
1	44.700	49.090	52.240	51.000
2	98.040	97.760	39.420	34.840
3	48.480	50.030	36.460	41.500
4	39.230	44.440	38.320	41.020
5	36.440	41.750	48.520	41.820
6	22.430	37.030	101.200	98.940
7	26.440	29.600	36.280	37.550
8	83.780	83.190	35.790	43.920
9	40.400	44.590	20.260	19.610
10	79.130	88.120	30.970	19.650
11	53.200	63.120	15.500	15.880
12	36.040	38.260	0	0
13	21.750	18.880	14.380	15.720
14	68.570	66.610	54.400	56.550
15	62.730	41.830	35.690	34.500
16	50.220	63.390	28.450	12.610
17	50.500	46.180	46.320	49.270
18	37.900	51.620	0	0

*Based on undifferentiated Civlian Labor Force (CLF) hiring
goals consistent with EEOC MD-702. (See Footnote 8)

$$\frac{A_{79}}{G_{79}} = \frac{\left(\begin{array}{l}\text{Actual \% representation of}\\ \text{a minority group for FY79}\end{array}\right)}{\left(\begin{array}{l}\text{Goal for \% representation of}\\ \text{a minority group set for FY79}\end{array}\right)}$$

The output indicator

$$R = \frac{A_{79}/G_{79}}{A_{78}/G_{78}}$$

then reflects progress in goal attainment over the 78-79 transition
period. $R > 1$ indicates progress, $R = 1$ reflects no change, and
$R < 1$ shows an increasing divergence between the EEO goal and the

Table 3. Efficiency Measures and Slack Variables

ID Number	Output Factor Increase (z_o)	Efficiency Rating ($1/z_o$)	Slack Variables			
			Black Rep.	Hispanic Rep.	EEO Staff	Opportunity Rate
1	1.000	1.000	0	0	0	0
2	1.294	0.773	0	0	0.050	0
3	1.000	1.000	0	0	0	0
4	1.094	0.914	0	0	0	0
5	1.000	1.000	0	0	0	0
6	1.000	1.000	0	0	0	0
7	1.099	0.910	0	0	0.098	0
8	1.000	1.000	0	0	0	0
9	1.002	0.998	0	0	0	0
10	1.000	1.000	0	0	0	0
11	1.464	0.683	0	0	0	0
12	1.020	0.981	0	0	0	0
13	1.068	0.936	0	0	0	0
14	1.350	0.741	0	0	0.268	0
15	1.000	1.000	0	0	0	0
16	1.535	0.651	0.449	0	0	0
17	1.269	0.788	0	0	0	0
18	1.060	0.944	0	0	0	0

actual representation of a minority group. The output indicators
(R) for Blacks and Hispanics are shown in columns 2 and 3 of
table 1. Observe that only six of the EEO programs have registered
progress for both Blacks and Hispanics; in two cases there was a
decrease in the representation of both groups, and in ten units
progress in one group was accompanied by diminished representation
of the other group.[10]

_____

[10]As David Sherman has shown in his analysis of the relative
efficiency of health service organizations, ratios should be used
with caution. Our analysis here is illustrative however, and in
actual application nonratio output quantities could be readily
incorporated because of the linear programming methodology.

Table 3 gives a summary of the results obtained from an application of program (1)-(2) to the data in table 1. For each EEO program, identified by an ID number in the first column, we show the proportional factor increase in output (z_o^*, in column 2), the input efficiency score ($1/z_o^*$, in column 3), and the slack variables associated with the outputs (column 4 and 5) and the inputs (columns 6 and 7). By scanning column 3 we note that there are seven efficient units (#1, 3, 5, 6, 8, 10, 15), all with $z_o^* = 1$ and zero slacks. Units #2, 11, 14, 16 and 17 have efficiency scores of less than 0.8, with the remaining six units nearly efficient (i.e., $0.9 \leq z_o^* < 1$). Consider for example unit 14 which posted an efficient score of 0.741. The associated $z_o^* = 1.35$ suggests that, in this case, the evidence from the operations by others indicates that a 35% increase in the representation of all minorities should have been attained with no more input resources. In fact, the slack of .268 associated with EEO staff implies that, in addition, this 35% increase in representation was attainable with a concomitant reduction of .268 from the present level of 2.386 EEO staff per 1000 employees.

Observe that in the definition of the output indicator R, three of the four terms (G_{78}, A_{78}, G_{79}) used to compute R can be regarded as fixed. It follows then that z_o constitutes the proportional factor increase in A_{79}, i.e., in the actual representation during the FY79 evaluation year. Therefore, by subtracting unity from the entries in the second column of table 3, we can obtain the growth rate in minority representation that has not been achieved but has been shown attainable by other units.

There is also a difference in the number of civilian employees among the Navy units. The first nine units have more than 10,000 civilian employees each, while the size of the civilian labor-force of the remaining nine units range from 1000 to 10,000 employees. The nine larger units are on the average more efficient (.96) than the nine smaller units (.86). This suggests the presence of some economics of scale in EEO programs. In order to obtain direct measures of return to scale, a recourse to the bi-extremal variant of the method developed by Banker, Charnes, Cooper and Schinnar [3] is required. However, this would necessitate a parametric characterization of the "EEO program production function," which is not required by the present analysis.

In closing, we should like to underscore the importance of proper selection of variables for the interpretation of results. By scanning the rows of table 2 we note that in all but one efficient unit, progress has been made in approaching the EEO goals of at least one minority group. Unit 15 has registered a decline for both minority groups, but because it has employed few inputs it is found efficient. Observe, however, that limited amounts of input can have, in this case, two interpretations: (i) a low

opportunity rate implies few opportunities for internal and
external recruitment irrespective of the level of EEO effort;
and (ii) a small EEO staff may not be able to provide the necessary
assistance to maintain, let alone increase, minority representation.
This may explain, at least partly,[11] the lack of progress in goal
attainment achieved by unit 15. Nonetheless, in the context of the
present analysis, "efficiency" suggests that, given the limited
opportunities for recruitment and the small EEO staff, the loss in
minority representation might have been even greater had the EEO
staff not used its resources efficiently.

4. COMPARISON GROUPS

The second objective of this paper is to provide guidance for
further evaluation of EEO programs in order to identify, e.g.,
organizational impediments that may be the source of inefficiency
in the EEO programs. Such studies will provide the necessary
information for effecting actual improvements in the operation of
EEO programs. In this section we introduce a criteria for choosing
a comparison group in order to evaluate a given unit.

We define a comparison group as a set of units which share the
same representative set of efficient units on the frontier; the
representative set may include "dummy units" associated with slack
variables. As will be shown below, the dual variables associated
with any member of such group are scalar multiples of the dual
variables of any other member of the group. Consequently, it is
assumed that for designated contours of efficiency levels, the rates
of tradeoff among inputs and outputs are the same for all members
of a comparison group. This suggests that, in addition to sharing
the same subset of representative efficient units on the frontier,[12]
the comparison group also shares a "technology" for converting
resource inputs into EEO program outcomes.

We focus now on identifying the members of the comparison group
for a unit whose inputs and outputs are given by $\{x_{io}\}$ and $\{y_{ro}\}$.
We assume that a solution of program (1)-(2) has been obtained so
that z_o^*, $\{\lambda_j^*\}$, the slacks, $\{u_r^*\}$ and $\{w_i^*\}$ are readily available.
In addition, we retain the "inverse of the optimal basis" in (1)-
(2) and denote by N the matrix containing its first m+s-1 rows.

---

[11]Alternatively, the initial set of goals may have been too
ambitious.
[12]Note, however, that the degree to which a particular
efficient unit is representative of an inefficient unit will vary
across members of the group.

Proposition 1: A unit $\underline{k}$ is a member of a comparison group defined with respect to unit $\underline{o}$ if, and only if,

$$\sum_{r=1}^{s} u^*_{ro} y_{rk} > 0 \tag{9}$$

$$z_k \sum_{r=1}^{s} N_{jr} y_{rk} - \sum_{i=s+1}^{s+m} N_{ji} x_{ik} \geq 0 \tag{10}$$

for all j (correspond to basic λ_j and slacks), where

$$z_k = \frac{\sum_{i=1}^{m} w^*_{io} x_{ik}}{\sum_{r=1}^{s} u^*_{ro} y_{rk}} \tag{11}$$

Proposition 2: If (9) and (10) hold, then (11) gives z^*_k; the left-hand side of inequality (1) gives the value for λ^*_j and the slacks; and

$$w^*_{ik} = w^*_{io} / (\sum_{r=1}^{s} u^*_{ro} y_{rk}) \tag{12}$$

$$u^*_{rk} = u^*_{ro} / (\sum_{r=1}^{n} u^*_{ro} y_{rk}) \tag{13}$$

The proofs for the above propositions follow from the derivation of an algorithm in Schinnar [14].[13] Equation (9) is derived from the optimality condition for the linear programming problem (1)-(2), in which the y_{ro} and the x_{io} have been replaced by y_{rk} and x_{ik}, respectively. It is also a sufficient condition for constructing a basic solution for $\{y_{rk}, x_{ik}\}$ from the optimal basis associated with an optimal solution of (1)-(2) for $\{y_{ro}, x_{io}\}$. Equation (10) follows from the feasibility test for the new basis. Also note that z_k in (11) coincides with f_k in (6), and that within a comparison group the dual evaluators are unique up to multiplication by a scalar; cf. equations (12)-(13).

The above simple tests provide a way by which to identify the entire comparison group for a given EEO program from a single linear programming solution. This comparison group will include

---

[13]These results are also suggested in an earlier draft of Charnes, Cooper and Rhodes [7].

both efficient and inefficient units. Several options are thus
available for proceeding with case-studies to identify the
determinants of inefficiency:

1. A comparison of inefficient EEO programs with other inefficient
    programs in the comparison group in order to observe common
    features that may help explain the observed inefficiency.
2. A comparison of efficient EEO programs among themselves in
    order to identify common factors that may help explain their
    efficiency.
3. A comparison of the inefficient EEO programs with the efficient
    subset of units in order to observe systematic differences
    between efficient and inefficient units.

We now continue with the illustrative example of section 3.
Table 4 identifies the membership in six comparison groups into
which the set of 18 EEO programs, evaluated in table 3, may be
divided. Each row in the table marks the inefficient and efficient
members of a group. Note that the inefficient units are members in
one group only, while the efficient units appear in several comparison groups when they provide the corners of the facets forming
the efficiency frontier. For example, in group D, two inefficient
units are associated with three efficient units that are their
representatives on the efficiency frontier. Thus, a case study
designed to assess the sources of inefficiency in units 12 and 13
should include comparisons of these units with units 3, 10, and 15.
Further specificity in the comparisons can be obtained by reference
to the weights associated with each of the efficient units; the
weights are the primal variables obtained from program (1)-(2) and
are displayed in table 5. Efficient units with larger weight enter
more "heavily" into the evaluation of an inefficient unit and are
therefore more suitable for effecting comparisons between inefficient and efficient program units. Thus, the inefficient unit 12
should be primarily compared with the efficient unit 3, and the
inefficient unit 13 should be primarily compared with the efficient
unit 15.

Note also that groups A and E have only two efficient units
each, whereas the other groups include three. This results from
the presence of a (positive) slack variable in the basis of
groups A and E and implies, in turn, that their associated facet
is not efficient. However, this need not preclude comparisons
between the inefficient and efficient units in these groups.

A final observation is made with reference to the tradeoff
between Black and Hispanic representation at the efficiency
frontier of each comparison. The last column of the table shows
the substitution possibilities between the two minority groups
while remaining on the efficiency frontier. These figures are
obtained from the ratios of the dual variables associated with

Table 4. Membership in Six Comparison Groups of Navy Units and
          Tradeoff Between Black and Hispanic Representation

Group	Inefficient Units											Efficient units							Tradeoffs between Blacks and Hispanics
	2	4	7	9	11	12	13	14	16	17	18	1	3	5	6	8	10	15	
A	X	X						X						X	X				-2.6
B	X		X									X		X	X				-2.4
C			X								X	X	X			X			-0.7
D				X	X							X				X	X		-2.9
E							X							X		X			0
F								X				X			X		X		-3.4

output indicators in program (3)-(4). For example, in group A
(units 2, 7, and 14), the tradeoff rate of 2.6 suggests that a loss
of two Black employees should be replaced by a gain of approximately
five Hispanics if the EEO program is being operated efficiently.
This indicates that the EEO programs in comparison group A are more
sensitive to progress in the representation of Blacks and Hispanics,
and may suggest a greater supply of Hispanics in the relevant
external labor market of qualified personnel. In comparison
group D, the situation is reversed: fewer Hispanics are required
to replace Blacks in order to maintain the efficiency rating of
the EEO programs. These tradeoff rates reflect the slopes of the
facets of the piece-wise linear frontier and are therefore meaningful in the context of a discussion of comparison groups. Any
attempt to attribute these tradeoff rates to specific efficient
units is erroneous because of their membership in several groups.
(See Schinnar [13] for further discussion.)

5. CONCLUSION

     We have presented a framework for monitoring the relative
efficiency of EEO programs in the Navy, and demonstrated how the
information can be used to guide further evaluation of the "cost
effectiveness" of these programs in order to obtain improvements
in EEO program operations. We have shown an example of an application of the method, using preliminary data on 18 organizational
units in the Navy. The reader should bear in mind, though, that
this application is intended for illustration only, and the figures
should not be used to evaluate any of the units involved. For this
purpose, a more complete data base is required, coupled with a
better specification of the input and output indicators. For

Table 5. Representation Weights for Efficient Units

Units Evaluated	Efficient Units						
	#1	#3	#5	#6	#8	#10	#15
1	1.000	0	0	0	0	0	0
2	0	0	0	0.425	0.594	0	0
3	0	1.000	0	0	0	0	0
4	0.437	0	0	0.183	0.461	0	0
5	0	0	1.000	0	0	0	0
6	0	0	0	1.000	0	0	0
7	0	0	0	0.360	0.640	0	0
8	0	0	0	0	1.000	0	0
9	0	0.444	0.456	0	0	0.112	0
10	0	0	0	0	0	1.000	
11	1.235	0	0	0.166	0.108	0	0
12	0	0.617	0	0	0	0.335	0.108
13	0	0.445	0	0	0	0.018	0.673
14	0	0	0	0.205	0.981	0	0
15	0	0	0	0	0	0	1.000
16	0	0	0	0.689	0	0.718	0
17	0.438	0	0	0	0.295	0	0.580
18	0	0.038	1.036	0	0	0.196	0

example, in subsequent applications, the output measures that relate to EEO goals will be computed using the more appropriate data which is based on relevant labor force supply statistics.

Further conceptual developments of the model would involve a more explicit linkage between the organizational structure of opportunities and the availability of qualified personnel in the relevant labor market, as well as the incorporation of data (such as the size of EEO programs) that is not only of an "input" or "output" variety.

REFERENCES

1. Atwater, D.M., R.J. Niehaus and J.A. Sheridan, "EEO Goals Development in the Naval Sea Systems Command," OASN(M,RA&L) Research Report No. 35 (Washington, D.C.: Office of the Assistant Secretary of the Navy [Manpower, Reserve Affairs and Logistics], 1979).

2.   Atwater, D.M., R.J. Niehaus and J.A. Sheridan, "EEO External
        Relevant Labor Force Analysis," OASN(M,RA&L) Research
        Report No. 37 (Washington, D.C.: Office of the Assistant
        Secretary of the Navy [Manpower, Reserve Affairs and
        Logistics], 1980).
3.   Banker, R., A. Charnes, W.W. Cooper and A.P. Schinnar,
        "A Bi-Extremal Principle for Frontier Estimation and
        Efficiency Evaluation," Management Science, forthcoming.
4.   Charnes, A. and W.W. Cooper, Management Models and Industrial
        Application of Linear Programming, (New York: Wiley, 1961).
5.   Charnes, A., W.W. Cooper, A. Lewin, R. Morey and J. Rousseau,
        "Efficiency analysis with non-discretionary resources,"
        Center for Cybernetic Studies Research Report 379,
        The University of Texas, Sept. 1980.
6.   Charnes, A., W.W. Cooper, K.A. Lewis and R.J. Niehaus, "Design
        and Development of Equal Employment Opportunity Human
        Resources Planning Models," NPDRC TR79-141 (San Diego:
        Navy Personnel Research and Development Center, 1979).
7.   Charnes, A., W.W. Cooper and E. Rhodes, "Measuring the Effi-
        ciency of Decision-Making Units," European Journal of
        Operational Research, Vol. 2, 1978, pp. 429-444.
8.   Charnes, A., W.W. Cooper, A.P. Schinnar and N.E. Terleckyj,
        "A Goal Focusing Approach to Analysis of Tradeoffs Among
        Household Production Outputs," American Statistical
        Association 1979 Proceedings of the Social Statistics
        Section, 1979, pp. 174-199.
9.   Färe, R. and C.A.K. Lovell, "Measuring the Technical Efficiency
        of Production," Journal of Economic Theory, 1978,
        pp. 150-162.
10. Farrell, M.J., "The Measurement of Productive Efficiency,"
        J.Roy, Statist. Soc. Ser. A, Vol. III, 1957, pp. 253-390.
11. Niehaus, R.J., Computer-Assisted Human Resources Planning,
        (New York: Wiley-Interscience, 1979).
12. Niehaus, R.J. and D. Nitterhouse, "Planning and Accountability
        Systems for EEO and Affirmative Action Policy," OASN
        (M,RA&L) Research Report No. 38 (Washington, D.C.:
        Office of the Assistant Secretary of the Navy [Manpower,
        Reserve Affairs and Logistics], 1980).
13. Schinnar, A.P., "Measuring Productive Efficiency of Public
        Service Provision," CDSE/HUD Project Report, Fels
        Discussion Paper No. 143, University of Pennsylvania,
        July 1980.
14. Schinnar, A.P., "An Algorithm for Measuring Relative
        Efficiency," Discussion Paper No. 144, University of
        Pennsylvania, August 1980.
15. Schinnar, A.P., "Constrained Isoquant Analysis of Efficiency
        Frontiers," Fels Discussion Paper No. 145, University
        of Pennsylvania, August 1980.

SECTION 6

STUDIES IN ORGANIZATIONAL EFFECTIVNESS

   This set of case studies illustrates the adoption of
management methods to meet the perceived need for more effective
utilization of human resources. Tweeddale reports on the U.S.
Navy Productivity Program, which resulted in such measurable
improvement as work load reductions and overtime decreases through
various productivity awareness instruments, quality circles, and
incentive schemes. In the same spirit, Wanders treats problems
of effectiveness in maintenance of the new generation of fighter
airplanes in the Royal Netherlands Air Force as a problem in
participative management and self-directed actions research.
Similiarly, the use of consultation in effectively securing
employee cooperation in British frozen food factories described
by Ursell and Blyton.

   The adoption of new management technologies is also driven
by the need to cope with greater environmental turbulance and
competition. As corporations attempt to gain degrees of freedom
via strategic positioning, managers must re-evaluate with their
own thought and decision processes. One case in point is the
paper of Tanimitsu, Yanagishita and Baba, who reflect on the R&D
organization in the Mitsubishi Electric Corporation, thus doing
research on research. Similarly, Lehnberg reports a cost-
effectivenss analysis of an effectiveness simulation tool: A
tank crew training system introduced into the West German Tank
Corps. This tool. of course, is a classical human resource
development instrument. Incidentally, the cost-effectiveness of
the U.S. Navy productivity program, according to Tweeddale, is
five dollars savings for every dollar invested in this effort.

PRODUCTIVITY ENHANCEMENT: AN APPROACH TO MANAGING BENEFICIAL

CHANGE IN A MILITARY-INDUSTRIAL WORK SETTING

James W. Tweeddale

Director, Productivity Management
Office of the Assistant Secretary of the Navy (M,RA&L)
Washington, D.C.   20360

INTRODUCTION

New technology, the rising cost of production, and expectations
that work should be enriching as well as profitable are causing
transformations in the Navy. Traditional tools of economic management are being reassessed as they demonstrate diminishing ability
to promote long-term economic improvement.

Created in 1978, as a reflection of growing management concern,
the Navy's Productivity Program establishes the framework for improvement. A network of productivity principals has been established in higher echelon Navy and Marine Corps commands with a
Director of Productivity Management located in the Navy Secretariat.

Under the direction of the Chief of Naval Material, Admiral
A. J. Whittle, Jr., the Naval Material Command (NMC) has established a Productivity Management Office with a cadre of professionals representing the various disciplines attendant to a functionally integrated Productivity Enhancement Program. Although
the Navy Productivity Program encompasses all military and civilian
organizations, the NMC organic industrial community is the focus
of current planning. Corporately, this community employs over
100,000 civilians and involves a budget of more than $6 billion.

The Navy Productivity Program, as presently structured,
explores three major areas of opportunity. These are technological
advancement, organizational development and process management.

TECHNOLOGY

Productivity improvements derive from changes in production methods, materials and machinery which in turn stem from the accumulation of scientific and technological knowledge. The technology factor has been credited with at least 40 percent of the growth in productivity over the past five decades of domestic industrial experience.

The thoughtful integration of beneficial technology to the Navy's industrial base represents a critical dimension of the Productivity Enhancement Program. To a very large degree, the process is limited by the capacity of the organization to accommodate innovation and to handle uncertainty.

To create a climate which encourages technological venture and innovation, a number of funding mechanisms have been introduced.

Cost of Ownership Reduction Investment (COORI) Program

Established as a part of the Navy's FY-82 budget planning, this program creates a funding base to support high payback capital investment opportunities. Candidate projects are placed in competition by operating managers during budget planning (approximately two years before budget approval). Ultimately, the amount of funding is determined by the quality of candidate projects and a subjective comparative assessment of the relative merits of COORI vis-a-vis other programs in competition for budget funds.

Fast Payback Program

The Fast Payback Program is designed to create a funding mechanism through which managers can fund high payback projects with short lead time provided the projects satisfy the following criteria:

1. The project costs less than $300,000 ($100,000 non-NIF).

2. The project has a payback of less than three years (two years non-NIF).

Funds are made available to support the Fast Payback Program through two funding mechanisms. These are:

Naval Industrial Fund (NIF). Naval Shipyards, Air Rework Facilities (NARFs), Public Works Centers (PWCs) are among many of the Navy's industrial organizations which are NIF funded. Under this funding concept, "earnings" are credited to an industrially funded activity by charging fleet customers for goods and services

rendered. Contractural relationships between these organizations
and weapons systems custodians (fleet customers) involve fixed price
bidding and cost accounting practices not uncommon to those found
in the private sector. NIF managers deciding to invest in beneficial technology which qualifies under "fast payback" criteria do so
using industrial funds. Costs associated with the investment are
amortized by the stream of dollar savings which accrue from the
investment. Additional savings, above and beyond the investment
cost, are reapplied to enhance the productivity of the organization.

     Procurement Funds (OPN). A productivity fund is created as
a part of the Navy budget to allow non-NIF activities (hospitals,
training agencies, headquarters, etc.) to have access to the "fast
payback" program.

Manufacturing Technology (MT) Program

     This program explores the application of emerging technology
to (1) reduce procurement and life cycle costs and (2) increase
productivity. The program reduces the risks associated with new
technology exploration by providing "seed money" to MT program
participants. The impact of technology on weapons system procurement costs is impressively demonstrated by a recently completed MT
project involving search radomes which are in common use on ship-
borne and land based radar systems. In this case, the MT project
found that radomes fabricated from foam filled fiberglass may be
substituted for search radomes conventionally manufactured from
honeycomb structures. The use of foam filled fiberglass reduced
the cost of the search radome from $6,000 to $450. Projected net
savings from this project thru FY 1985 exceeds 5.2 million dollars.
(The total cost of the MT project was $116,000.)

Office of the Secretary of Defense (OSD) Sponsored Productivity Enhancing Capital Investment (PECI) Program

.   In addition to the above Navy funded programs, OSD, starting
in FY 1981, established a "grant" fund to underwrite the cost of
certain productivity enhancing capital investments. To compete
for these funds, Services respond to an OSD PECI "project call"
by submitting projects meeting the following basic criteria:

     1. Minimum investment cost $1 million (modified to
$300,000 for FY 1982).

     2. At least 50 percent of economic return of invested
funds must accrue from labor savings.

     3. Maximum return on investment--4 years.

     4. Internal rate of return--at least 10 percent.

The Decision Model contained in figure 1 depicts the relationship which exists between the various funding mechanisms which provide access to productivity enhancing technology. It must be kept
in mind that technology transfer differs from ordinary scientific
information transfer in the fact that to be transferred, technology
must be embodied in an actual operation of some kind. Decisions to
adopt new technologies are basically investment decisions; they involve elements of risk and uncertainty. Organizations vary in their
capability to accommodate uncertainty as a reflection of their value
systems, social structures and/or culture. These elements are
brought into focus in a second major target of opportunity of the
Navy's Productivity Program.

ORGANIZATIONAL DEVELOPMENT (OD)

The importance of the internal organization environment for the
success of any productivity enhancement endeavor is emphasized in
the literature and attested to by successful managers. The Navy's
OD program provides the structure to recognize and pursue opportunities which strengthen the organization and influence the quality of
member behavior. A number of specific programmatic initiatives and
accomplishments follow.

Performance Contingent Reward System (PCRS)

An incentive program designed to improve individual productivity
was developed by the Naval Personnel Research and Development Center
and implemented in the data entry section of a data processing center at the Long Beach Naval Shipyard. The employees participating
in the study were Navy civilian key entry operators. Production
standards were developed based upon keying speed and the amount of
time spent working. A Performance Contingent Reward System was
designed in accordance with behavioral principles and federal guidelines such that a monetary bonus was awarded for high individual
productivity. The amount of the reward was directly proportional
to the amount of work exceeding a production standard.

Production for the 12-month period improved substantially. Ex-
cessive overtime and a heretofore perpetual backlog were virtually
eliminated (figs. 2 and 3). The workforce decreased in size but not
in productivity as a few employees left the organization through natural attribtion and were not replaced. A rigorous cost-effectiveness
analysis showed that the set-up costs of the program were recovered
in the first three months of operation. A similar follow-on study
was conducted at the Mare Island Naval Shipyard with comparable results.

Implementation of PCRSs at other Naval Shipyards has shown that
where the basic tenets of a PCRS exist (viz. regular performance
feedback, timely reward, and reasonable time-on-the-job standards)
significant increases in productivity may be expected.

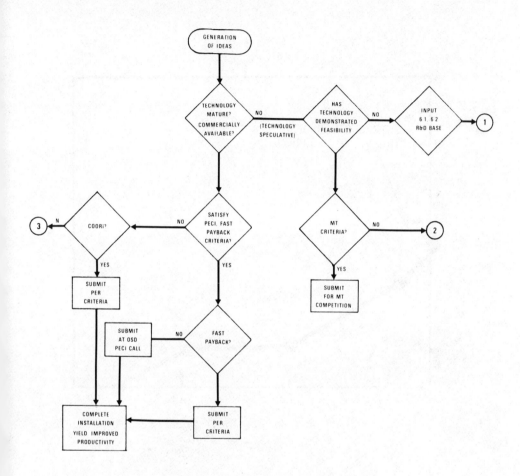

1 HOLD IN BACKLOG FOR FUTURE CONSIDERATION IN APPLIED SCIENCES OR ABANDON
2 EXPLORE ALTERNATIVE SUPPORT, NEED ASSESSMENT, ETC.
3 CONSIDER OTHER FUNDING MECHANISMS (ENERGY, OSHA, ETC.)

Figure 1.   Capital Investment: Technology Decision Model

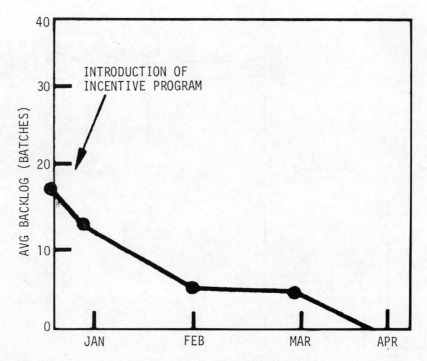

Figure 2. Productivity and Motivation: Work Backlog Reduction

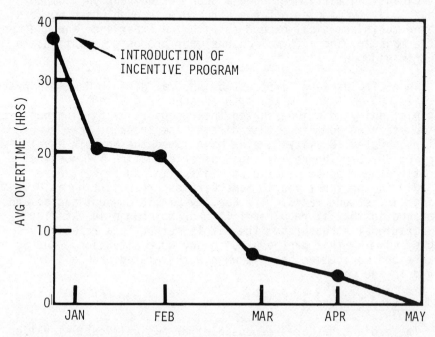

Figure 3. Productivity and Motivation: Overtime Reduction

Quality Circles (QCs)

Quality Circles are small groups of workers who get together voluntarily on a regular basis to solve everyday work problems that cause frustration, product defectiveness and hinder productivity. At this time, there are more than one hundred Quality Circles in the Naval Material Command; the number is growing weekly. Circle participants normally meet once a week for an hour on "company time." Quality Circles offer a structure through which workers combine their practical expertise with the experience and judgment of the manager, and a formal channel for proposing solutions to higher management.

When Circles were first established at Norfolk Naval Shipyard, whose 12,000 employees make it the second largest employer in Virginia, shipyard unions showed their support by filling positions on the steering committee that directs the program. For every dollar invested, Quality Circles have saved the Norfolk Naval Shipyard $5.00. The net savings, after deducting all costs of operating the program, including staff time and travel, were $150,000 in the first year. Some Circles are reporting an 8:1 savings:investment ratio. The Quality Circle concept has an added advantage in that it requires little or no change in existing organizational structures. The Circle method is a relatively simple technique that can be used in any work situation. And both workers and management benefit when work-related problems are solved in the Circle.

Productivity Awareness

The media today are responsible for a significant elevation in the level of social consciousness associated with productivity. Officials in labor, management, academia and government appear to agree that productivity must improve if our nation is to remain economically and socially sound. These factors when coupled with the many emerging productivity issues coming from the world of work plus ever-present changes in management methods, structures, priorities and personalities establish the basis for a productivity awareness program.

Under the sponsorship of the Director of Productivity Management and the Naval Material Command Productivity Management Office, the following initiatives are being pursued:

Prospective Commanding Officers (PCO) Course. Commanding Officers (COs) of naval field activities are subject to rotation at regular intervals. The Prospective Commanding Officers Course is structured to equip all expectant COs with information on programs of contemporary interest prior to their assignment. The PCO

Course curriculum includes segments providing the critical information which the PCO will need to effect decisions and implement
productivity programs at the field activity level. It is expected
that by the end of calendar year 1981, eighty percent of naval
industrial activity COs will have been indoctrinated in the Navy's
Productivity Enhancement Program. By 1982, and thereafter, at
least 90 percent of the industrial activity commanding officers
should have a working knowledge of the program, its accomplishments
and objectives.

        Distinguished Lecture Series. This initiative, started in
November 1980, involves a host of distinguished lecturers from
government, academia, industry and labor. Attendance is by nomina-
tion, involving a select group of senior Navy line managers. These
face-to-face encounter sessions with the leaders in the world of
productivity provide for a continuing infusion of new ideas for
Navy's senior managers.

        Innovations in Productivity. Published quarterly by the Naval
Material Command, "Innovations in Productivity" provides a mechanism
to keep Navy personnel up-to-date on the status of programs, new
initiatives and matters of general interest in the field of productivity.

        Seminars and Workshops. The Naval Material Command and its
constituent Systems Commands regularly host seminars and workshops
in the field and at headquarters to keep managers at all levels
involved in the dissemination of information and the formulation
of plans for future action. During FY 1980, more than a dozen such
seminars and workshops were hosted.

Disincentives to Productivity

        In addition to the pursuit of systems and programs to improve
productivity, the Naval Material Command sponsored a study to isolate
impediments to productivity within the Navy's industrial community
(Shipyards, Naval Air Rework Facilities, Public Works Centers,
Supply Centers and Weapons Stations). The purpose of the study
was to identify factors which impede organizational achievement and
to recommend management actions to reduce their effect on productivity.

        The study was completed in October 1980. A detailed report
of findings now serves to provide direction to a corporate team
which is chaired by the Director of Productivity Management with
membership from each major headquarters command representing the
study participants.

PROCESS MANAGEMENT

The ultimate objective of the Navy's Productivity Program is to improve military preparedness. Within the industrial segment of the Navy, military preparedness equates to the delivery of goods and services as required by fleet operators. To accomplish this objective, management controls must be established. With the advent of modern computer technology, a major new tool has been made available--information, huge quantities of it, quickly and at a low cost. There is now a means to establish meaningful objectives and measure performance, accurately and on a timely basis. Information technology makes available to management the means to manage the organization's resource capacity heuristically. A properly developed data base system of management, however, introduces a source of personal intrusion into the manager's work life as well as greater interdependence and visibility within the organization, all sources of turbulence to its members. Yet, by pointing to problems and successes in real time, management is presented with a powerful tool through which correct and timely action may be taken. The data base provides the manager with accessable information about the performance of his work group.

In the Pearl Harbor Naval Shipyard, the Planning Director has a small data base system to appraise employee performance based on work count and product quality. The system has been thoughtfully integrated into the planning department in a way which strengthens the linkage between the employee and the supervisor. The data is used as an instrument to explore actual performance in the light of potential for performance; it is not made a part of the formal personnel record. In the few years that the system has been operative, performance of the planning department has increased over 20 percent, and the cost of the planning function is 20 percent below the "unit norm" established by other shipyards. The success of the system is a reflection of the manager's style and sensitivity to the behavioral principles which cause the organization to focus on the achievement of mission objectives.

In the broader sense, the Navy is devoting considerable attention toward the development of statistically supportable measures of productivity. These measures involve headquarters and field activities and at this point focus on the individual, a group of individuals, and a major command (aggregate measures). Work toward the development of aggregate measures of productivity is in progress. These measures are expected to provide a useful tool to restructure systems of incentives and management, and for the first time provide a realistic portrait of productivity in a military industrial work setting.

PRODUCTION CONTROL, PARTICIPATIVE MANAGEMENT AND

ACTION RESEARCH: A Case Study

Heinz J.D. Wanders

Directorate of Economic Management RNLAF
The Hague
The Netherlands

ABSTRACT

The introduction of the third generation of fighter aircraft into the air forces of the western hemisphere will soon
be completed, involving a gigantic increase in operational capabilities and the consequent necessity to develop new concepts
of tactics, operational command and control systems, etc. Apart
from these problems, however, the air forces concerned were
confronted with fundamental changes in the maintenance concept
for their new aircraft. The Royal Netherlands Air Force (RNLAF) *)
is right now in the process of introducing the General Dynamics
F-16 aircraft and encounters various problems with regard to
the above-mentioned changes. An attempt is being made to cope
with these problems at each maintenance level. The case to be
dealt with by this paper illustrates the difficulties experienced in connection with the adaptation of the organization
and the production control system of a major RNLAF overhaul
facility as a result of the innovations in fighter aircraft
technology.

LEVELS OF PREVENTIVE AND CORRECTIVE MAINTENANCE

Preventive and corrective maintenance activities are performed at four levels:

1. First level maintenance. Preventive activities include
   pre-, post- and through-flight inspections, lubrication,
   refuelling, etc. Corrective maintenance takes place by way
   of the so-called "repair by replacement". Malfunctions of
   systems are corrected by a simple exchange of defective

system components. This type of maintenance is performed
by the operational units.

2. Second level maintenance. Preventive maintenance consists
   of extensive inspection and overhaul of the airframe and
   major components, such as the power plant, the fuel and hy-
   draulic systems components, the brake systems, the navigation and fire-control equipment, etc. These inspections are
   scheduled after 200 hours of elapsed flying time. Corrective maintenance includes the repair of complicated system
   malfunctions, necessitating the use of test equipment
   available only in workshops. Personnel employed at second
   level maintenance are highly specialized and trained to
   repair specific subsystems of the aircraft.

   Second level maintenance is performed by the technical and
   logistic branch at air base level. For reasons of
   efficiency, expensive equipment has been centralized in a
   separate branch. The same applies to the employment of
   skilled personnel.

3. Third level maintenance. More complicated maintenance
   activities are performed at central level for the air
   force as a whole. Power plants are overhauled after 1,000
   hours; complete dismantlement of RADAR and navigation
   systems takes place every year, etc. This type of
   maintenance for aircraft may be compared with the docking
   of a ship after a collision or for a major overhaul. As far
   as the air force is concerned this type of maintenance
   takes place in so-called "depots".

4. Fourth level maintenance is carried out by the industry
   and becomes imperative as soon as an aircraft needs to be
   demagnetized after having been struck by lightning or in
   case of any major structural modification. The air force
   occasionally has recourse to this level of maintenance,
   while preventive maintenance is in essence excluded from
   fourth level maintenance.

SOME CONSEQUENCES OF TECHNOLOGICAL CHANGE FOR THE MAINTENANCE
OF FIGHTER AIRCRAFT

In the past aircraft were designed mainly with a view to
parameters of performance and aerodynamics. Things, however
have changed. Problems of ergodynamics and maintainability are
for instance, dealt with and taken into account from the very
first phase of the aircraft design: Modern aircraft possess a
so-called "built-in maintainability" [1,2]. The built-in
maintainability has been accomplished in the following ways.

1. Every piece of electronic or avionics equipment
   incorporates BITE (built-in test equipment). It means
   basically that this kind of systems checks itself prior to
   becoming operative and consequently facilitates the
   detection of failures and malfunctions at the very moment
   of occurrence.

2. Electronic equipment is of the solid state type, resulting
   in a dramatic decrease in the failure rate and facilitating
   modularization as well as, as a result, repair-by-
   replacement.

3. OCAMS (on-board check and monitoring systems) check and
   monitor all vital and complex aircraft systems and parameters during the flight. Detection, diagnosis and correction of malfunctions after landing have thus become much
   easier than with the traditional "rudder kicks pilot" type
   of failure description, so often given by pilots who
   experienced difficulties during the flight and intended to
   leave the aircraft as soon as possible after touch-down.

4. Scheduled preventive maintenance is replaced by "on
   condition maintenance", which means that only faulty,
   defective or malfunctioning system components or parts are
   repaired or replaced and that extensive overhauls, during
   which many parts had to be replaced and many systems,
   whether of not faulty, had to be checked, has become something of the past.

5. Nearly all systems, including the power plant, are built in
   accordance with the modular concept. In former aircraft
   types complete systems had to be removed for overhaul,
   modification, upgrading or repair. Nowadays only system
   modules have to be exchanged and the accessibility of
   primary systems and system components is such that the
   exchange can be performed within a few minutes.

With regard to the maintenance concept described in the previous
paragraph, these developments have the following consequences.

1. The repair-by-replacement philosophy has gained a new
   dimension as a result of the modular conception of aircraft systems. In consequence of this development, the
   workload of the second maintenance level has decreased
   considerably. The same applies to the quantitative as well
   as the qualitative aspects of activities executed at this
   level. On the one hand less failures occur and extensive
   inspections are no longer carried out. On the other hand
   the failure diagnosis has been largely automated and the
   technician has only to replace a part or a module.

2. Maintenance polarizes, so to speak, at first and third
   level. At first level it comprises simple inspections,
   functional checks and exchange of modules; at third level
   the maintenance activities are mainly of the on-condition
   type and therefore, by definition, corrective.

3. As far as the depots are concerned the development can be
   said to be such that scheduled preventive major overhauls
   are replaced by unscheduled repair activities of modules.
   This means that the predictability of input, workload and
   workflow of the depots decreases and the planning of these
   parameters becomes increasingly difficult.

THE ORGANIZATION SETTING OF THE CASE

The organizational problem to be described below presented
itself in a third level major overhaul facility of the RNLAF.

THE PROBLEM

The changes in aircraft technology and their impact on
maintenance compelled the management of the depot to adopt new
production planning techniques. The management decided to
implement a computer-based planning system. The main features of
this system, which will be called PERT, are:

1. All overhaul/repair activities are looked upon as network
   elements.

2. All network elements are sequenced in accordance with a
   PERT network analysis technique [3,4,5] , implying the
   critical path method.

3. All activities defined as network elements constitute
   workorders for the specific workshops of the depot and
   specify the moment of start and finish of the activities,
   facilitating a precise workflow control as well as an
   optimal utilization of machine and personnel capacities.

4. In order to construct standard networks it is necessary
   to standardize as accurately as possible the duration of
   each network element.

Approximately a year after the introduction of PERT at the
depot the management had a feeling that the system had not been
accepted yet by the workfloor and particularly not by the first
line management level, such as the workshop chiefs and their
assistants. The planning quality did not improve as expected and
the commanding officer decided to call in the assistance of two
industrial psychologists of the Applied Social Sciences Division

of the RNLAF Personnel Department in order to promote the
acceptance of the new production planning system especially by
first line managers.

THE PROBLEM REDEFINED

From a thorough investigation of the problem by the social
scientists involved the conclusion was drawn, first of all, that
the management had defined the problem in a far too simple manner.
All consultants on organization and management problems, however,
have the same experience. The disfunctioning of the planning
system appeared to be just one of a complex of interrelated
subproblems and developments. In the last few years prior to the
introduction of PERT more than ten organizational changes had
been implemented, ranging from new intrastructure, tooling and
machinery to a total restructuring of the production department
and the introduction of a fundamentally redesigned quality
control system. The advisers observed that a certain degree of
tiredness caused by continual changes was widespread. Regarding
the original definition of the problem the psychologists reached
the following tentative conclusions:

1. PERT in its present form was considered, to a certain
   extent, inadequate to meet the expectations with regard to
   an increased planning quality.

2. The introduction of PERT had been carried out in such a way
   as to create a low level of commitment. This applied not
   only to all personnel in general but also held good
   especially for the workshop chiefs and their assistants.

So PERT had to be adjusted to the problems inherent in the
specific planning situation of the depot, its organization and
processes. The commitment of the workshop chiefs and their
assistants had to be improved.

ADJUSTMENT OF THE PLANNING SYSTEM

According to one of the important principles of the modern
management, planning and organization theory the paramount
problem of management, planning and organization is to develop
ways of reducing the complexity [6,7,8] . In seeking ways to
accomplish the reduction of the managerial complexity managers
often generate more complexity by means of the very instruments
developed to reduce the complexity. This phenomenon is
illustrated by the present case.

1. As noticed, it is necessary to standardize the duration of
   each activity, defined as a network element. Without such
   standards it will not be possible to build a network. Due

to the on-condition maintenance activities it is, however,
extremely difficult to standardize the activities in
advance. Consequently the planning will most likely be
frustrated, especially, in those shops where standardizatior
is difficult or even impossible. Planning targets are
reached easily where standardization can be achieved, e.g.
in assembly and disassembly workshops.

2. There exists an optimum level of splitting up an entity of
work for planning purposes. The required capacity for
controlling and monitoring the workflow and workload of the
specific workshops is directly related to the total amount
of network elements defined. In our depot the definition of
the planning network was carried through in great detail.

3. Network planning is highly effective as far as project
management or the planning of standardized cyclical
activities is concerned. With regard to the overhaul
facility the conclusion must be that PERT would have fitted
the control demands of scheduled maintenance perfectly, but
appeared to be more or less unfit for planning the unscheduled corrective on-condition maintenance activities.

4. The centralizing tendency of computer-based planning
systems is self-evident. This tendency increased by the
growing need for a control and monitoring capacity at staff
level of the depot as a result of the too detailed network
definition. The lack of a real-time, on-line availability
of planning information at workshop level created the
inevitable feeling of being kept in the dark at this level,
paralyzing the self-regulating capacity at workshop level.

The conclusion with regard to PERT must be that the system
introduced more planning-complexity instead of reducing the
complexity and that system changes should be realized so as to
improve the use of the planning-capacity at a lower level of the
depot hierarchy. Decentralization and relief of redundant information at the depot planning level should therefore be
contemplated.

IMPROVING COMMITMENT

The introduction of PERT had been ordered by the depot
commander. Study, preparation and selection of a planning system
had taken place at unit staff level. Until the introduction in
the line and the actual implementation of the PERT procedures
distribution of information to the lower levels had been margina
First line management had not been informed of the configuration
and the procedures and had no idea of the impact PERT might hav
on the quality of their work until they were actually confronted

with the system. In other words: participation in the decision
making process by the group that would be affected most by the
introduction of PERT had been minimal. First line management was
affected most because PERT involved mainly a short circuiting of
workshop chiefs. Prior to the introduction of PERT the chief had
acted as mediator between the workload control office at staff
level and the workshop. He was fairly well informed about the
input of work to be carried out in his shop and about the deadline for turning out his output and he executed the direct
production control in his shop. In the new situation he
considered himself to be merely a "policeman" forcing his
workers to meet the demands of the pre-established standards
with regard to the allowed duration of their activities. As
far as the perception of the participation is concerned the
first-line managers were affected by the introduction of PERT
in two ways, namely:

1. The planning system had been developed and introduced
   without any consultation, let alone participation, of
   the first-line management.

2. The system as such reduced the authority of the workshop
   chiefs with respect to leadership and production control.

These two aspects inevitably caused a degree of "resistance to change" that cannot be removed simply by "soft words".

Another conclusion is that the problems regarding the
role, the position and the task of the workshop chief are
closely intertwined with the problems of adaptation of the
planning system as described in the previous paragraph.
Improvement of the quality of work done by the workshop chiefs
could be identical to a reduction in the managerial complexity.

REMEDY: PARTICIPATIVE MANAGEMENT

Participation in decision making processes is one of the
main determinants of job satisfaction [9,10] . I should like to
emphasize that it reflects two basic concepts; one is the
possibility of influencing the decisions of higher management
levels, while the other is synonymous with the autonomy in the
performance of a job. Within the context of the present case
the usage of the term participative management embraces both
concepts. The psychologists who acted as consultants of the
depot management team set themselves two goals, namely:

1. To increase the autonomy of the workshop chiefs with
   regard to the planning and control of the workload and
   workflow of their specific shop. A basic condition for
   such an increase would be the adjustment of PERT.

2. To improve the overall participation of all levels of the organization, but in particular that of the first-line managers, in organizational changes and in the development of policies on this matter.

   The consultants strongly believed that solving organizational problems is a responsibility of all organization members and not only of the (top) management. Participative management in the sense as described above could serve as an instrument to improve the organizational flexibility and to reduce the managerial complexity. It may also be a means to improve commitment because an actual change in structure or processes is achieved by means of a decision making process, in which each concerned member of the organization takes part.

STRATEGY: ACTION RESEARCH

   Action research is a paradigm for solving (organizational) problems. The main steps in this paradigm are: [11,12]

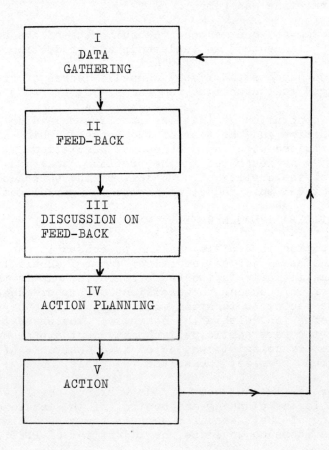

In the paradigm it is of crucial importance that all steps are done by the consultants together with the involved members of the organization. The only exception sometimes made is in the the first data gathering phase.

At this particular moment the project at the depot is in phase four of the first cycle:

1. Data gathering took place by means of some sixty interviews as well as by observation.

2. Feed-back on the sampled information was provided to the top management team of the depot and to all levels of the organization, down to the first-line managers and the workfloor.

3. Exhaustive discussions on the feed-back took place at all levels, with the exception of the workfloor. This discussion phase lasted some eight months and the outcome of it was that:

   + The top management team had been convinced that participation of the lower levels in the decision-making processes is of paramount importance for effectively coping with problems arising from organizational changes.

   + The top management team had been convinced that adjustment of PERT so as to realize a more decentralized system of production-planning and control is vital for the improvement of the quality of planning and control. It was also acknowledged that first-line managers should have a substantial say in the development of the change direction and in the expansion of their autonomy.

4. All levels wish to establish a formalized communications network in order to:

   + Guarantee an effective top to bottom flow of information with regard to possible future changes in organizational processes and structures.

   + Guarantee an effective flow of ideas from bottom to top

   + Improve job satisfaction, motivation and commitment.

5. Action planning takes place in workgroups and discussion teams, composed of all levels of the hierarchy, especially the workshop chiefs and their assistants. Three teams are right now developing solutions for the problems in question.

One team has the major responsibility to integrate and co-
ordinate all suggested solutions and actions. Another team
adjusts PERT to the new insights, while a third team tries
to design a formal communications network. This team will
at the same time formulate demands regarding aspects of
the development and the improvement of social skills.
Meeting these demands will be an indispensable condition
for an effective communications network.

SUMMARY AND CONCLUSIONS

One of the main management problems nowadays is to reduce
the managerial complexity. In seeking ways of accomplishing this
reduction, however, managers often introduce more complexity by
the very instruments developed to reduce the complexity. The present case constitutes an illustration of this mechanism.

There is a tendency in such bureaucratic organizations as
the military, to develop increasingly more centralized command
and control systems, as is illustrated by the introduction of a
computerized system of production-control into the overhaul
facility discussed above. This centralization meant the creation
of more managerial complexity at unit staff level.

Centralized control results, among other things, in a loss
of commitment at the lower levels of the organization. The
management tends to neglect the negative impact that inadequate
information about plans and policies has on the lower levels. It
also tends to ignore the possible contributions by the lower
levels to the problem solving processes in the organization.

Participative management will elicit the following positive
results:

1. Use of the self regulating potential at first-line management level will reduce the need for centralized control.
   Adjustment of the planning system in order to achieve
   participation will be necessary.

2. Personal involvement of lower levels will be improved,
   thus enhancing job satisfaction and motivation.

Action research is a suitable strategy to implement the
necessary changes.

NOTES

1. An excellent illustration of the point made here, was
   found in: Klaus Lewandowski, Wartbarkeit beim Tornado
   gross geschrieben, in: "Jahrbuch der Luftwaffe 13",

Verlag Wehr und Wissen, Bonn-Duisdorf (1976).

2. The on-condition maintenance concept for guided missile systems is exemplified in: Rodin Bethke, Moderne Material-erhaltung für Flugkörpersysteme, <u>Wehrtechnik 3/81</u>, Verlag Wehr und Wissen, Bonn-Duisdorf (1981).

3. Albert Battersby, "Network Analysis for Planning and Scheduling", MacMillan & Co, London (1965).

4. K.G. Lockyer, "Netwerkplanning in het kort", Samson, Alphen aan den Rijn (1969).

5. H. Poolman and Th. M. Femer, "Netwerkplanning volgens PERT", Universitaire Pers Rotterdam, Rotterdam (1973).

6. A philosophical base for the concept of reduction of complexity is elaborated by Luhmann in: Niklas Luhmann, "Zweckbegriff und Systemrationalität", Suhrkamp Taschenbuch Verlag, Ulm (1973).

7. Deutsch makes the same point for governmental steering systems in the development of his thesis that theory-construction and modelling are essentially ways for reducing complexity in: Karl W. Deutsch, "The Nerves of Government", The Free Press, New York (1966).

8. The concept of reduction of complexity as basic managerial problem is developed in: Stafford Beer, "The Brain of the Firm. The managerial cybernetics of organization", The Penguin Press, Harmondsworth Middlesex (1972).

9. There exists an abundant amount of literature on the determinants of job satisfaction. Blumberg summarizes some classics on this subject, and especially on the importance of participation, in: Paul Blumberg, "Industrial Democracy", Constable, London (1968).

10. Dean Champion states that: "Probably no other variable has received so much attention as a prerequisite and determinant of job satisfaction and motivation. The concept of "psychology of participation" is that the more a person participates in a decision, the more likely he is to stick with it and to enjoy his work. A more recent phrase which expresses this notion is "participative management"." Dean J. Champion, "The Sociology of Organizations", McGraw-Hill, New York (1975).

11. A.J.G.M. Bekke, "Organisatieontwikkeling: Confrontatie van individu, organisatie en maatschappij", Universitaire Pers Rotterdam, Rotterdam (1976).

12. Wendell L. French and Cecil H. Bell jr, "Organization Development: behavioral science interventions for organization improvement", Prentice-Hall, Englewood Cliffs N.J. (1978).

CONSULTATION: A METHOD TO SECURE EMPLOYEE COOPERATION IN

ORGANIZATIONAL AND TECHNOLOGICAL CHANGE PROGRAMMES?

Gillian Ursell and Paul Blyton

Trinity and All Saints College, Leeds
and
University of Wales
Institute of Science and Technology
Cardiff, United Kingdom

ABSTRACT

This paper examines the proposition that joint consultation can
be a successful technique for securing employee co-operation and commitment to improvements in organizational efficiency, even where these
improvements involve organizational and technological change programmes which include redundancies and job modifications. Empirically the study investigates the operation of a consultative system
under the different conditions prevailing in 3 factories of a major
British frozen food company. Theoretically these data are considered
against two competing hypotheses: one, that joint consultation represents a bureaucratic endeavour to secure greater information-
management and is dysfunctional to encouraging co-operative
attitudes among employees; the other, that consultation is both
meaningful to employees as an information-exchange venue and has
important relational consequences reducing social distance, the net
effect of the two aspects being to encourage co-operative employee
attitudes.

INTRODUCTION

To the casual observer the British economy presently offers a
picture of almost unrelieved gloom. Manufacturing output and our
share of world trade have fallen as surely as unemployment and inflation have persisted. Questions about causes and cures are put by
politicians, economists and citizens alike. And not just in Britain,
for it is apparent that the British plight is shared to different
degrees by many other industrialized nations. Whatever the causes
(and analyses range from too much public spending, e.g. Bacon and
Eltis, 1976, to a crisis of capitalism, e.g. Heilbronner, 1979,

Fitt et al, 1980), all roads to recovery indicate attendant political
instabilities and the need for radical reshapings of major political
and economic institutions if public consensus is to be achieved.

Necessarily, much of this reshaping will take place within the
individual workplace. This will be the context for fundamental
changes as increasingly corporatist-minded governments seek greater
"demand and supply" management (Friedman, 1973), and managements
chase profitability in the face of rising costs of basic commodities
(Rostow, 1979), intensifying competition (Lewis, 1978), and the profound implications of new electronic technologies for organizational
forms and manpower needs (Jenkins and Sherman, 1979).

Previously in Britain crisis conditions have occasioned a shift
towards more consultative strategies by managements. In the First
and particularly Second World Wars, for example, government-
encouraged joint consultation committees (JCCs) proliferated. Indeed
the enthusiasm for this form of labour-management committee led to
the formal embodiment of joint consultation in the constitutions of
industries taken under public control during the late 1940s (e.g.
coal, steel and railways). By the early 50's JCCs were to be found
operating in 73% of companies surveyed (NIIP, 1952). But the crisis
conditions of war gave way to full employment and prosperity, and the
interest in consultation waned as the economic basis for shopfloor
independence and negotiating power grew. By the late 60's the number
of companies operating JCCs had fallen to below one-third (Clarke et
al, 1972), and politicians of all main parties were viewing this rank
and file independence with some concern (Donovan, 1968) - one eye
being on their inability to exercise authority over the British
workers, the other on the accumulating evidence that Britain's
economic strength was deteriorating and persistently so.

Earlier parliamentary efforts to relocate shopfloor power to
more formal, centralized mechanisms were largely unsuccessful (e.g.
Industrial Relations Act, 1971; Bullock recommendations, 1976), but
they were heeded. Perhaps persuaded that "better the devil you know
than the one designed by Parliament", many British managements in the
mid-70's introduced consultative machinery. According to a recent
survey (Beaumont and Deaton, 1981), JCCs now operate in 62% of establishments, over half of these having been introduced within the previous 5 years.

Undoubtedly the general economic downturn in Britain has been a
major contributor to this revival of consultation, creating on the
one hand managerial needs for greater information about and control
over organizational subunits, so as to co-ordinate and speed up
adaptive responses to environmental uncertainty (Lawler and Rhode,
1976) and, on the other, union needs for early indications of likely
changes in manpower, and terms and conditions of employment (Clegg,
1979).

This union response must be viewed against the background of reduced political leverage and lowered expectations as a result of high levels of unemployment. No longer in a position to demand radical forms of industrial democracy, the British unions have tended to accept management offers of joint consultation even though this represents only a very limited form of industrial democracy in that it does not redistribute formal decision-making rights to lower strata. Indeed it could be argued that consultation is less a step towards industrial democracy and more a form of managerial information control system. If, e.g., we accept Burns and Stalker's (1968) evidence that the typical bureaucratic response to environmental uncertainty is to create new formal functionaries and committees for the purpose of liaising with and co-ordinating subordinates, then joint consultation is just such a formal bureaucratic response. The prediction of Burns and Stalker is, however, that these bureaucratic responses are largely dysfunctional in that, failing to decentralise decision-making powers, they improve neither the capacity of sub-units to respond to environmental change nor do they raise the sense of involvement and commitment of subunits to overall goals. In short, bureaucratic responses perpetuate mechanistic forms of organization rather than promote the organismic forms which, arguably, are better able to cope with environmental change.

It is within this context - on the one hand the practical difficulties currently faced by British managers and workers and, on the other, the theoretical considerations surrounding our understanding of consultation - that we wish to discuss our case study. On the basis of this we hope to demonstrate that even when understood as a formal bureaucratic mechanism for information-management, consultation is not necessarily dysfunctional in the ways implied by Burns and Stalker. We will argue that, under certain conditions which it will be our task to identify, consultation may go a long way to establishing that degree of trust which has been identified as lacking in British industrial relations (Fox, 1974). In addition, where it is successful in this manner, we shall contend that it secures for management a greater degree of flexibility in manpower management and labour relations.

What follows is firstly a brief description of the British company under investigation and one aspect of its response to various market, supply and technological problems - namely, an 8-point deal with the unions which included but was not confined to the introduction of a comprehensive consultation system. Secondly, attention will be focussed on 3 of the company's factories, geographically distanced and in many respects different from each other. These 3 factories have facilitated comparative analysis of how the consultation system has operated under different conditions. Thirdly, though this research is still in its infancy, we shall use currently available data to support - with due caveats - the contentions indicated above.

THE COMPANY
     Our research site is a major food-processing and freezing com-
pany, enjoying over the last three decades the market leader position
in British domestic sales of freezer foods but experiencing over
recent years intensifying competition (there are now over 350 frozen
food companies in Britain), a recession-induced fall in demand for
some of its "luxury" products, and the effects of inflation.
Operating in 9 sites, the company employed about 12,000 people in
1979, of whom 4,500 were part-timers. Union membership stands at
about 80%, the main unions being TGWU (Transport and General Workers)
and GMWU (General and Municipal Workers) organizing process and distribution workers; the AUEW (Amalgamated Union of Engineering
Workers) and the EETPU (Electrical, Electronic, Telecommunications
and Plumbing Union) representing craft workers; and the ASTMS
(Association of Scientific, Technical and Managerial Staffs)
representing roughly 50% of lower managers.

     Market changes in the early 70's prompted the company, by 1975,
to embark on a radical 5-year plan to reshape the entire organization, the 3-fold goal being i) product specialisation in each
factory, ii) tighter manning levels, especially but not only at
managerial levels, and iii) technical modernisation, including
company-wide computerisation. In terms of employment the implications of the plan were substantial - in some factories, e.g., two
tiers of management and 50% of process jobs were to go over a 5 year
period.

     In their own words, top management wanted "to discuss the plan,
warts and all, with representatives of the whole workforce ... but
the existing consultation structures were quite inadequate." Inspite
of a highly developed union negotiating structure, some categories of
employee had no representation at all, and there was no representative structure extending to company level where the most important
decisions were taken. The company made no pretense of altruism in
its desire for consultation but spoke publicly of it as the means for
getting "you (the employee) to see things our (the employer) way."
To facilitate this goal the company proposed a system of linked consultative committees extending from company level through factory and
departmental levels to workgroup level, and open to all employees
regardless of union status.

     Union responses to the proposal ranged from outright hostility
to indifference. However consultation was offered not on its own but
as part of a wide-ranging, 8-point package of changes (namely, manpower planning; organization development; workgroup development and
workstyle; personnel development; training; remuneration; employee
relations and consultation) many aspects of which were highly attractive to the unions. For example, in return for their co-operation in
the introduction of consultation, lower management sections (represented by the ASTMS) were offered, at factory level, their own

Management Consultative Committees. This deal reflected not just the special position of management but also the significance of the organizational change programme for managerial redundancies and alterations in the pattern of executive command. Similarly, rationalisation of the company included a shift from local to central pay negotiations, from multiple settlement dates to a single date, and from various pay structures to a single, banded system. The appeal of this to the unions was 3-fold; firstly, it guaranteed for local shop stewards new and regular access to company-level management; second, it laid the basis for new lines of contact and communication between stewards in different factories; and third, somewhat paradoxically, the company's stress on the distinctiveness and role of consultation had the effect of further legitimating the unions' prerogatives in negotiations. Such aspects as these had the effect of winning, albeit after protracted discussions, union acceptance of the consultative scheme.

Clearly the company's multi-dimensional approach here has contributed substantially to its successes - 6 years later, without any industrial disputes attending the large numbers of managerial and operative redundancies, the greater part of the change programme has been carried out. Elsewhere we have argued the importance of a multi-dimensional approach: too many consultative committees in the past have suffered from being mere appendages to existing structures rather than being built into wider programmes of organizational redesign (Blyton, 1981), a consequence of which is that frequently consultation is seen and rejected as an impotent and meaningless activity. Data from the 3 factories indicate, with relatively few exceptions and to the extent of winning converts, consultation in this company is by no means regarded as meaningless. It is operating regularly and fully at all levels (with only workgroup levels yielding a patchy performance) and it is clear that much of its appeal derives from its being the venue where longer terms plans are raised and discussed.

It is however also evident from differences among the 3 sites that factors other than the company's 8-point deal are influencing the efficacy of consultation. Certain of these factors are again aspects of broader company policy; namely, a successful marketing strategy, a tight rein on manning levels, agreed procedures with regard to the handling of redundancies and extensive welfare provisions. Thus, by focussing more on basic rather than luxury food products and by operating relatively tight manning levels, the company has largely ridden out the worst effects of recession. Meanwhile, any redundancies necessitated by reorganization have, via the consultative system, been declared well in advance, attracted substantial compensation and favoured volunteers. Supported by extensive welfare provisions, these policies have helped create the ethos of a caring -and successful - employer with whom the majority of workers seem satisfied.

There do however remain exceptions and problem areas, and it is
by reference to these that we can see the limits of management policies plus the importance of other factors which are, in a sense,
fortuitous. A closer look at the 3 factories should help demonstrate
these complexly related factors.

THE THREE FACTORIES

At this early point in our research we cannot speak directly of
employee attitudes but only of their indirect reflection through the
eyes of managers and those worker representatives serving in the
negotiating or consultative systems. Whilst this represents a void
which our subsequent inquiries will seek to fill, we do not believe
it destroys our ability to drawn meaningful, if tentative, conclusions from present data. This belief rests on the recognition that
for both managers and worker representatives the attitudes of the
workforce are a major concern and therefore something demanding their
close and continuous attention. Whilst allowing for some degree of
error and subjectivity, management and worker representative descriptions of worker characteristics should be reasonably accurate.

All 3 factories, A, B and C, have experienced substantial change
over the past 6 years and are continuing to do so in the areas of job
content, changes in top management and job losses among assistant
managers and production operatives. However, the conditions under
which the changes have been effected substantially differ between
each factory. Factories A and B share a similar product base of fish
and vegetable processing, though A (the smaller of the two, in 1979
employing 827 people compared to 2,335 in B) specializes in the more
expensive fish products. Factory C (1,139 employees) deals solely in
meat products. All 3 factories have enjoyed a steady demand for
their products but this demand represents less than that anticipated
in the company's investment plans, resulting in excess capacity at
both factories A and C.

Excess capacity has generated problems of surplus manpower in
addition to that envisaged in the 1975 plan. Yet again the 3 factories differ in their capacity for absorbing these problems. At
factories A and B the majority of process operatives (by far the
largest section of the total workforce) are part-time or temporary
female employees. At factory A a 50% reduction among these has
already been achieved since 1977, and at factory B management anticipate no difficulty in similarly reducing their operative workforce
by 33-40% over the next period. Because their employment does not
constitute a major life interest, these women employees are not
militant; many are accustomed to regarding this occupation as season-
al (though technological change has made this less the case) and
among them there are generally high turnover rates (anything from 15%
to 50% depending on age and stage in the family life cycle). Such
high rates of natural wastage, plus quiescent attitudes, provide
factories A and B with an enviable margin of flexibility in man-

power management - at least at operative levels.

For these 2 factories the principal area of difficulty with
regard to manpower plans has been the shedding of two levels of
management, and here consultation has been utilized to the full. It
has however taken the form of individual rather than collective discussions (though the latter were also utilized at a later stage).
The reasons for this approach were, firstly, that senior management
feared a collective announcement of managerial redundancies "might
have lost us the best" and, secondly, that individual performance
was the criterion for continued employment. Utilizing the routine
annual staff appraisal system, senior management duly notified
other managers - from six months to two years in advance - whether
or not, and in what capacity, they had a future with the company.
One direct consequence of this was to create informal differentiation amongst managers according to whether they were staying and
assuming new responsibilities, staying but not progressing, or
leaving the company. This differentiation substantially undermined
the ability of the managerial union, ASTMS, to mobilize resistance.
Additionally, however, resistance was undermined because these same
managers continued to be responsible for holding consultations with
their subordinates at which company performance and the rationale
for change were major topics which they personally had to detail and
plead. Thus, while isolated pockets of resentment can be found
among managers adversely affected, overall managerial redundancies
have been accepted with surprisingly little acrimony - and much of
the credit for this achievement must accrue to the consultative
approach.

A similar "success story" is apparent at Factory C, though the
practical problems have been somewhat different. Here redundancies
are anticipated but not yet achieved and the task for consultation
so far has been general re-education of the workforce aimed at
improving attitudes to work. Factory C operates in an area where
10% male unemployment is the norm, and 30% that reached during the
recession. The area is noted on the one hand for militant trade
unionism (e.g. a very high number of worker occupation of factories
during the late 60's, early 70's) and, on the other, for the
demoralising effects of trans-generational unemployment. Many of
the newly employed, with minimal family employment experience, have
little sense of work discipline - a fact which manifests itself in
high levels of absenteeism, unarticulated insubordination and
pilfering. The task over the last 3 years (following a strike in
1977 which caused massive lay-offs and brought the plant near to
closure) has been, in the eyes of management, "monumental, involving
almost suicidal levels of management control and complete dedication
to the task of changing attitudes, norms, custom and practice.
There was an absence of guilt and now there is some sense of it -
we've got absenteeism down to average levels, and the level of theft
has been reduced." A principal vehicle for this management effort

has been consultation. Here - to the detriment of the time given to
their production functions, say some managers - they have hammered
home the rationale behind company and factory decisions, and the
importance of the individual's contribution to company success.

Managerial and trade union attitudes towards consultation at
factory C remain wary however: the stewards continue to see consultation as a management ploy aimed at putting off the achievement of
genuine industrial democracy, while the managers fear the stewards'
proclivity to turn consultation into negotiation. Few stewards are
however willing to deny their fellow workers the right to be consult-
ed. They also recognize two other positive features of consultation;
firstly, it secures a substantial degree of managerial accountability
in that, having publicly detailed past and present performance and
future plans, management can then be called upon to explain any
delays, alterations and failures; and secondly, trade union organization just as much as manufacturing enterprise benefits from disciplined and conscientious workers - the stewards therefore see
themselves as having something to gain from a re-educated workforce.
For these reasons and also because, inspite of the recession and
high levels of unemployment locally, the stewards remain confident of
their ability to take an independent stance on negotiable issues,
they have acceded to the management policy of consultation.
Currently at factory C, management regards consultation as "vital"
and both stewards and ordinary workers are prepared to agree that it
is "valuable".

Possibly the real test of consultation at factory C has yet to
come. The 1977 strike (not related to the re-organization programme)
lasted 18 weeks and occurred as the factory was in the first stage of
a massive investment project. The loss of market which the strike
brought has not yet fully been recovered, partly because the factory
specializes in relatively higher cost meat products which have been
to a degree adversely affected by the recession. Ironically the
investment project was intended to expand capacity and this will come
on line in 1982. Not due to re-organization (which has in fact
temporarily expanded the factory's demand for managers and engineers)
so much as to excess capacity, the factory in 1982 must make redundant substantial numbers among its process operatives. The nature of
the factory's processes, plus cultural features of the area, however,
mean that process operatives here are primarily full-time males and
females rather than part-time women. They do not offer the same flexibility in manpower management as is enjoyed at factories A and B. The
margins for managerial manoeuvring so as to avoid the worst clashes
of interest between employer and employee would seem, at factory C,
to be very limited. It is doubtless for this reason that factory C
managers are investing so heavily in "endless consultation" as their
only promising route for heading off strife.

Will they succeed in this use of consultation? It would perhaps

be foolhardy to predict an outcome. Certainly though, at all three
factories, managers - even those originally antagonistic to the idea
of consultation - have found it not to be a waste of time but
actually to improve their authority. This it does in two ways which
reflect our earlier theoretical distinctions. Viewed as a bureaucratic mechanism for information-management, consultation affords a
new or additional venue in which managers, because of their greater
command of functionally relevant information, can appear in the role
of "experts" and so are better able to influence proceedings (see
also Obradovic, 1975, and Marchington, 1980, in this connection).
It is also clear that workers involved in these consultative proceedings (even the minority who view askance the evidence of
managerial domination) find the information thereby provided as
valuable in itself and as not likely to be forthcoming at all or so
readily through any other venue. This aspect ties up with the second
way in which consultation improves managerial authority, namely, in
its relational consequences. Any act of communication involves not
just the exchange of information but also the negotiation of a
relationship (Danziger, 1976). It is apparent from interviews that,
in this company, the willingness of managers to consult regularly
and to be influenced to some degree by workers' contributions has
raised the workers' self-esteem or sense of status, has reduced
social distance and the sense of hierarchy, and has simultaneously
personalised relations and broadened familiarity. The phrase, "its
like a family here" fell frequently from the lips of interviewees
and was given greater articulation by some of the stewards. Thus
one commented, "Management are willing to give away as much as they
dare without undermining their own position. I'm inclined to let
this rub off on me - straight away we're aiming for the middle
position;" another said, "...here we all sit together, lunch
together, talk like nextdoor neighbours ... Consultation allows us
to meet ... it stops us becoming mere numbers"; and a third opined,
"Our management are employees, just like us, and sackable ... it
takes a basic trust for it to work well ... I don't believe in
taking to the trenches. The better they get on, the better we'll
get on."

CONCLUSIONS
     This affirmative tone should not of course allow us to lose
sight of the fact that the intransigent conditions under which
factory C must in the near future achieve substantial lay-offs may
stretch consultation beyond its ability to secure co-operation.

     Certain of these conditions - notably, local labour market
characteristics, a predominantly full-time workforce and excess
capacity - are not readily open to managerial intervention.
Neither however are the general recession and high unemployment
levels helpful to trade union organization, nor to such pressures as
there may be to take a firm, negotiating position against managerial
ambitions. Within these limits to their powers set by factors

largely beyond their control, both managers and shop stewards can be
expected to make the most of consultation. As to the employees more
generally, it is apparent in all three factories that consultation
has contributed substantially to the emergence of more co-operative
attitudes. It has done this for a number of reasons which can be
summarised as follows:-
firstly, company policies in other fields (e.g. marketing, welfare)
have contributed to the creation of an ethos of a caring and also a
successful employer;
secondly, the meaningfulness and acceptability of consultation have
been underwritten by a wider-ranging package of related policies,
and,
thirdly, even where employees regard consultation as a vehicle for
bureaucratic information-management which fails to decentralize
formal decision-making rights, they nonetheless value it both as a
forum for information-exchange and for its relational consequences.
In short, given a supportive organizational environment, consultation
can contribute positively to meeting the needs and expectations of
managers, union representatives and ordinary workers.

    As a contribution to further considerations though, let us
finally indicate that there is the suggestion in our argument that
before consultation can make this positive contribution these
needs, and more especially expectations, may have to be rendered
more compatible by forces deriving from broader economic and social
conditions. The ratcheting up of expectations in periods of prosperity has been generally acknowledged, and the possibility of
worker demands for decentralized decision-making in place of consultation will likely grow in proportion as the economy recovers. Such
demands are likely to be particularly acute where workers' views
have been too long neglected, where they feel themselves to be "mere
numbers", and where they have little understanding of what the
running of a business entails. If, however, a consultative system
has already operated successfully to relieve these complaints, then
demands for decentralized decision-making should be either less
urgent or, if granted, should present fewer problems of polarization
and incompatibility. In short, companies who consult well today may
have little to fear from shared decision-making tomorrow.

REFERENCES

Bacon, R., and Eltis, W., (1976) Britain's Economic Problems, 2nd
    edition, Macmillan, London.
Beaumont, P., and Deaton, D., (1981), "The extent and determinants
    of joint consultative arrangements in Britain", Journal of
    Management Studies, 18, p.49-71
Blyton, P., (1981, in press), "Cross-national currents in joint
    consultation" in Mansfield, R., and Poole, M., (eds)
    International Perspectives on Management and Organization,
    Gower, Farnborough.

Burns, T., and Stalker, G., (1968), The Management of Innovation, 2nd edition, Tavistock, London.

Clarke, R., Fatchett, D., and Roberts, B., (1972), Workers' Participation in Management in Britain, Heinemann, London.

Clegg, H. A., (1979), The Changing System of Industrial Relations in Great Britain, Blackwell, Oxford.

Danziger, K., (1976), Interpersonal Communication, Pergamon, New York.

(Donovan) Royal Commission on Trade Unions and Employers' Associations, Report, Cmnd 3523, H.M.S.O., London.

Fitt, Y., Faire, Z., and Vigier, J., (1980), The World Economic Crisis, Zed Press, London.

Fox, A., (1974), Beyond Contract: Power, Work and Trust Relations, Faber, London.

Friedman, I., (1973), Inflation: a world-wide disaster, Hamilton, London.

Heilbronner, R., (1979), Beyond Boom and Crash, Boyars, London.

Jenkins, C., and Sherman, B., (1979), The Collapse of Work, Methuen, London.

Lawler, E. E., and Rhode, J. Grant, (1976), Information and Control in Organizations, Goodyear Publication Co., Pacific Palisades, California.

Lewis, W. A., (1978), The Evolution of the International Order, Princeton University Press.

Marchington, M., (1980), Responses to Participation at Work, Gower, Farnborough.

NIIP (National Institute of Industrial Psychology), (1952), Joint Consultation in British Industry, Staples, London.

Obradovic, J., (1975), "Workers' Participation: who participates?", Industrial Relations, 14, p.32-44.

Rostow, W., (1979), Getting from Here to There: a policy for the post-Keynesian age, Macmillan, London.

ENVIRONMENTAL CHANGES AND R&D ORGANIZATION IN MITSUBISHI ELECTRIC CORPORATION

Taro Tanimitsu, Kazuo Yanagishita, and Junichi Baba

ABSTRACT

The organization of research and development (R&D) in the Mitsubishi Electric Corporation (Melco) has progressed as follows:

- Original staff development.

- Progression from technology to market-oriented organization.

- Dual ladder career system of technological specialists and R&D managers.

- Alternate introduction of a management and a research oriented organization in keeping with environmental change.

- Development of research groups in plants to make a direct return to Melco.

- Transfer of R&D managers to various plants as Melco's products became more sophisticated.

I. ORIGIN OF RESEARCH ORGANIZATION

The Mitsubishi Electric Corporation (Melco) was established in 1920 from the Electric Machine Division of the Mitsubishi Ship Building Company. Because Melco had engineering license contracts with Westinghouse Electric Corporation, in its early years Melco was dependent upon Westinghouse for technology. As the firm realized the importance of engineering research, Melco established its own research group in 1925.

The research group originally was part of the Manufacturing
Section, located in Kobe Works near the city of Osaka, and studied
techniques of manufacturing and materials. The research group
grew to 10 individuals (1 head officer, 8 engineers, and 1 operator)
and was shifted to Headquarters in 1935, where it became the
Research Department. By 1940, the Research Department had 80
personnel, and in 1944 it became the Research Laboratory, with a
staff of 100 people.

The original organization of the Research Laboratory was
according to traditional sections, and until its reorganization
in 1951, consisted of:

General Manager
        Office Section
        License Section
                Electric Research Section
                Wireless Research Section
                Physics Section
                Chemistry 1st Section
                Chemistry 2nd Section
                Workshop

II. SECTION SYSTEM TO ROOM SYSTEM (1951)

In the early 1950s, Japanese industry entered a boom period
because of the Korean War which broke out in 1950. The demand
for products was high. Many business companies invested capital
in the modernization of their production facilities, and many
licenses were introduced, mainly from the United States.

Japanese industries had been completely ruined by World War II,
but the Korean War enabled reconstruction to begin, and to surpass
the old days. We regard the 1950s as the age of reconstruction,
modernization of industry and the introduction of foreign technology.

The following were necessary goals for Melco in this period:

- Use the fruits of its research to develop new products and
  get new business.
- Improve and expand into new research fields.
- Mobilize and develop research personnel.

For these reasons a large reorganization took place in 1951,
called Section System to Room System.

This reorganization aimed to:

- Change the traditional organization to a soft organization
  system and to diversify the research fields.
- Establish a Survey Room.
- Introduce senior researchers of equal status with group
  chiefs.

This was the first step in the dual ladder system of manager and
professional careers.

General Manager
     General Affairs Dept.
     Survey Room
            Electric Research 1st Room
            Electric Research 2nd Room
            Chemistry 1st Room
            Chemistry 2nd Room
            Materials Research Room
            Machine Research Room
            Workshop
            Semiconductor Research Room

In 1959, Melco founded a Consumer Products Research Laboratory
(CPRL). CPRL was charged with developing consumer products, and
aimed to cope with the future home appliance boom which many
people foresaw. One of the CPRL's important duties was the design
of home appliances, and many industrial designers have been employed
at Melco since then.

This industrial design group became the Industrial Design
Center in 1977.

III. ROOM SYSTEM TO DEPARTMENT SYSTEM (1963)

Japan's rapid economic growth began in the late 1950s, and
the 1960s became, so to speak, the age of high economic growth.
As for R&D activities, we could say that this period was the age
of application of foreign technology.

The industrial business world of Japan in the early 1960s
was characterized by:

- Improvement of technological innovations.
- An active productivity development campaign.
- Development of heavy and chemical industries.
- Large scale mass production of home appliances.

To cope with this, and especially with technological innovation, many corporations established research laboratories to reinforce the R&D activities of the early 1960s. In certain industries, such as electronics and pharmacy, the importance of R&D equalled that of the traditional sectors of sales, production and finance. The effects of laboratory activity had become so great that laboratory management was spotlighted in Melco, and research activities were required to be conscious of their return to the enterprise.

"Research Policy at Melco" was written in 1962. The policy statement declared that research is the most important managerial function, which supports the very existence of the enterprise. All research operations in our corporation must be aimed at profits and should be carried out for the purpose of private enterprises.

There are three stages in research activities---basic, applied and development---and each stage has differences in the time lag between the completion of research and the realization of returns or the possibility of success. However, the aim of research shall never change. The policy statement stressed that continuous prosperity of the enterprise depends upon its research operations.

The staff of Melco's laboratory grew to more than 1,000, and the name was changed to the Central Research Laboratory in 1963. At this time the Research Department System was introduced.

This reorganization aimed to:

- Establish clear cut lines of authority and
  responsibility for research management.
- Emphasize research management. (For this
  reason the Research Management Department
  was introduced.)
- Improve and expand branch laboratories in
  several plants.
- Introduce a system of research fellows and senior
  researchers (In this system, a research fellow was
  of equal status with a section head. A senior researcher was of equal status with a subsection chief.)

Branch laboratories were established in the mid 1960s and were located in the plants. Each manager of a branch laboratory was under the supervision of the general plant manager, but the Central Research Laboratory's general manager coordinated the functions of the laboratories. These branch laboratories were as follows:

- Electroncs R&D Department   (established at Kamakura Works in 1966)
- Semiconductor R&D Department   (established at Kita Itami Works in 1967)
- Material R&D Department   (established at Sagami Works in 1965)

There were three main streams in Melco's research activities at that time: (1) basic and future technology, (2) research for manufacturing development, (3) product development. Since there were many differences in planning, conducting, and evaluating the research in these three streams, it was necessary to let these activities be independent.

The Manufacturing Development Laboratory and the Product Development Laboratory were established in 1970 and 1974, respectively. The Manufacturing Development Laboratory had three main functions:

- Development of manufacturing facilities and process-engineering technology.
- Evaluation of materials and components.
- Development of insulation materials.

The main purpose was to improve productivity in Melco.

In the early 1970s, labour costs had jumped because of a labour shortage, so labour saving machines were seriously demanded.

One of the most important strategies in Melco since that time has been the development of high productivity, so the establishment of the Manufacturing Development Laboratory was in accordance with Melco's strategy and environment in Japan.

The human side of productivity development has been the establishment of quality control circles and zero defect campaigns. The machine side has been the development of labour saving machines, and especially many types of sophisticated robots.

The Product Development Laboratory was founded with the intention of strengthening the organization for new product development and promoting new products to meet the needs of the times.

General Manager
    General Affairs Department
    Research Management Department
        1st Research Department   (Physics)
        2nd Research Department   (Electric & Machine)
        3rd Research Department   (Chemistry)
        Workshop
        - Electronics R&D Department (1966)
        - Semiconductor Research Department (1967)
        - Metal Materials Research Department (1965)
        - Branch Laboratories Coordination

IV. RESEARCH PLANNING BOARD, AND GROUP SYSTEM (1973)

In the mid 1960s, mainly because of the energy crisis, Japan's economic environment shifted from a rapidly growing economy to a stable one.

The fruits of technological innovation, introduced mainly from the United States, made Japan's economy flourish. Melco renewed its technological contract with Westinghouse, changing from a one way to a mutual technology transfer.

After 1970, Japan entered the age of self-made technology. At this time the scientific and engineering environment was as follows:

- Science and engineering became large scale.
- Information processing technology was highly developed. (From the mid 1960s, computer development had become full-scale.)
- Industrial pollutions had increased which caused serious public discussion about the relationship between pollutions and engineering.

From the business view point the following problems occurred:

- The necessity for new product development and new engineering development increased. To cope with these necessities, it became important to establish a research strategy.
- The rapidly changing R&D environment, such as the requirement that R&D activities speed up, necessitated making the best of highly talented young people.
- The problem of a "wall" between different departments occurred.

Melco's technological challenges at this time was as follows:

- Large-scale technological development, nuclear reactors, energy saving, computers, LSI (large-scale integrated circuit) technology.
- Development of technology connected with social and human life saving materials, anti-pollution devices, systems of social life.
- Development of new products and improvement in efficiency, security and reliability of products.

Because of these problems, in 1973 Melco established the Research Planning Board, introduced the Group System, and abolished the Department System.

R&D key tasks such as setting up research themes, following up research program, and connecting research fruits with business, were primarily the tasks of the R&D general manager. He managed research activities with the help of his managers in the early days. If an R&D organization had a personnel size of 100 to 200, the general manager could manage the key tasks effectively and speedily. But as an organization grew bigger, the separation into R&D management and R&D research became necessary.

The first administrative staff organizations were the Office Section and the License Section, both established in 1943. The first research staff organization was the Survey Room, which was established in 1951. Its functions were to:

- Survey and select research themes.
- Edit research plans.
- Follow up on research programs.

In 1963, the Research Administration Department was established to strengthen research staff functions. From this period, the format, rules, and operational system of research affairs became set.

In 1970, the Research Planning Department was established to pursue the following functions:

- Grasp research needs.
- Evaluate research themes.
- Organize research projects.
- Allocate research resources.

In 1973, the Research Planning Board was established, composed mainly of former department managers. This was the general staff or executive staff organization for the R&D general manager, formed to strengthen the planning and executive functions.

To cope with both the speed of R&D activities and the kaleidoscopic of changes in the R&D environment, it became necessary to make the most of the highly talented young researchers to strengthen the basic research field.

Under the Department System, the hierarchical structure of the R&D organization was complex, including researchers, senior researchers, fellow researchers, vice department managers, department managers and a general manager. To sharpen authority and responsibility, it was necessary to simplify the structure.

In the ten years since the Department System had been established in 1963, "walls" between departments had developed, because of the various views of department managers. Since this was a big problem the Group System was introduced. This system consisted of Groups, each with about 20 members. The Group head was under the direct control of the general manager.

Group heads were highly talented young researchers in their own right. Under the Japanese seniority system, it is difficult for young men to be in a management position with elder subordinates, so former department managers were appointed to the Research Planning Board.

At this time, the Research Director System was introduced, with directors equal in status to a department manager.

```
General Manager
      Research Planning Board
            General Affairs Department
            Technical Affairs Department
                  Unit Group
                  Unit Group

                        .
                        .
                        .

                  Consumer Electronics Development Department
                  Products Development Department
                  Coordination
```

V. MARKET ORIENTED RESEARCH DEPARTMENT SYSTEM (1978)

In the environment of the late 1970s, keen R&D competition emerged in the business world of Japan. One of the biggest requirements at that time was to make R&D activities effective, and to minimize risk in the low growth economic environment. Though the Group System of 1973 made the most of highly talented young men, the absence of coordination between Groups and the lack of an integrated long-term R&D view were evident.

To make the Group System efficient and to cope with the environment of the late 1970s, a Market Oriented Department System was introduced.

This system had three research departments: Materials Science & Device Department, Energy Science & Electro Mechanical Technology Department, System Control & Information Science Department, and some project groups.

There were several research groups under these departments, and the Group heads appointed were competent young men.

General Manager
    Planning Board
    General Affairs Department
    Technical Affairs Department
        Materials Science & Devices Department
        Energy Science & Electro-Mechanical Technology
            Department
        System Control & Information Science Department
        Micro-Computer Education Center

VI. CONCLUSION

R&D organization in Melco became large scale after its origin in the 1950s with about 1,000 members. After the 1950s, many R&D groups were shifted to related plants to build up new technically oriented products, or were assigned the task of founding new technically based units, such as the wireless technology works (Wireless Works), Braun tub works (Kyoto Works), and the transistor technology works (Kita-Itami works).

One of the recent Melco strategic products is semi-conductors. Melco has four plants and will have five. But if we examine the past, we can see that the origin of the semi-conductor works was

in a small R&D group. The research on transistors began in 1952,
with a few people at a research laboratory; after two years Melco
contracted transistor technology licenses with Western Electric
Corporation.

The transistor workshop was established at a research
laboratory in 1955, and Kita-Itami Works, the main semi-conductor
plant in Melco was founded in 1959. Many scientists and engineers
were shifted from the laboratory to Kita-Itami Works.

The 1950s, from a technological viewpoint was a period of
introducing foreign technology, and it was necessary for the R&D
organization to be highly research oriented. The Room System of
1951 was geared to flexible R&D activities. We can see it as a
research oriented organization.

In the 1960s, R&D activities were strongly urged to generate
returns from the technology which had been introduced from foreign
countries during 1950s. To be more effective in generating returns,
we had to be more management oriented. The Department system of
1963 established branch laboratories at related works.

In the 1970s, many Japanese enterprises perceived that they
had to create their own technology; this has been the most important
strategy in many Japanese enterprises since that time. It was
important to decide what research to do, and how to entrust highly
talented young men with these research themes. Therefore, Melco
established the Research Planning Board and Group system in 1973.

In the 1950s, we observed the technology in advanced foreign
countries and endeavored to introduce it. In the 1960s we digested
and applied these technologies. Since it became difficult to rely
upon advanced foreign technology, we decided to find new technology
in the 1970s.

Our recent strategy has been to produce our own technology,
to be practical, and to survey the market.

From the 1970s, R&D management and activities have been
required to be market oriented, and the reorganization of 1978
was aimed at remodeling the organization for a market orientation.

No organization is optimal and fresh forever. On the contrary,
too frequent reorganizing can be worse than having a bad but stable
organization. Reorganization should take place to meet changes in
the research environment and the demography of researchers. It is
often carried out coincident with the appointment of a new general
manager of the research laboratory and is effective in refreshing
researchers' minds.

We have traced the above history of our research activity with emphasis on its relation to environmental changes. The Melco organization will be changed every few years to meet the needs of the business environment.

COST-EFFECTIVENESS ANALYSIS FOR A TANK CREW TRAINING SYSTEM

Siegfried Lehnberg

Industrieanlagen-Betriebsgesellschaft mbH
Einsteinstraße
8012 Ottobrunn
Germany

ABSTRACT

For the gunnery training of tank gunners and commanders a complete simulation system has been developed encompassing the entire turret crew compartement, the battle terrain, and the observations of the crew during the process of system handling and target engagement.

The evaluation of the cost-effectiveness of the tank crew training system has shown that the annual operating cost of the tank corps training system may be reduced, and that the simulator is an adequate device for training tank gunners and commanders in preliminary gunnery training and firing practice.

BACKGROUND

In general, trainers are used for one or more - sometimes for all - of the following purposes: to reduce cost, save time, symplify training procedures, and minimize the amount of material necessary to attain the required proficiency in performing designated tasks with the operational equipment.

In training tank crews, the German tank corps has been faced with the following problems for some time:
o Wear of tanks and guns
o Reduced availability for emergency case
o High ammunition cost
o Limited availability of training ranges
o Limitations imposed by safety requirements
o Firing range constraints (e.g. range effect)

o   Weather influences
o   Environment protection conflicts
o   Short period of service (15 months)

For these reasons, the Federal Ministry of Defense, Army Staff, in the early seventies contracted HONEYWELL to develop the prototype of a training system for main battle tank crews.

TECHNOLOGICAL CONCEPT (SYSTEM DESCRIPTION)

For the gunnery training of Leopard MBT 1, A4, gunners and commanders, a complete simulation system has been developed encompassing the entire turret crew compartement, the battle terrain, and the observations of the crew during the process of system handling and target engagement.

The tank turret crew training system consists of the following major sub-systems (see Fig. 1):

o   Training turret with optical system
o   Model terrain and targets
o   Process control computer and interface for firing and tracer simulation
o   Instructor's console with audio and motion simulation controls
o   Illumination system

The system is installed in an air conditioned building.

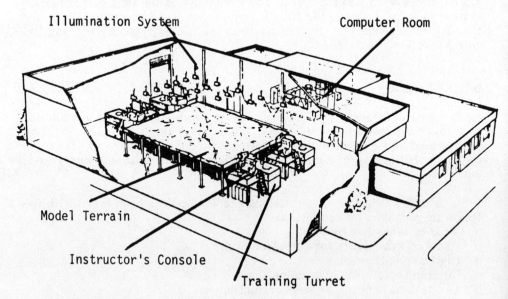

FIGURE 1 Tank Crew Training Center (Developed by HONEYWELL)

TRAINING CAPABILITIES

For engaging stationary or moving targets from a stationary position or a moving battle tank with the main gun, the following subtasks/functions can be performed with the simulation system:

o   Battlefield observation
o   Target detection, recognition, and identification
o   Target designation
o   Selection of ammunition
o   Stereoscopic ranging
o   Aiming
o   Firing
o   Shot sensing
o   Shot correction

PROBLEM

The ministerial directive on armaments management (1971) formalized the sequence time cycle of the planning process and how the organizational elements were to execute the process. According to the directive several GO/NO-GO decisions are to be made during the process of planning and developing military equipment.

The problem to be solved by the study was to provide decision aid by evaluating the cost-effectiveness of the training system at a very early stage of development. At that time (1973) only little or no information was available on the cost and effectiveness of the simulaton system.

APPROACH

For the purposes of this study, cost-effectiveness is defined as that procedure by which
o   the cost of alternative means of achieving a stated effectiveness,
     or, conversely,
o   the effectiveness of alternative means for a given cost,
are expressed in a series of numerical indices and compared.

The objective of the analysis is to isolate the alternative, or combination of alternatives, that produces either the maximum expected effectiveness for a given expected cost or the given expected effectiveness for the minimum expected cost.

Cost Analysis

The expected operating cost of alternative training systems, as the expected R&D and investment cost of the (planned) simulator and a range of potential savings were estimated under the assumption of constant tank crew proficiency.

Effectiveness Analysis

The decision-maker's objective was to obtain information whether and to what extent the actual equipment can be replaced by the simulation system in tank gunnery training or, possibly, firing practice.Used as an appropriate measure of effectiveness was the proficiency of alternatively trained crewmen.

The structures of the cost and effectiveness models are described below.

COST MODEL

The objective was to compare the annual operating costs of the operational training system and a simulator-aided training system. The firing proficiency of the tank crews was assumed to be constant.

Steps of Cost Analysis

a. Estimating the annual operating cost of the operational training system (variable cost).
Cost elements included:
o   Personnel cost (travel allowances for attending training courses)
o   Maintenance
o   POL
o   Main-armament ammunition
o   Subcaliber ammunition
o   Training area lease
o   Transport of tanks and material
o   Transport of personnel

The distribution of the operating cost per year and crew to the cost elements is shown in Fig. 2.

b. Estimating total cost of the simulation system

o   Research and Development
    -   Management
    -   Prototype
    Total R&D

o   Investment
    -   System cost (1 training center including 4 turrets)
    -   Infrastructure
    Total Investment

o   Operating
    -   Per center and month
    Total Operating (1 year)

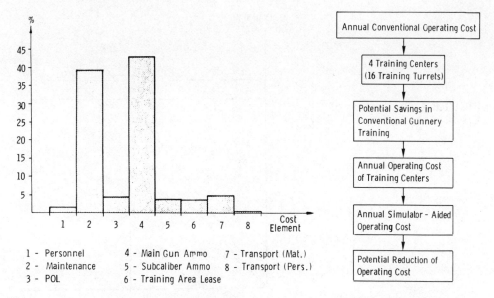

1 - Personnel   4 - Main Gun Ammo   7 - Transport (Mat.)
2 - Maintenance   5 - Subcaliber Ammo   8 - Transport (Pers.)
3 - POL   6 - Training Area Lease

FIGURE 2   Distribution of Operating Cost per   FIGURE 3 Structure
Year and Crew to Cost Elements   of Cost Model

c. Estimating a range of potential savings achievable by replacing
    alternative parts of the existing training system by the simulation
    system.

    Our estimates were made to show the significance of potential
    savings rather than to establish goals that should be reached by the tank
    corps. The estimates are based on the assumptions of replacing
    preliminary gunnery training, firing practice, and operation of tanks.
    The greatest potential for savings is in main-armament ammunition
    because of its high cost and because a major percentage of the annual
    ammunition cost is spent on practice firing.

d. Estimating the range of annual operating cost to be incurred by the
    simulator-aided training system.

e. Estimating a break-even interval, i.e. a period of time in which the
    expenditures for the simulator (R&D, Investment) can be compensated
    by a significant reductiuon in operating cost.

The structure of the cost model is shown in Fig. 3 and 4.

The current decision process relating to potential savings is still in
progress.

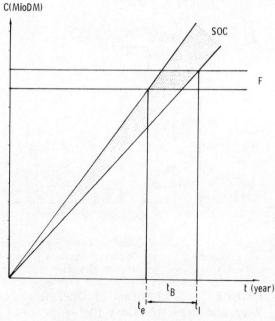

FIGURE 4 Estimation of
Break-even Points

F   – Estimated Interval of Fixed Cost (R&D, Investment)
SOC – Estimated Interval of Saving Operating Cost
t_B – Estimated Break-even Interval
t_e – Estimated Earliest Break-even Point
t_l – Estimated Latest Break-even Point

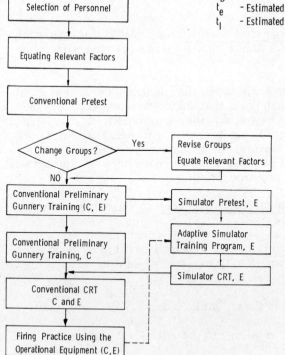

FIGURE 5 Test Procedure

EFFECTIVENESS MODEL

a. Objective

As mentioned earlier, the objective of the study was to provide information on whether and to what extent the existing equipment can be replaced by the simulation system in training tank gunners and commanders. In other words, the hypothesis of constant effectiveness under which the cost analysis had been conducted was to be tested.

b. Background

Testing the overall adequacy of the tank crew training system is not much different from testing any other thing - it must be tested with the end use in mind. Therefore, the suitability of the simulator is tested under simulated or actual training conditions using trainees and instructors representative of those who will operate the training system. Furthermore, the only control unit possible, other than the criteria established for the specific device, is the actual item of equipment which is to be operated by the students after simulator training. This involves the selection of two equivalent groups of trainees - one is called the control group the other the experimental group.

c. Measure of Effectiveness

As an appropriate measure of effectiveness, the gunner and commander proficiency firing the main gun at various types of practice targets was chosen.
The following main criteria were used to test the suitability of the simulator:
o   Time
o   Accuracy
o   Consumption of ammunition

d. Test Procedure

A series of tests to determine the advantages and disadvantages in terms of safety hazards, pre-operational inspection, physical characteristics, human factor engineering, and availability were conducted by the tank corps.

The study was focussed on testing the overall adequacy of the simulator as a substitute for the actual equipment in training tank gunners and commanders. The procedure used in the study is shown in Fig. 5. It is described below.

o   Selection of personnel

The   selection   of   personnel   is   of   vital   importance   to   the validity of the results of the effectiveness analysis, so it was done as systematically as possible.

Two groups of tank crews were selected as trainees, one called the control group (n=39), the other the experimental group (n=69). The following factors which influence student performance were equated:
- Aptitude-test scores
- Stereopsis test
- Term of service
- Educational and occupational level attained (civilian)
- Previous period of service
- Previous training and experience

To ensure a valid test, the instructors for the control and experimental groups had equivalent abilities, too. Additionally the following factors were considered:

- Experience as an instructor
- Past performance
- Supervisory ratings

o Conventional pretest

To ensure equivalent abilities, the control and experimental groups took part in a conventional pretest. In this test, the groups were to engage 5 stationary targets from a stationary position at the station drill-ground. The final selection of the groups was based on the pretest results.

o Conventional preliminary gunnery training

Both, the control and the experimental group were given standard instructions for the first part of preliminary gunnery training with the actual equipment. The second part of the conventional program was only performed with the control group.

o Simulator preliminary gunnery training

In parallel to the second part of the conventional preliminary training, the experimental group took part in an adaptive simulator training program. Each crew (gunner and commander) was trained

- 4 hours in engaging stationary and moving targets from a stationary position,
- 1 hour in engaging stationary and moving targets from a moving battle tank, and
- 1 hour in engaging stationary and moving targets alternately from a stationary and a moving battle tank.

For internal control purposes the simulator training was preceded by a pretest and followed by a criterion-referenced test.

o Conventional criterion referenced test

This test corresponded to the conventional pretest and provided initial information on the effectiveness of simulator-aided training.

o   Firing practice using the operational equipment

1. The first part of testing the adequacy of the simulator was to provide information on whether the system may be used as a substitute for the operational equipment in preliminary gunnery training.In line with the current tank corps regulation for firing the main gun, each crew of the control and experimental groups engaged 16 targets (gunners 13 targets, commanders 3 targets).

The performance of the alternatively trained crewmen was measured by 7 criteria as to time, accuracy, and consumption of ammunition. Performance was compared for each target and criterion by suitable statistical methods.

The following hypothesis was to be tested:
The alternatively trained groups achieve equivalent proficiency in firing the main gun at the described set of targets.

2. The second part of testing the adequacy of the system had to provide information on whether the simulator may be used as a substitute for firing practice.In a first step, it was decided to diminish the set of targets to be engaged by the experimental group. The firing program was reduced from 16 to 14 targets. The gunners of 3o additional experimental crews were to engage only 5 (instead of 7) stationary targets from a stationary position. All other conditions and parameters - except the weather -were kept constant.

The second part of the effectiveness analysis was to test the following hypothesis:
In spite of firing less, the experimental group, in regard to the remaining targets, attains a proficiency equivalent to that of the control group in the first run of the effectiveness analysis.

e.   Results of Effectiveness Analysis

The hypotheses were rejected. The experimental groups attained a significantly higher proficiency level than the control groups.

The effectiveness analysis indicated that the tank crew training system is an adequate device for training tank gunners and commanders in preliminary gunnery training and firing practice.

CONCLUSION AND PROSPECT

The results of cost-effectiveness analysis should be considered an important but only initial step in the process of integrating the simulator into the tank corps training system. Based on the results of the study the decision maker developed a simulator aided training system that is going to be adopted.

To find the optimum mix (ratio) of conventional and simulator training, the evaluation of simulator-aided trained crewmen should be carried on as described. After identifying the combination that promises optimum effectiveness the cost estimates will have to be revised.

This will enable the decision maker to take the right choice between the two alternatives for a simulator-aided training system:

a. take that system that ensures maximum effectiveness of the tank corps training system for a given budget, i.e. keeping operating cost constant, or

b. minimize operating cost while keeping performance constant.

During the development of the simulator the available ammunition per tank and year has been reduced by 18%. Therefore the decision maker intends to reassign the ammunition that can be saved by simulator training from service practice to field firing exercises. The objective is to raise the proficiency level of tank crews by integrating the simulator into the tank corps training system.

Whatever the final decision may be, the evaluation of a technically advanced, sophisticated training device has shown that, by duplicating the complex process of battle tank handling and target fighting, the annual training costs may be reduced. Although the capital expenditure on the training facility is high, it will pay off in a relatively short time depending on the final design of the tank corps simulator aided training system. In addition, projected fuel and ammunition shortages may be alleviated.

BIBLIOGRAPHY

1. Corcoran, A.W., "Costs - Accounting, Analysis, and Control", John Wiley & Sons, Santa Barbara (1978)
2. Goldman, Th.A., ed., "Cost-Effectiveness Analysis", F.A. Praeger, 2nd printing, New York (1968)
3. Honeywell "Sondertechnik", "Training System for Tank Crews of Main Battle Tanks - System Description", BRO 1003, 2nd ed., Maintal
4. Lehnberg, S., "Kosten-Wirksamkeitsschätzung eines Panzerturm-Simulators unter Berücksichtigung der Fahrerbeteiligung", IABG - Bericht B-SZ 1029/01, Ottobrunn (1975)
5. Lehnberg, S., "Auswertung des Truppenversuches Kampfraum-Trainer", IABG - Berichte M-SZ 1030/05-07 und B-SZ 1030/08, Ottobrunn (1980)
6. Seiler, K., "Introduction to Systems Cost-Effectiveness", John Wiley & Sons, New York (1969)

NA

SECTION 7

METHOD FOR MOBILITY ANALYSIS

There are many social, as well as economic consequences of
accelerated technological change. These consequences may seem
paradoxical when they reflect underlying tradeoffs between
different social and economic aims. The mobility of individuals
and groups arising from technological change provides an example of this situation. The tradeoffs between the desirable rate
of economic growth and restructuring, on one end, the desirable
or undesirable rate of social mobility and social structure
change, on the other end, are little understood. This section
provides mobility analysis in a level of depth and detail not
available in welfare economics or the sociology of change.

Schinnar proposes a manpower flow model which relates the
composition of the population to conditions of the external
labor market, to advancement opportunities in internal labor
markets, and to desired manpower flows. According to Schinnar,
different population groups show different tendencies to take up
opportunities. Likewise, Gaimon found that cohort groups within organizations (persons with similar organizational age) are
affected differently by career opportunities which the market
provides.

Within a fairly narrowly defined labor market category,
such as university academic personnel, or within a subset thereof, such as instructors in economics, one may view the mobility
problems from the point of view of the employer or the employee.
Feuer constructs a manpower planning framework for university
faculties who view personnel mobility targets, such as equality
or stability of promotion opportunities, as subjec to budgetary
and procedural constraints. Beckmann, on the other hand, constructs a personnel flow model which addresses such interesting
questions as the relationship between salary level and personnel
flow between universities of various ranks for personnel of
different levels. This is the kind of model that the scholar
needs if he wants to explicitly assess his career possibilities
and probabilities.

8330
8210

MICRO CONDITIONS FOR MACRO MODELS OF MANPOWER MOBILITY[1]

Arie P. Schinnar

School of Public and Urban Policy
University of Pennsylvania
Philadelphia, Pennsylvania 19104

1. INTRODUCTION

This paper combines systems of micro and macro manpower accounts to link heterogeneous labor groups and employment opportunities within and across internal and external labor markets. The resulting framework is then used to construct a time-inhomogeneous model of social mobility in organizations in order to show how conditions in the external labor market affect advancement opportunities in the internal labor market and, vice-versa, the possible effect of conditions in the internal labor market on flows into and out of the external labor market. The model is then shown to converge to an equilibrium in which the aggregate manpower flow parameters become time-homogeneous.

The formulation is based on the premise that the transition rates between the states of a human resources system need be a function of three considerations: (i) The composition of the population in the origin state is heterogeneous, e.g., sex, age, education, family background, work experience; (ii) the availability of opportunities in the destination state affects movement rates between states, e.g., job opportunities for college students, retirement plans for retirees; and (iii) different population groups show different tendencies to take up opportunities leading to a move from one state to another state.

The framework is intended to serve three objectives: To help circumvent distortions in Markov type models resulting from

[1]This research was supported in part under ONR project NR 047-228.

to provide a general social accounting framework for formulating
specific manpower models throgh the varying supply of job opportunities in the destination grades; and to show how the framework might
be used to develop explicit linkages between external and internal
labor markets.

2. A SYSTEM OF MACRO AND MICRO MANPOWER ACCOUNTS

This section develops the relationship between two systems of
manpower accounts: A macro system that traces the patterns of manpower flows within and without the organizational boundaries, and
a micro system that relates the classification of personnel to job
categories within and across organizational grades.

A Macro Manpower Account

Manpower mobility in the organization is characterized by
patterns of personnel flows between states within the organization
(the internal labor market), and by patterns of personnel flows in
and out of the organization, or between the internal and the
external labor markets. To set the background for the developments
in this paper we adapt Richard Stone's (1970) system of demographic
accounts to connect the opening and closing stocks of year t with
the manpower flows of year t. The symbols of table 1 have the
following definitions:

α = a scalar; denotes the total number of individuals who both
    enter and leave the internal labor market in the course of
    year t and are thus not recorded in either the opening or
    the closing stock of that year. An example is a person
    hired during the year who leaves before the end of the year.
b = a row vector; denotes the new entrants into the internal
    labor market, namely the recruits of year t, who survive to
    the end of the year. Individuals in this group are recorded
    in the closing stock but not in the opening stock.
d^T = a column vector; denotes the leavers from the internal labor
    market, namely the attrition or terminations of year t.
    Individuals in this group appear in the opening stock but not
    in the closing stock.
S = a square matrix; denotes the 'survivors' in the internal labor
    market through year t who are recorded in both the opening and
    the closing stock. They are classified by their opening states
    in the rows and by their closing states in the columns.
$\mu(t)$ = a row vector; denotes the opening stock in each state of
    the internal labor market.
Note that the closing stock of year t is the opening stock of
year t + 1, i.e., $\mu(t+1)$.

Table 1. Scheme for a System of Manpower Accounts

State at year t \ State at year t + 1	External labor market	Internal labor market: closing states	Opening stock
External labor market	α	b	
Internal labor market: Opening states	d^T	S	$\mu(t)^T$
Closing stock		$\mu(t+1)$	

From the second column in table 1 we have the column sums

$$\mu(t+1) = e^T S + b \qquad (2.1)$$

where e is the unit vector, so that pre-multiplication of S by the transpose of e, $e^T S$, gives the column sums of S. Let

$$P \equiv \hat{\mu}(t)^{-1} S \qquad \text{or} \quad S = \hat{\mu}(t)P \qquad (2.2)$$

be a matrix of outflow (transition) coefficients and where $\hat{\mu}(t)$ = diag $\{\mu(t)\}$, a diagonal matrix constructed from the components of $\mu(t)$. Next, by substituting (2.2) into (2.1) we obtain a set of linear difference equations

$$\mu(t+1) = \mu(t)P + b \qquad (2.3)$$

connecting the opening and closing stocks of the account.

The relation (2.3) constitutes the conceptual basis for the development of most macro manpower models. The internal labor market of the organization is partitioned into n-grades (states) with promotion following a hierarchical pattern. The nxn matrix P in (2.3) is the usually stationary promotion matrix containing rates of promotion and retention in the manpower system with $Pe \leq e$. A typical element p_{ij} denotes the expected promotion rate from grade i to j. $q = e - Pe$ is the nxl vector of quit rates from the various grades; b is the lxn vector of recruits entering the n grades, and $\mu(t)$ is the lxn grade-size vector of personnel in the organization.

With these definitions, $\mu(t)P$ provides the personnel in the organization in period $t + 1$ who were also present in the organization in period t; $\mu(t)q$ the portion of the manpower system that quits the organization between time t and $t + 1$; and b is the allocation of new recruits to the manpower system at the beginning of period $t + 1$. The intertemporal realignment of the grade-size distribution is thus portrayed by (2.3). See, e.g., Bartholomew (1973), Schinnar (1979), or Charnes, Cooper and Niehaus (1972).

Table 2 gives a schematic arrangement of the promotion, retention, attrition and recruitment parameters of a macro model of manpower mobility derived from the account in table 1. The p_{ij} and p_{ii} give the promotion and retention rates, and the p_{io} are the elements of the q vector of grade specific attrition rates usually found in most Markov type manpower models. The new parameter p_{oj} is not commonly found in these models; it denotes the probability that a person in the 'recruitment pool' will take a job in grade j. Unlike the first n rows in the table where

$$\sum_{j=0}^{n} p_{ij} = 1 \qquad\qquad (2.4)$$

the p_{oj} do not sum to unity. A recruitment pool is defined here, similar to the definition of a promotion pool in Lewis and Schinnar (1978). It consists of the group of people in the external labor market who are being actively considered for a position in grade j; it is a subset of the group of applicants which the organization is prepared to hire. The parameters p_{oj} thus reflect the rates of recruitment from the recruitment pool. A simple measure of the recruitment pool is the total number of job offers that an organization makes annually in the external labor market.

A Micro Manpower Account

In many macro manpower models the flow rates p_{ij} are assumed constant. Next, we make use of a framework developed in Schinnar (1980) to show how the heterogeneity of the population in the origin grades and job opportunities in the destination grades affect the flows between grades in the manpower system; i.e., we expect that as the composition of the population and job opportunities in various grades change, the flow parameters between grades will change also. We refer to such characteristics of the internal labor market as micro conditions.

We start by considering a single parameter in table 2, say p_{ij}, which is the expected flow rate from state i to state j. Consider h heterogeneous population groups in grade i and k job opportunities in grade j. Let $w_{i\ell}$ denote the number of people of

Table 2.   Promotion, Retention, Attrition and Recruitment
           Parameters of a Macro Manpower Model

Promotion from Grade (t)	Promotion to Grade (t+1)				Attrition
	Grade 1	Grade 2	...	Grade n	
Grade 1	P_{11}				P_{1o}
Grade 2	P_{21}	P_{22}			P_{2o}
⋮					⋮
Grade n			$P_{n-1,n}$	P_{nn}	P_{no}
Recruitment	P_{o1}	P_{o2}	...	P_{on}	

population group type ℓ in grade i. Then, the aggregate flow rate
P_{ij} is expressible in terms of the expected flow rates of each
population group;

$$P_{ij} = \sum_{\ell=1}^{h} \frac{w_{i\ell} p_{ij}^{\ell}}{\sum_{\ell=1}^{h} w_{i\ell}} = \sum_{\ell=1}^{h} \left(\frac{w_{i\ell}}{\sum_{\ell=1}^{h} w_{i\ell}}\right) p_{ij}^{\ell}$$

$$= \sum_{\ell=1}^{h} \omega_{i\ell} p_{ij}^{\ell} \tag{2.5}$$

where $\omega_{i\ell} \geq 0$, $\sum_{\ell=1}^{h} \omega_{\ell} = 1$, gives the relative size of population
group ℓ in grade i, and p_{ij}^{ℓ} is the population specific transition
probability between grades i and j.

     Let $R_{\ell r}^{ij}$ denote the conditional probability that a person
type ℓ in grade i will take up opportunity type r in grade j
provided it is made available, and let γ_{rj} be the probability of
(or relative frequency with which) opportunity type r being
available at grade j. Then $R_{\ell r}^{ij}\gamma_{rj}$ gives the joint probability
that a person type ℓ in i will capitalize on opportunity r in j.
By summing these probabilities over all the k opportunities
we obtain

$$P_{ij}^{\ell} = \sum_{r=1}^{k} R_{\ell r}^{ij} \gamma_{rj} \qquad (2.6)$$

the population specific transition probability. Now, by inserting (2.6) into (2.5) we obtain the bilinear relation

$$P_{ij} = \sum_{\ell=1}^{h} \sum_{r=1}^{k} \omega_{\ell i} R_{\ell r}^{ij} \gamma_{rj} \qquad (2.7)$$

where $\omega_{\ell i}$ and γ_{rj} are in the unit simplexes

$$\{\omega_{\ell i} : \sum_{\ell=1}^{h} \omega_{\ell i} = 1, \ \omega_{\ell i} \geq 0\} \qquad (2.8)$$

$$\{\gamma_{rj} : \sum_{r=1}^{k} \gamma_{rj} = 1, \ \gamma_{rj} \geq 0\} \qquad (2.9)$$

The parameters $R_{\ell r}^{ij}$ reflect the propensities for mobility by diverse population groups; they need not be constant and, in fact, will most probably vary as, e.g., the job opportunities γ_{rj} change. Even if the $R_{\ell r}^{ij}$ were assumed to be fixed parameters, these different propensities for mobility will alter the composition of personnel in the grades ($\omega_{\ell i}$), and lead to adjustments in the aggregate flow P_{ij}. There are certain strict conditions under which the P_{ij} may remain stable however. For example, when the mobility propensities $R_{\ell r}^{ij}$ are equal across all job opportunities, the P_{ij} remain invariant to variations in the distribution of job opportunities $\{\gamma_r\}$. Similarly, when the mobility propensities $R_{\ell r}^{ij}$ are equal for all population groups, the P_{ij} remain invariant under variations in the population composition $\{\omega_\ell\}$. These conditions will be rarely met however, which, in turn, calls for making allowance for variations in mobility rates when these are used to effect forecasts of realignment in grade size distributions via (2.3). In the remaining sections of the paper we outline the development of such a model, using the relationships in (2.3)-(2.9). In Schinnar (1980), the framework (2.7)-(2.9) is used to analyse the effect of population heterogeneity on the aggregations of parameters in social accounts; see also Kemeny and Snell (1960) for related conditions of lumpability in Markov type models.

In closing, I should like to clarify several notations used in the next section. The number of job and personnel classifications used in (2.7) will differ among grades and should therefore be denoted by k_j and h_i; we dropped the subscripts in (2.7) for notational convenience. We also make use of a matrix form to represent (2.7):

$$p_{ij} = \omega^i R^{ij} \gamma^j \qquad\qquad (2.10)$$

where ω^i is a row vector with elements $\omega_{\ell i}$, γ^j is a column vector
with elements γ_{rj} and R^{ij} is the $h_i x k_j$ matrix with entries $R^{ij}_{\ell r}$.
In addition, $\hat{\gamma}j = \text{diag} (\gamma^j)$ will be used to denote a diagonal
matrix construction from the elements of the vector γ^i.

3. A TIME-INHOMOGENEOUS MODEL OF MANPOWER MOBILITY

Those who are familiar with the many difficulties of applying
aggregate manpower models (e.g., of a Markov variety) would
undoubtedly suggest extra caution when considering application of
the framework proposed in section 2. The increase in the number
of parameters places a heavy burden on data bases, and the many
possible inter-grade relationships between the heterogeneous labor
groups and the various job opportunities appear intractable, at
least at first glance. In this section we apply the social accounting framework (2.3)-(2.9) to illustrate the formulation of a simple
time-inhomogeneous model of manpower mobility. This model is only
one of many possible developments however ; it is intended primarily
to serve as an example of how the linkage between micro and macro
manpower accounts might be operationalized.

Classification of Jobs and Personnel

To facilitate tracking the relations between labor groups and
job opportunities, we use job classifications to describe the
various groups of labor. Thus, for a given grade, we have a one-
to-one correspondence between job opportunities and manpower groups,
i.e., $h_j = k_j$ all j. Table 3 gives a schematic representation of
the dimension of the framework (2.3)-(2.9) for a three-grade manpower system. The classification of jobs (in the columns) and of
manpower (in the rows) are commensurate within each grade:
C_1 classes in grade 1, C_2 classes in grade 2, and C_3 classes in
grade 3. The formulae within the 'boxes' designate the aggregate
flow rates between the grades in accordance with table 2.

The dimensions of the R^{ij} matrices are defined by the row and
column classification of manpower and jobs. Thus, for example, R^{22}
is a $C_2 x C_2$ matrix of conditional retention rates within the 2nd
grade; it is predominantly diagonally dominant with off-diagonal
components reflecting (lateral) job mobility within the same grade.
The matrix R^{21} is a $C_2 x C_1$ matrix of conditional promotion rates
from C_2 manpower groups in grade 2 to C_1 jobs in grade 1.

Table 3.   Schematic Representative of a Three-Grade Manpower System
           With Commensurate Job and Manpower Classifications

Classification of Jobs

	C_1	C_2	C_3
Classification of Manpower C_1	$\omega^1 R^{11} \gamma^1$	(shaded)	(shaded)
C_2	$\omega^2 R^{21} \gamma^1$	$\omega^2 R^{22} \gamma^2$	(shaded)
C_3	(shaded)	$\omega^3 R^{32} \gamma^2$	$\omega^3 R^{33} \gamma^3$

The commensurate job and manpower classification system
facilitates the linking of the ω's and γ's across grades. Note,
for example, that γ^1 is a $C_1 \times 1$ vector of job opportunities in
grade 1 for the C_2 groups of workers in grade 2; its classification
corresponds to the $1 \times C_1$ vector ω^1 of worker group sizes in grade 1.
This correspondence in classification thus helps to compute how,
e.g., attrition from grade 1 alters job opportunities for employees
in grade 2, and how the propensities for upward mobility in grade 2
affect the heterogeneity of the worker population in grade 1.

Two further assumptions are evident from the notations in
table 3. First, the distribution of population in the origin
grade, ω^i, is considered a 'pool' for all patterns of mobility:
promotion, retention and attrition. E.g., we presently do not make
allowance for the possibility that certain employees may be inherent
'movers' or 'stayers' as was suggested by Blumen, Kogan and
McCarthy (1955), or that the attrition pool may be predominantly
a subset of the promotion pool, consisting of those employees who
were considered but not actually promoted. The second assumption
applies to the job opportunities in the destination grades, γ^j,
which are taken to be equal for all origin grades, including the
recruitment pool. Given the set of job opportunities in a grade,
we assume that irrespective of seniority in the grade and the
organization, all manpower shares equal opportunities for filling
these jobs. In reality, however, there exists a sequence of almost
preemptive priorities that follows seniority: the first choice of

jobs is usually given to those with seniority in the grade (i.e., retentions), the remaining job opportunities are then offered to workers with seniority in the organization (i.e., promotions from a lower grade), and the unfilled jobs are lastly available to recruits from the external labor market.

Derivation of Job Opportunities

From (2.4) and (2.6), the vector of promotion, retention and attrition rates is expressible as $R^{ij}\gamma^j$, and because R^{ij} has C_i rows for all j, we can write

$$R^{ij}\gamma^j + R^{ii}\gamma^i + R^{io}\gamma^{io} = e_i, \qquad \text{all } i \qquad (3.1)$$

where e_i is the $C_i \times 1$ vector with unity for all its components; $i = j+1$. We next assume that γ^{io} is determined by conditions in the external labor market and is thus exogenous to the model. While γ^j and γ^i might be regarded as employment goals and hence offer some discretion about their choice, the systemic form of (3.1) requires that the $\sum_{i=1}^{n} C_i$ unknowns $\{\gamma^j\}$ satisfy the following system of $\sum_{i=1}^{n} C_i$ equations

$$
\begin{pmatrix}
R^{11} & & & \\
R^{21} & R^{22} & & \\
& \cdot & \cdot & \\
& & \cdot & \cdot \\
& & R^{n-1,n} & R^{nn}
\end{pmatrix}
\begin{pmatrix}
\gamma^{1*} \\
\gamma^{2*} \\
\cdot \\
\cdot \\
\gamma^{n*}
\end{pmatrix}
=
\begin{pmatrix}
e_1 - R^{1o}\gamma^{1o} \\
e_2 - R^{2o}\gamma^{2o} \\
\cdot \\
\cdot \\
e_n - R^{no}\gamma^{no}
\end{pmatrix}
\qquad (3.2)
$$

where

$$e_j^T \gamma^{j*} = 1, \quad \gamma^{j*} \geq 0, \qquad \text{all } j \qquad (3.3)$$

For a given set $\{R^{ij}\}$ a solution $\{\gamma^{j*}\}$ of (3.2) that also satisfies (3.3) may not exist.[2] This is in part due to the R^{ii} being diagonally dominant and thus nonsingular (see Varga, 1962). Therefore, for a unique $\{\gamma^{i*}\}$ in (3.2) to also satisfy (3.3), adjustments in the propensities for internal $\{R^{ij}\}$ and external $\{R^{io}\}$ mobility will usually be required.[3] Clearly, R^{ij} can be easily adjusted

---

[2]Additional constraints will be added with eqns. (3.5) and (3.6).

[3]This may be computed by using the nonsurvey methods discussed in Bacharach (1970).

to obtain the desired γ^{j*} when needed. For the purpose of the present paper, we assume these to reflect distributional goals of jobs in each grade.

Derivation of Manpower Distributions

The availability of the $\{\gamma^{j*}\}$ provides immediate access to the aggregate flow rate p_{ij} in (2.7). However, manpower flows produce a realignment of the ω's, and, consequently, the aggregate flows will remain stable only in the unlikely event that all components of the $\{R^{ij}\gamma^{j*}\}$ vectors are equal. To compute the realignment of personnel across and within grades, recall that $R^{ij}\hat{\gamma}^{j*}$ is a matrix of joint probabilities: each entry denotes the probability that a person in a certain job (row) in grade i will fill a certain job opening (column) in grade j. It is a joint probability of matching a job in grade j with a person from grade i. Similar to the way $R^{ij}\gamma^{j*}e_j$ gives the vector promotion probabilities to job j for personnel in grade i, the vector

$$\eta_j = \sum_{i=o}^{n} e_i^T R^{ij} \hat{\gamma}^{j*} \tag{3.4}$$

gives the probabilities of promotion into jobs in grade j, $\eta_j e_j = 1$, $\eta_j \geq 0$. In case of a hierarchical pattern of mobility, this is expressible as

$$\eta_j = e_j^T R^{jj} \hat{\gamma}^{j*} + e_i^T R^{ij} \hat{\gamma}^{j*} + e_o^T R^{oj} \hat{\gamma}^{j*} \leq e_j^T, \qquad \text{all } j \tag{3.5}$$

which, in turn, suggests two additional constraints for the solution of (3.2) : (3.5) and

$$(e_j^T R^{jj} + e_i^T R^{ij} + e_o^T R^{oj}) \gamma^{j*} = 1, \qquad \text{all } j. \tag{3.6}$$

The matrix $R^{ij}\gamma^{j*}$ may also be interpreted as a matrix of conditional flow rates; that is, given a person of certain characteristics in grade i, and given a certain job in grade j. However, the flow of personnel into C_j jobs in grade j also depends on the ω^i distribution of personnel in the origin grade i. The expected manpower flow from grade i into C_j jobs in grade j is thus expressible via $\omega^i R^{ij} \hat{\gamma}^{j*}$ with total flows given by $N_i \omega^i R^{ij} \hat{\gamma}^{j*}$, where N_i denotes the size of grade i, $\sum_{\ell=1}^{h} w_{\ell i}$ in (2.5). Thus,

the total number of recruits from the internal labor market to jobs in grade j can be written as

$$N_j \omega^j R^{jj} \hat{\gamma}^{j*} + N_i \omega^i R^{ij} \hat{\gamma}^{j*} \tag{3.7}$$

where $i = j + 1$.

Recruitment from the external labor market is defined here by the propensities of mobility into the organization, R^{oj}, the distribution of labor in the relevant applicant pool, ω^{oj}, and the size of the applicant pool, N_{oj}. The expected number of recruits into the organization thus consists of $N_{oj} P_{oj} = N_{oj} \omega^{oj} R^{oj} \hat{\gamma}^{j*}$.

Since the $p_{jj} = \omega^j R^{jj} \gamma^{j*}$ and the $p_{ij} = \omega^i R^{ij} \gamma^{j*}$ are available, the size of grade j grows, is stable or contracted depending on whether

$$N_{oj} P_{oj} > N_j - N_j P_{jj} - N_i P_{ij} \quad (\Rightarrow \text{ growth})$$

$$N_{oj} P_{oj} = N_j - N_j P_{jj} - N_i P_{ij} \quad (\Rightarrow \text{ constant grade size}) \tag{3.8}$$

$$N_{oj} P_{oj} < N_j - N_j P_{jj} - N_i P_{ij} \quad (\Rightarrow \text{ contraction})$$

respectively. Therefore, for a particular choice of $\{N_{oj}\}$ we may forecast the realignment in grade sizes and the realignment of the manpower distribution within each grade via the following system of difference equations

$$N_j(t+1) = N_j(t) p_{jj}(t) + N_i(t) p_{ij}(t) + N_{oj}(t) p_{oj}(t) \tag{3.9}$$

$$\omega^j(t+1) = \left(\frac{1}{N_j(t+1)}\right) [N_j(t) \omega^j(t) R^{jj} + N_i(t) \omega^i(t) R^{ij}$$

$$+ N_{oj} \omega^{oj} R^{oj}] \hat{\gamma}^{j*} \tag{3.10}$$

where the

$$p_{ij}(t) = \omega^i(t) R^{ij} \gamma^{j*} \tag{3.11}$$

give rise to a time-inhomogeneous flow process. In this special case of the model we assume that R^{ij}, γ^{j*} and ω^{oj} remain stable over time. This can be easily relaxed by introducing intermediate solutions of (3.2). Next we examine the temporal properties of (3.10).

Weak Equilibrium[4]

    A manpower model attains equilibrium when the realignment of
grade sizes stops; i.e., when the relative distribution of grade
sizes and the distribution of personnel within jobs categories
stabilize . We shall call such an equilibrium a _weak_ equilibrium
condition, and shall say that a _strong_ equilibrium condition has
been attained when all the job vacancies have been filled, or when
the distribution of manpower in jobs (ω^i) matches the distribution
of job opportunities (γ^i). We limit the present discussion to
weak equilibrium in the context of the (3.9)-(3.11) model
formulation.

    To determine whether the process (3.10) converges to equilib-
rium we introduce the matrix form

$$w(t+1) = w(t)A + r \tag{3.12}$$

where

$$w(t) = [N_1(t)\omega^1(t), N_2(t)\omega^2(t), \ldots, N_n(t)\omega^n(t)]$$

$$A = \begin{pmatrix} R^{11}\hat{\gamma}^{1*} & & & \\ R^{21}\hat{\gamma}^{1*} & R^{22}\hat{\gamma}^{2*} & & \\ & \cdot & \cdot & \\ & & \cdot & \cdot \\ & & & \cdot \\ & & R^{n,n-1}\hat{\gamma}^{n-1*} & R^{nn}\hat{\gamma}^{n*} \end{pmatrix}, \quad r^T = \begin{pmatrix} N_{o1}P_{o1} \\ N_{o2}P_{o2} \\ \cdot \\ \cdot \\ \cdot \\ N_{on}P_{on} \end{pmatrix}$$

Because A is a non-negative matrix and because at least one entry
in $R^{oj}\hat{\gamma}^{j*}$ is positive, at least one inequality in (3.5) is strict;
it then follows that the dominant eigenvalue of A is less than
unity, giving rise to the following τ-period projection equation

$$w(t+\tau) = w(t)A^\tau + r(I-A)^{-1}(I+A^{\tau-1}) \tag{3.13}$$

with

$$\lim_{\tau \to \infty} w(t+\tau) = r(I-A)^{-1} \tag{3.14}$$

because A^τ approaches the null matrix as $\tau \to \infty$.

---

[4]Our distinction between strong and weak equilibrium follows
from the substantive application of the model. For a general
discussion of equilibrium conditions in time-inhomogeneous
Markov chains see Seneta (1973).

The existence of a weak equilibrium (3.14) is secured by the non-singularity of (I-A); and, indeed, by inserting (3.14) in (3.12) we obtain

$$w(t+1) = w(t)A + r = r(I-A)^{-1}A + r$$

$$= r[(I+A+A^2+...)A + I] = r(I-A)^{-1} \qquad (3.15)$$

the desired equilibrium. We have thus shown how a stationary micro flow model (3.13) gives rise to a non-stationary macro flow model (3.9); the aggregate flows stabilize when the micro model attains weak equilibrium. However, because a weak equilibrium does not satisfy $\omega^i = (\gamma^i)^T$, these equilibria remain unstable as certain job vacancies remain unfilled. We shall address the treatment of strong equilibrium in a sequel to this paper.

External Labor Market

While the main focus of this paper has been on the workings of the internal labor market, we have also provided an explicit linkage with the external labor market. R^{io} describes propensities of mobility from the internal to the external labor market. The actual flow rates $R^{io}\gamma^{io}$ will also depend on the opportunities in the external labor market such as early retirement, employment opportunities in other organizations, continuing education, etc. It is evident from (3.2) that as these opportunities are changed by economic fluctuations, business cycles, and supply and demand conditions in the relevant labor markets, both job opportunities and mobility propensities within the organization are altered as well.

A second linkage with the external labor market is provided through recruitment. The size and structure of the applicant pool and its propensities for entry into the organization will undoubtedly be affected by job opportunities within the organization as well as by conditions in the external labor markets. We have thus sketched how, e.g., conditions in the external labor markets, γ^{jo}, affect opportunities in the organization, γ^{j*}, which, in turn, has a feedback effect on conditions in the external labor market through its impact on the recruitment pool. What remains to be developed is an explicit model describing the workings of the external labor market in order to complete the system.

4. CONCLUSIONS

In this paper we have shown how a system of micro accounts
could be used to model aggregate patterns of mobility in a manpower system. The framework also provides an explicit linkage
between employment (or other) opportunities in external and internal labor markets. The simple model developed in section 3
can be extended in a number of ways (already suggested in the body
of the paper); but perhaps most significantly, by articulating
more clearly the subsets of personnel and job opportunities which
are relevant to a particular flow pattern. Even though we have
neglected to discuss the importance of lateral mobility within
grades, it is fully defined by the bilinear representation
of retention rates $(\omega^j R^{jj} \gamma^j)$. This is an integral component of
the model which has received but scant attention in the general
literature of manpower mobility.

Lastly, a technical remark. The very simple variant of the
model in (3.12) bears resemblance to the two-class Markov mobility
models discussed by Boudon (1973), Henry, McGinnis and Tegtmeyer
(1971) and Schinnar and Stewman (1978). The resemblance is only
in form, however, since the two-class models use conditional
probabilities, whereas the matrix A in (3.12) contains joint
probabilities. While both approaches try to account for population heterogeneity in the origin states, the bilinear framework
(2.7) makes also allowance for varying job opportunities in the
destination states.

A numerical example using the model developed in section 3
was presented at the NATO symposium and is available from
the author.

REFERENCES

Bacharach, M., (1970) Biproportional Matrices and Input-Output
    Change, Cambridge University Press: London.
Bartholomew, D.J., (1973) Stochastic Models for Social Processes,
    2nd edition, Wiley: New York.
Blumen, I., M. Kogan and P.J. McCarthy, (1955) The Industrial
    Mobility of Labor as a Probability Process, Cornell Studies
    of Industrial and Labor Relations, Vol. 6, New York.
Boudon, R., (1973) Mathematical Structures of Social Mobility,
    Elsevier, New York.
Charnes, A., W.W. Cooper and R.J. Niehaus, (1972) Studies in
    Manpower Planning, Office of Civilian Manpower Management,
    Department of the Navy, Washington, D.C.
Feuer, M.J. and A.P. Schinnar, (1978) "Exchange of Advancement
    Opportunities in Hierarchical Organizations," Fels Discussion
    Paper No. 136, University of Pennsylvania, Philadelphia.

Henry, N.W., R. McGinnis and H.W. Tegtmeyer, (1971) "A finite
     model of mobility," Journal of Mathematical Sociology,
     Vol. 1, pp. 107-118.
Kemeny, J.G., and L. Snell, (1960) Finite Markov Chains,
     Van Nostrand: New York.
Lewis, K.A. and A.P. Schinnar, (1978) "Analysis of Job-Related
     Promotion Pools," presented at the ORSA/TIMS joint national
     meeting, Los Angeles, November 1978.
Niehaus, R.J., (1979) Computer Assisted Human Resources Planning,
     Wiley-Interscience: New York.
Schinnar, A.P., (1979) "Organizational growth and realignment of
     manpower grade-size distributions," presented at the joint
     national ORSA/TIMS meeting, Milwaukee, October 1979.
Schinnar, A.P., (1980) "Frameworks for social accounting and
     monitoring of invariance, efficiency and heterogeneity,"
     presented at the Seminar on Models for Alternative Develop-
     ment Strategies, October 1980, The Hague, Netherlands.
Schinnar, A.P. and S. Stewman, (1978) "A class of Markov models
     of social mobility with duration memory patterns," Journal
     of Mathematical Sociology, Vol. 6, pp. 61-86.
Seneta, E. (1973) Non-Negative Matrices, Wiley: New York.
Varga, R.S., (1962) Matrix Iterative Analysis, Prentice-Hall:
     New Jersey.

LATERAL SKILL PROGRESSION/RANK RESTRUCTURING STUDY

Don Slimman

Canadian Forces

National Defence Headquarters, Ottawa, Ont, Canada

BACKGROUND

1.   Like most other military organizations, the Canadian
Forces use promotion in rank, with its attendant rewards in
compensation, social status and perquisites, as the primary
means of recognizing superior performance and marking the
visible steps in career advancement   (Figure 1).

If, however, rank is seen as an instrument and symbol of
authority, as it traditionally functioned in the pre-technical
armed forces throughout preceding centuries, its use as a
reward for superior performance in a purely technical/trade
sense may be inappropriate and even dysfunctional. Although the
personal skills required for leadership/supervision/management
are different from those required for technical or trade
proficiency, the CF have virtually no other means available,
other than promotion in rank, for motivating servicemen and
women to increase their technical knowledge and performance
levels and to reward them when they do. From many discussions
with CF servicemembers, it appeared that a substantial
proportion of men promoted to higher rank solely on the basis
of their technical skills turned into poor leaders and
supervisors. In cases such as these, the CF can be said to
have lost a good tradesman and gained a poor leader.

Figure 1:
1. Each trade is as-
    signed to one of
    three groups - Stan-
    dard, Specialist 1
    or Specialist 2 -
    depending on its
    technical complexity.
2. Approximately 80% of
    the CF noncommis-
    sioned population is
    placed in the Stan-
    dard Trade Group.

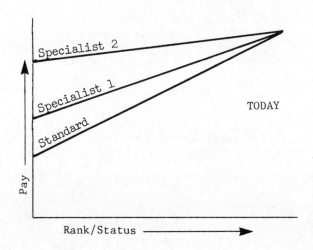

Figure 2:
1. TPL - Trade Progres-
    sion Level.
2. In most, if not in
    all trades supervi-
    sors would be required
    to be at least "com-
    petent technicians,"
    with TPLs in excess
    of the basic levels.

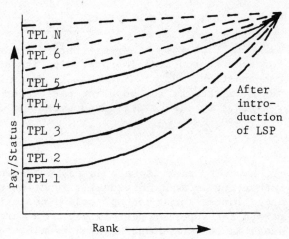

2.   The concept of Lateral Skill Progression (LSP) is
accordingly being studied by the CF in order to move beyond a
unidimensional reward system. LSP can be defined as a career
advancement system for measuring, recognizing and rewarding
technical/trade proficiency concurrently with, but
independently from, supervisory/leadership ability   (Figure
2).

3.   The study to determine the feasibility of applying LSP
to CF other rank trades has been underway for over six months,
and it appears likely that the results will be positive.
Because LSP is in effect divorcing technical knowledge and
skill from the interpersonal skills associated with leadership,
management and supervision, the study team was also asked to
reaffirm the status of rank in accordance with this concept and
to suggest a revised ideal rank structure for each CF trade
that would represent a minimum consistent with the need to
permit supervisors and leaders to exercise proper command and
control. Finally, the team was to determine a mechanism which
would permit an orderly return to this ideal structure. The
model being proposed appears to go well beyond measures used by
other armed forces to separate leadership from technical skill.

DISCUSSION

4.   The problems arising from the present system were
identified several years ago in the report of the Other Ranks
Career Development Program (ORCDP). It was recognized that the
personnel system of the Canadian Forces which preceded
unification had evolved slowly for the Navy and the Army, and
rapidly, in an industrial/managerial context, for the Air
Force. The single personnel management system which was forged
at the time of unification was structurally affected by the
management theories in vogue at the time, and by a greatly
enhanced role of the central agencies of Government in the
internal policies of the armed forces. Not surprisingly, many
managerial concepts which had been developed to control the
civilian public service along substantially private sector
employment concepts were embedded within the conceptual base
and the subsequent thought patterns of the military personnel
system. Great benefits can be cited from this major
readjustment, but no one should be surprised if some
assumptions and the resultant policies proved over time to have
inadequacies in the military context.

5.   The Ministers Manpower Study (Men) (MMSM) had earlier
stated in June 1966 "... in our view it is an artificiality, to
say the least, to try and separate rank and skill. They are,
in the final essence, one and the same thing." The assumption

contained in this statement is characteristic of the literature
of the management science and business administration dogma of
the era, which was concerned with the profitability of
competitive corporations and focussed almost exclusively on the
managerial caste within the corporate structure. In this
model, executive rank level could well be expected to be
closely correlated with managerial skills, so that the
assumption was satisfied in practice. Workers, in this model,
were regarded as a commodity - "labour" - which, when combined
with machines and raw materials, produced a marketable
product. The shop foreman was a worker of superior work skills
who could teach novices, assign tasks, and keep the output up.
Thus in the worker group also the assumption seemed valid. The
Public Service approach to manpower was predominantly that of
defining a job to be done and hiring a person who already had
skills to do that job, preferably to continue doing it until
age 65. In the atmosphere created by the pervasiveness of this
model of the workforce it was reasonable that the MMSM should
go on to state:

> "We consider that any technicians who can plan,
> organize, direct, and control the work of his junior
> technicians with all the technological complexities
> involved is quite capable of discharging any general
> military duties that may fall to his lot."

6.   The application of this industrial model to the
Canadian Forces has led to several major problems. The ORCDP
found that:

> a.   Service personnel support the "team concept" that
>   all members of the Canadian Forces are of equal
>   importance to the capability and effectiveness of
>   the team but they are also aware that individuals
>   differ in ability.

> b.   They are quite willing to accept pay differentials
>   based on demonstrated individual qualifications
>   and performance.

      c.   They are not satisfied with a pay system that rewards trade groups rather than individual initiative and achievement.

      d.   They felt that many personnel in specialist groups are better paid simply because they were fortunate enough to be selected into those trades as recruits.

      e.   In trade progression, there is little emphasis on personal achievement or initiative.

      f.   In many cases, selection of personnel into a trade is dictated by the existing requirement of the Forces, and not by the preference and ability of the individual.

7.   Subsequent to the submission of the original ORCDP LSP recommendation, several more problem areas that might be solved by an LSP system were determined:

      a.   <u>Inadequacies of the Promotion System</u>. The CF requires a comprehensive span of trade skills. Recognizing and rewarding these skills through rank promotion tends to degrade rank as an instrument and sign of authority.

      b.   <u>Rank Escalation</u>. Because the only tangible reward for superior performance is promotion, there is a tendency to upgrade the ranks of positions in an attempt to provide an opportunity to pay personnel for their performance by increasing promotion opportunity.

      c.   <u>Loss of technical "hands on" personnel through promotion</u>: Lacking any other reward system, there is a tendency to try to promote superior technicians on their technical ability even though they may lack the desire or talent to be leaders.

The man is thus promoted "out of his tool box". A
valuable technician in many cases is made into a
below-average and unhappy NCO. The CF required a
system which could reward superior technical
performance without promoting the tradesman away
from the very aspect of his employment in which he
excelled.

d. Decline in the Quality of Leadership. A number of
observations on the loss of respect for NCO ranks
were made by servicemembers during visits and
interviews. Some NCOs lost the respect of their
subordinates because they did not perform as NCO's
are expected to. They were not given the
opportunity to lead, and in some instances the
supervisory ranks outnumbered the personnel
available to be supervised.

e. Responsiveness to a Lateral Entry Scheme. The
present system does not provide a satisfactory
method to compensate or recognize technical
training received by recruits prior to joining the
CF. The CF must be able to compete with civilian
industry for technical school graduates in
recruiting for Other Rank positions. We must be
able to offer pay and status to a skilled recruit
commensurate with his level of training obtained
prior to entry. The projected savings in training
especially in small, highly skilled trades, would
be substantial.

8.   In order to devise a system that might find solutions
for these identified problems, a twelve-man team has been
tasked with determining the feasibility of applying LSP to each
of the 100 other ranks trades, known as Military Occupation
Classifications (MOCs). The task is formidable. Six MOCs,
selected to form a representative sample across environments,
technical complexity, leadership requirements, homogeneity and
diversity of employments, and are being examined during the
initial year-long stage of the study. Through formal

organizational analysis (OA), the spectrum of jobs performed by
each trade is being determined. This will provide the basis
for identifying the skills required to perform these jobs.
What we are calling a TPL is simply the discrete group of
skills required to perform a particular job. The second
dimension of that job description is the requirement for
leadership or supervision i.e. a rank. The requirement for
rank will no longer be based on skill or longevity, or on the
perceived remuneration level required. Rather it will be
determined through a "bottom up" analysis. In other words,
when a certain number of workers are together in a specific
group, a supervisor (or worker/supervisor) with one rank higher
will be needed. The number will be determined primarily by the
appropriate span of control and may differ from one type of
employment to another, depending on circumstances. Rank, then
will be allotted on the basis of the requirement for legal
authority over others in the group. This will represent a
substantial change from current CF trends, where some trades
show a rank-to-rank ratio of 1:1 at certain levels. In
colloquial terms, we have too many chiefs and not enough
Indians. Moreover, many of these chiefs are in fact highly
skilled Indians and what we propose to do is to remove their
headdress and give them instead different insignia to mark
their technical skills, different in nature to the
interpersonal skills required by the true chief but equally
essential to the efficient functioning of the organization.
Removing headdress is a touchy business, however; and an
important part of the LSP study will be to discover whether
personnel can be motivated to seek the rewards and status of
the master craftsman as eagerly as they now seek those of
increased rank. This behavioural science aspect is being
investigated in the CF Personnel Applied Research Unit and the
conclusions will be important in making the final
recommendations.

TRADE PROGRESSION LEVELS

9.   As suggested earlier, delineation of the various TPLs
will rely to a great extent on occupational analysis (OA)
techniques developed for CF use by our colleagues in the

Directorate of Military Occupational Structure. The process of
sorting the entire range of skills and knowledge required to
perform the various tasks called for within each MOC into a
hierarchical TPL structure will be complicated and
time-consuming. Needless to say, the end results, both within
and between trades, must be seen to be demonstrably fair, in
particular within the context of the wide rank of occupations
represented in our unified services. As an example of one of
the problems to be faced is the question of whether the number
of TPLs should be the same for each MOC. (Figure 2, Note 3).
Representatives of the technical trades tend to argue that the
number of TPLs should reflect the technical complexity of each
trade, so that the Radar Systems Technician trade might, for
example, have double the number of levels as the Infantryman
trade, where the leadership component predominates over the
technical. Spokesman for the Combat Arms, however, would argue
that in the same way that each private can aspire to the rank
of Chief Warrant Officer, he should also be able, theoretically
at least, to attain the highest TPL in the CF. The number of
TPLs, they maintain, should be the same for each trade, with of
course a greater proportion of positions in the technical
trades being established for higher TPLs than in the less
technical trades. Since substantial pay increases are
envisioned for each TPL increase, the question is not merely
academic.

10.   One of our underlying concerns is that under the
current progression system where financial records are tied
basically to promotion in rank, there is and will continue to
be pressure to upgrade more positions for no other reason than
to provide more competitive pay. With inflation in rank comes
a corresponding deflation in the understanding of rank for what
it was meant to be in a military context - i.e. to ensure the
effective accomplishment of missions by groups of individuals
over a wide spectrum of peacetime and wartime scenarios. There
is little question, in the CF at least, that the perception of
rank as a symbol and instrument of authority has been seriously
eroded. The trend must be reversed.

11.   This paper has so far looked as LSP/RR solely in the
context of an established system. Despite the fact that the CF
will probably be going further than any other military force in
creating a career advancement system with independent, and yet
concurrent, dimensions of recognition and reward, there appears
to be general acceptance that LSP will represent a substantial
improvement over the current unidimensional system. The
introduction of such a radically new system creates its
own problems, particularly in the initial transition phases,
and the study team will have to demonstrate that these problems
will not outweigh the apparent advantages of the LSP concept.
What may be required is a balancing act; bringing LSP into
being quickly enough to overcome the problems inherent in the
present system, and yet gradually enough to permit individuals
to adjust to a new set of values and to a new system of career
progression.

THE EFFECTS OF FLUCTUATIONS IN THE ECONOMY OR TECHNOLOGY ON COHORT CAREERS

Cheryl Gaimon

Carnegie-Mellon University
Graduate School of Industrial Administration
Pittsburgh, PA 15213

ABSTRACT

We have examined the effects on the careers of cohorts in response to the interaction between goal levels of manpower corresponding to a period of contraction followed by a period of expansion, and (1) the effectiveness versus salary relationship, (2) the initial size of the workforce, and (3) the rate of change in the goal levels of manpower. The mathematical model that we have employed in this paper is the normative, longitudinal (or cohort), 'pull' formulation presented by Gaimon and Thompson in [4] that is solved using optimal control theory.

1. INTRODUCTION

Changes in the economy or technological advancement can have a significant impact on the employee structure of an organization. In this paper, we focus on the manner in which such variations affect the careers of cohort groups within an organization, assuming that the organization behaves optimally to achieve a given objective. (By cohort group, we mean those persons sharing the same organizational age.) It is assumed that we can translate the impact of expected changes in the economy or technology into quantifiable changes in the future goal levels of manpower desired by the organization. Therefore, when we speak of increasing or decreasing goal levels of manpower in an organization it should be understood that we are implicitly referring to changes in the economy or technology.

We show that the effects of variations in the goal levels of manpower on the careers of cohort groups can only be ascertained when analyzed in conjunction with a variety of exogenous factors. Specifically, we examine the effects which changing the goal levels of manpower have on the optimal hiring, separation, promotion, and retirement policies. We solve a normative, longitudinal, 'pull' formulation under several hypotheses concerning (a) the relationship between the salary and effectiveness levels of personnel, (Section 3), (b) the initial size of the workforce present in various grades of the organization, (Section 4), and (c) the rate of growth or decline in the goal levels, (Section 5).

Since we characterize the effects that growth or decline have on an organization by determining the resulting changes in the optimal personnel policies, our approach is normative. Descriptive models have been employed to analyze the effect that size has on the internal structure of an organization experiencing growth or decline, as well. The cross-sectional studies in [5],[6],[11],and [14], and the longitudinal studies in [2],[7],[8], and [10] focus on explaining changes in the A/P ratio, (the ratio of administrative personnel to production personnel), as a function of changes in the size of the organization. In addition, in [7], it is suggested that the rate of growth or decline affects the A/P ratio. In this paper, using our normative approach, we derive conclusions that are similar to those in the descriptive literature cited above, except that we focus on explaining changes in the careers of cohort groups rather than changes in the A/P ratio.

The mathematical model that we have employed in this paper is the normative, longitudinal (or cohort), 'pull' formulation presented by Gaimon and Thompson in [4] that is solved using optimal control theory. In a comparison of the accuracy of 'push' versus 'pull' descriptive manpower models in imitating the actual flows of personnel, Stewman [12] and Konda and Stewman [9] have shown that 'pull' models (such as the control theory formulation in [4] that is used here) outperform the 'push' models in both intermediate and long-term testing. A summary of the results of [4] is presented in Section 2.

2. THE OPTIMAL POLICIES

In this section, we present a simplified version of the cohort personnel planning model in [4] that is used in this paper. The objective function is interpreted as the maximization of the effectiveness of personnel minus the costs associated with attaining the goal levels of manpower over the planning period. Costs are incurred due to separation, promotion, hiring, retiring, and salary of personnel. In addition, a cost is assigned to the squared deviation of the difference between the goal level and the actual level of personnel.

The state constraints of the model consist of a system of first order partial differential equations that depict the change in the number of persons in the organization (state variable) over time. Each equation describes the flow of personnel into a grade (through promotion-into and hiring), and out of a grade (through separation, promotion-out, voluntary departure, and retirement). We assume that promotion can occur only into the next successive grade, and we note that promotion cannot occur into the lowest grade nor out of the highest grade. In addition to the state constraints, we require that the control variables satisfy a constraint prohibiting promotion into a grade and separation from that same grade from occurring simultaneously for persons of the same organizational age. Lastly, the initial level of manpower is given for all organizational ages and grades.

Given the objective and constraints described above, Gaimon and Thompson derive the optimal separation, promotion, and hiring policies as expressed in Equations (1), (2), and (3) using the following notation, (see [3], and [4]).

t=time, $t \epsilon [0,T]$, T is the terminal time of the planning period.
y=a person's organizational age, $y \epsilon [0,Y]$, $y=0$ for persons being hired, $y=Y$ for persons being retired.
g=grade in the organization, $g=1,2,\ldots,G$, where G is the highest grade.
$x(t,y,g)$=number of persons of organizational age y, in grade g, at time t, (state variable).
$u(t,y,g)$=number of persons of organizational age y, at time t, being promoted from grade g into grade g+1, (control variable).
$v(t,y,g)$=number of persons of organizational age y, at time t, being separated from grade g, (control variable).
$k(t,y,g)$=number of persons of organizational age y who voluntarily depart from grade g at time t, (exogenous variable).
$E(y,g)$=effectiveness of a person of organizational age y, in grade g, (exogenous variable).
$\bar{x}(t,y,g)$=the goal level of manpower at time t, for persons of organizational age y, and in grade g, (exogenous variable).
$b(y,g)$=the salary cost of a person of organizational age y, in grade g, (exogenous variable).
$c(t,g)$=the value of a person retiring from grade g, at time t, (y=Y).
$d(y,g)$=the value of a person of organizational age y, in grade g, at the terminal time, (t=T).

p(g)=the cost of promoting a person from grade g into grade g+1, per unit squared promotion.

q(g)=the cost of separating a person from grade g, per unit squared separation.

w(g)=a positive constant representing the weight or cost of the squared deviation between x(t,y,g) and $\bar{x}$(t,y,g).

λ(t,y,g)=marginal value of a person of organizational age y, in grade g, at time t, (adjoint variable), derived to be a function of E(y,g), b(y,b), w(g), and $\bar{x}$(t,y,g).

$$v(t,y,g)= \begin{cases} -\lambda(t,y,g)/(2q), & \lambda(t,y,g)<0, \\ 0, & \text{otherwise}, \end{cases} \tag{1}$$

$$u(t,y,g)= \begin{cases} [\lambda(t,y,g+1)-\lambda(t,y,g)]/(2p), & \lambda(t,y,g+1)>\lambda(t,y,g) \\ & \text{and } \lambda(t,y,g+1)>0, \\ 0, & \text{otherwise}, \end{cases} \tag{2}$$

$$h(t,g)= \bar{x}(t,0,g)+(E-b-\lambda_t-\lambda_y)/(2w), \quad t\varepsilon[0,T], \; y=0, \; g\varepsilon[1,G]. \tag{3}$$

From Equation (1), it is clear that separation of personnel with organizational age y, from grade g, at time t occurs if the marginal value of personnel is negative. The optimal promotion policy described in Equation (2) can be interpreted as follows. A person of organizational age y at time t is promoted from grade g into grade g+1 if the marginal value of a person in grade g+1 is positive and exceeds the marginal value of a person in grade g. The expression in Equation (3) depicting the optimal hiring policy states that we hire an amount equal to the goal level of manpower plus adjustments made for the cost/value of personnel.

Given the optimal separation, promotion and hiring policies, the number of persons in the organization (state variable) is determined over all time and for persons of all organizational ages and grades. Furthermore, the number of persons reaching retirement from each grade is known over all time since it is simply the value of the state variable at the terminal organizational age. (For a more detailed discussion of these results see [3], [4].)

3. THE EFFECTIVENESS VERSUS COST OF PERSONNEL

In this Section, we examine the careers of cohort groups which result from changes in the goal levels of manpower for various effectiveness versus cost relationships of personnel. We implement the numerical solution technique of [4] to solve the discrete version of the model presented in Section 2. For simplicity, we assume that two grades exist, the retirement organizational age is ten years, and the planning horizon is five years. Furthermore, we

assume that the goal levels of manpower in grades one and two
decrease during the first half of the planning horizon, (from time 0
to time 2.5), and increase during the remainder of the planning
period, (from time 2.5 to time 5). Assuming that fewer persons are
desired in grade two (the higher grade) than in grade one, we have
the following.

$$
\bar{x}= \begin{cases}
225-25t, & t \in [0,2.5], \ y \in [0,10], \ g=1, \\
\bar{x}(2.5,y,1)+25(t-2.5), & t \in (2.5,5], \ y \in [0,10], \ g=1, \\
150-15t, & t \in [0,2.5], \ y \in [0,10], \ g=2, \\
\bar{x}(2.5,y,2)+15(t-2.5), & t \in (2.5,5], \ y \in [0,10], \ g=2.
\end{cases} \tag{4}
$$

Since decreases (increases) in the goal levels of manpower are
meant to reflect contraction (expansion) in the economy, the
functional representation of voluntary departure corresponding to
equation (4) is assumed to be

$$
k= \begin{cases}
16-2t, & t \in [0,2.5], \ y \in [0,10], \ g=1, \\
k(2.5,y,1)+2(t-2.5), & t \in (2.5,5], \ y \in [0,10], \ g=1, \\
10-t, & t \in [0,2.5], \ y \in [0,10], \ g=2, \\
k(2.5,y,2)+(t-2.5), & t \in (2.5,5], \ y \in [0,10], \ g=2.
\end{cases} \tag{5}
$$

Therefore, during an economic contraction (expansion) we expect
fewer (more) persons to voluntarily leave the organization. The
remainder of the exogenous variables, with the exception of the
effectiveness and salary, are defined below.

$$
x(0,y,g)= \begin{cases} 275, \ \text{for } g=1, \ y \in [0,10], \\ 175, \ \text{for } g=2, \ y \in [0,10], \end{cases} \tag{6}
$$

$$
d = w[\bar{x}(5.0,y,g)-\bar{x}(4.5,y,g)]+E-b, \ \ t=5, \ y \in [0,10], \ g=1,2, \tag{7}
$$

$$
c = \begin{cases} w[\bar{x}(t+.5,10,g)-\bar{x}(t,10,g)]+E-b, & t \in [0,4.5], \ y=10, \ g=1,2, \\ w[\bar{x}(5.0,10,g)-\bar{x}(4.5,10,g)]+E-b, & t=5, \ y=10, \ g=1,2, \end{cases} \tag{8}
$$

$$
w(1)=.125, \ w(2)=.200, \tag{9}
$$

$$
q(1)=1.0, \ q(2)=1.5, \tag{10}
$$

$$
p(1)=0.5. \tag{11}
$$

In addition, we define two Cases. In Case (1), the effectiveness
of personnel is differentiated by grade but not by organizational
age, whereas salary is a linearly increasing function of
organizational age and higher in grade two than in grade one.
Therefore, in Case (1), salary is the dominant factor in the
determination of a person's marginal value. In Case (2), the salary
of personnel is differentiated by grade but not by organizational
age, whereas effectiveness is a linearly increasing function of
organizational age and higher in grade two than in grade one. It

follows that a person's marginal value is dominated by the
effectiveness function in Case (2). Equations corresponding to Cases
(1) and (2) are written below.

Case (1):
$E(y,1)=20,$ $E(y,2)=20,$ $y\epsilon[0,10],$
$b(y,1)=.05+1.8y,$ $b(y,2)=.20+2.1y,$ $y\epsilon[0,10],$

Case (2):
$E(y,1)=.00+1.3y,$ $E(y,2)=.35+1.4y,$ $y\epsilon[0,10],$
$b(y,1)=3.5,$ $b(y,2)=4.5,$ $y\epsilon[0,10].$

To analyze the way in which the relationship between the
effectiveness and salary of personnel affects the success or failure
of the careers of cohorts in an organization whose exogenous
functions are expressed by Equations (4)-(11), we compare the
optimal solutions derived for Cases (1) and (2). A cohort group has
a successful career if its members achieve high levels of hiring,
promotion, and retirement, and low levels of separation. An
unsuccessful career is identified by the opposite behavior of the
optimal policies.

The optimal numbers of separations, promotions and the
subsequent numbers of remaining personnel are displayed in Tables
(1.1)-(1.3) and (2.1)-(2.3) corresponding to Cases (1) and (2),
respectively. In each Table, time is measured horizontally and
organizational age is measured vertically in increments of one year.
Since both time and organizational age vary together, to examine the
career of a particular cohort group, we simply follow the diagonal
entries (upward and to the right) in the Tables.

We first examine the careers of senior cohort groups, (persons
having organizational age exceeding four at the initial time). In
Case (1) where salary dominates the optimal policies, these cohorts
experience unsuccessful careers identified by high levels of
separation and no promotion due to high salaries. In Case (2) where
effectiveness is the dominant criteria, careers are mixed with
relatively small levels of initial separation followed by promotion.

Next, we examine the careers of junior cohorts (persons having
organizational age less than five at the initial time). We
characterize the careers of junior cohorts as successful in Case (1)
due to the lack of separation and the high degree of promotion, and
as unsuccessful in Case (2) due to the extensive separation in the
contraction period. The contrast between the careers of junior
cohorts in Cases (1) and (2) occurs because persons of lower
organizational age are relatively desirable in Case (1) due to low
salaries, and undesirable in Case (2) due to low effectiveness.

TABLE 1

OPTIMAL SOLUTION OF CASE (1) WHERE
SALARY IS DOMINANT FACTOR

TIME ——▶
| ORGANIZATIONAL AGE
▼

SEPARATIONS   PROMOTIONS
GRADE 1   GRADE 2

	0	1	2	3	4	5	0	1	2	3	4	5
10	0	0	0	0	0	0	0	1	1	0	0	0
9	5	6	3	0	0	0	4	4	3	0	0	0
8	9	8	4	0	0	0	6	5	2	0	0	0
7	11	8	0	0	0	0	6	5	0	0	0	0
6	9	4	0	0	0	0	5	2	0	0	0	0
5	6	1	0	0	0	0	3	0	0	0	0	0
4	3	0	0	0	0	0	1	0	0	0	0	0
3	0	0	0	0	0	0	0	0	0	0	0	0
2	0	0	0	0	0	0	0	0	0	0	0	0
1	0	0	0	0	0	0	0	0	0	0	0	0
0	0	0	0	0	0	0	0	0	0	0	0	0

Table (1.1)

	0	1	2	3	4	5
10	0	0	0	0	0	0
9	0	0	0	0	0	0
8	0	0	0	0	0	0
7	0	0	0	0	0	0
6	0	0	0	0	0	0
5	0	0	0	0	0	0
4	0	1	0	0	0	0
3	2	1	0	0	0	0
2	3	1	0	0	0	0
1	3	0	0	0	0	0
0	0	0	0	0	0	0

Table (1.2)

NUMBER OF PERSONS
GRADE 1   GRADE 2

	0	1	2	3	4	5	0	1	2	3	4	5
10	275	255	233	215	204	199	175	162	147	136	130	128
9	275	251	229	216	214	203	175	159	146	138	137	131
8	275	249	230	226	218	205	175	159	147	145	140	133
7	275	250	238	230	220	204	175	160	153	149	143	134
6	275	254	242	232	219	204	175	163	156	151	144	135
5	275	257	244	231	218	231	175	165	159	152	144	154
4	275	258	243	231	245	239	175	167	159	152	163	160
3	275	257	243	258	254	254	175	167	160	171	169	170
2	275	257	269	266	269	286	175	168	179	177	179	191
1	275	283	278	281	301	352	175	188	185	187	200	233
0	299	291	293	313	366	456	198	194	195	208	242	300

Table (1.3)

TABLE 2

OPTIMAL SOLUTION OF CASE (2) WHERE
EFFECTIVENESS FUNCTION IS DOMINANT FACTOR

TIME ⟶
| ORGANIZATIONAL AGE
▼

SEPARATIONS

	GRADE 1						GRADE 2					
	0	1	2	3	4	5	0	1	2	3	4	5
10	0	0	0	0	0	0	0	0	0	0	0	0
9	0	0	0	0	0	0	0	0	0	0	0	0
8	2	1	0	0	0	0	0	0	0	0	0	0
7	5	1	0	0	0	0	1	0	0	0	0	0
6	5	0	0	0	0	0	2	0	0	0	0	0
5	3	0	0	0	0	0	1	0	0	0	0	0
4	6	1	0	0	0	0	2	0	0	0	0	0
3	8	3	0	0	0	0	3	1	0	0	0	0
2	10	5	0	0	0	0	5	2	0	0	0	0
1	12	0	0	0	0	0	6	0	0	0	0	0
0	0	0	0	0	0	0	0	0	0	0	0	0

Table (2.1)

PROMOTIONS

	0	1	2	3	4	5
10	0	0	0	0	0	0
9	2	2	1	1	0	0
8	0	3	1	1	0	0
7	0	3	1	1	0	0
6	0	2	1	0	0	0
5	0	3	1	1	0	0
4	0	3	2	1	0	0
3	0	0	2	0	0	0
2	0	0	0	0	0	0
1	0	0	0	0	0	0
0	0	0	0	0	0	0

Table (2.2)

NUMBERS OF PERSONS

	GRADE 1						GRADE 2					
	0	1	2	3	4	5	0	1	2	3	4	5
10	275	258	242	227	213	201	175	167	159	151	144	136
9	275	257	239	227	216	199	175	166	158	151	145	134
8	275	256	239	229	214	197	175	164	158	152	143	132
7	275	256	241	227	211	194	175	164	159	151	142	130
6	275	257	240	224	209	192	175	166	158	150	139	127
5	275	256	237	222	207	158	175	164	156	147	136	108
4	275	253	235	220	173	156	175	163	153	144	117	106
3	275	251	233	185	170	157	175	161	150	125	115	106
2	275	249	197	183	171	167	175	160	132	123	116	112
1	275	211	195	184	182	180	175	141	131	124	122	119
0	227	208	195	194	194	187	150	140	131	130	128	122

Table (2.3)

Lastly, we examine the careers of cohorts hired into the organization during the current planning period. (Refer to Tables (1.3) and (2.3) with organizational age equal to zero.) These cohorts are not separated or promoted in either Cases (1) or (2) since hiring decisions are made with consideration given to the anticipated changes in the goal levels. In addition, we observe that the levels of hiring in Case (1) greatly exceed the corresponding levels in Case (2) due to the Case (1) emphasis on low salaries and the Case (2) concern for high effectiveness.

From the detailed analysis presented above, we can conclude that the effect of changing goal levels of manpower on the success or failure of a cohort's career is dependent upon the way in which an organization defines the relationship between the effectiveness and salary of personnel. In particular, we have shown the following.

1. In organizations which characterize effectiveness as the dominant factor, cohorts of higher organizational age are the most valuable and experience the more successful careers.

2. Conversely, when salary is identified as dominant, cohorts of lower organizational age are the most valuable and experience greater success in their careers.

4. THE INITIAL SIZE OF THE WORKFORCE

The effect that the size of the workforce has in determining the employee structure of an organization, (specifically the A/P ratio), has been investigated a good deal using descriptive approaches, [2],[5]-[8],[10], [11],and [14]. In this Section, to illustrate the impact that the initial size of the workforce has on the careers of cohorts, we analyze the optimal solutions derived for our normative model that is again characterized by goal levels of manpower that first decrease and then increase in both grades one and two. We keep the values of the exogenous functions defined in Equations (4),(5), and (7)-(11) unchanged, except for redefining the rate of change in the goal levels of manpower to be twenty in grade one (instead of twenty-five) and and ten in grade two (instead of fifteen). We define the effectiveness and salary of personnel as increasing functions of organizational age and higher in grade two as written below.

$$E(y,g)= \begin{cases} 15+.25y, & y\epsilon[0,10], \ g=1, \\ 27.5+.25y, & y\epsilon[0,10], \ g=2, \end{cases}$$

$$b(y,g)= \begin{cases} 0+2y, & y\epsilon[0,10], \ g=1, \\ 3+3y, & y\epsilon[0,10], \ g=2. \end{cases}$$

In addition, we introduce three Cases, each Case corresponding to a different combination of initial levels of manpower in grades one and two.

Case (1): $x(0,y,g) = \begin{cases} 230, & y\epsilon[0,10], & g=1, \\ 175, & y\epsilon[0,10], & g=2, \end{cases}$

Case (2): $x(0,y,g) = \begin{cases} 265, & y\epsilon[0,10], & g=1, \\ 175, & y\epsilon[0,10], & g=2, \end{cases}$

Case (3): $x(0,y,g) = \begin{cases} 265, & y\epsilon[0,10], & g=1, \\ 150, & y\epsilon[0,10], & g=2. \end{cases}$

While the initial size of the workforce in grade two is identically defined in Cases (1) and (2), the initial level of grade one manpower in Case (2) exceeds the corresponding level in Case (1). Therefore, we explore the effects on the careers of cohorts caused by different initial levels of grade one manpower by examining the optimal solutions derived for Cases (1) and (2). In addition, while the grade one levels of manpower are identically defined in Cases (2) and (3), the initial level of grade two manpower in Case (2) exceeds the corresponding level in Case (3). A comparison of the optimal solutions derived for Cases (2) and (3) illustrates the effects of the initial size of the workforce in grade two on the careers of cohorts.

The optimal solutions derived for Cases (1), (2), and (3), show that cohorts of senior organizational age experience separation, whereas cohorts of junior organizational age experience promotion. This occurs because the consideration of salary (not effectiveness) is the dominant factor in the optimization of these Cases, (see Section 3).

First, we compare the solutions derived for Cases (1) and (2) to examine the effect that the initial level of grade one manpower has on cohort groups. During the period of contraction, senior cohorts experience separation from grade two at equal levels in Cases (1) and (2) since all exogenous functions related to grade two are identically defined. Senior cohorts experience more separation from grade one in Case (2) than in Case (1) because although the goal levels of manpower are identical in both Cases, the initial level of grade one manpower is greater in Case (2). Senior cohorts do not meet with promotion in either Cases (1) or (2). We conclude that senior cohorts do not have successful careers in either Cases (1) or (2).

Next, in examining the careers of junior cohorts we observe that separation from grades one or two does not occur in either Case (1) or (2). Although promotion occurs only at time five in Case (1), in

Case (2), junior cohorts experience promotion even during the period of contraction. The promotion in Case (2) occurs because the marginal value of an additional person in grade two exceeds the corresponding value in grade one due to the large initial stock of personnel in grade one. We can thus characterize the careers of junior cohorts as moderately successful in Case (1), and as successful in Case (2).

The optimal solutions derived for Cases (1) and (2) are identical for all cohorts hired into the organization during the current planning period since changing the initial level of manpower in the organization can have no effect on cohorts hired into the organization during the current planning period.

So far, we have analyzed the effects of changing the initial level of the workforce in grade one in conjunction with goal levels of manpower that first decrease and then increase in both grades one and two, (Cases (1) and (2)). A comparison of the solutions derived for Cases (2) and (3), illustrates the effect that different initial levels of personnel in grade two have under identical changes in the goal levels of manpower. The detailed analysis of the careers of cohort groups in Cases (2) and (3) can be found in [3].

To conclude this Section, we can make the following observations regarding the careers of cohorts in organization whose goal levels of manpower decrease and then increase in both grades one and two.

1.   If the initial level of manpower in grade one is relatively large, then the optimal solutions require the movement of personnel out of grade one and either into grade two (promotion) or out of the organization (separation).

2.   If the initial level of manpower in grade two is relatively small, then the optimal policies require less separation from grade two and more promotion into grade two.

The converse of both 1. and 2. above holds, as well. In addition, the actual decision as to which cohort groups meet with promotion or separation is determined by the salary versus effectiveness relationship.

5. THE RATE OF GROWTH AND DECLINE

In Section 4, we examined the effects on the careers of cohorts caused by the interaction between changing goal levels of manpower and various initial values of the workforce. In [7], the descriptive approach of Hendershoot and James suggests that changes in the personnel structure of an organization (specifically the A/P ratio) cannot be fully explained by size, and that the rate of growth in an organization must also be taken into consideration. In this Section, we derive results in agreement with those of Hendershoot and James.

Our normative model is solved for different rates of change in the goal levels of manpower in grade two, and the subsequent effects on the careers of cohorts are examined. The exogenous functions defined in Case (1) of Section 3 remain unchanged with the following exceptions which characterize Cases (1) and (2) of this Section.

Case (1): $\bar{x}(t,y,2) = \begin{cases} 150-30t, & t_\epsilon[0,2.5], \ y_\epsilon[0,10], \\ x(2.5,y,2)+30t, & t_\epsilon(2.5,5], \ y_\epsilon[0,10]. \end{cases}$

Case (2): $\bar{x}(t,y,2) = \begin{cases} 150-5t, & t_\epsilon[0,2.5], \ y_\epsilon[0,10], \\ x(2.5,y,2)+5t, & t_\epsilon(2.5,5], \ y_\epsilon[0,10]. \end{cases}$

Therefore, the Case (1) rates of decline and growth exceed the Case (2) values, while the goal levels of manpower at the initial and terminal times are identical in both Cases.

In the optimal solutions derived for Cases (1) and (2), respectively, we note that senior cohorts are the first to be separated, and junior cohorts are the first to be promoted since our exogenous functions characterize salary as dominating effectiveness in the process of optimization, (see Section 3).

The separation of senior cohorts from grade one occurs at almost identical levels in Cases (1) and (2) since the goal levels in grade one are the same in both Cases. However, while separation from grade two is restricted to senior cohorts and occurs in small amounts in Case (2), extensive separation from grade two is experienced by senior and junior cohorts in the Case (1) solution. Clearly, the severe separation from grade two in Case (1) is a response to the rapidly decreasing goal levels of manpower in that grade.

The most marked difference between the Case (1) and (2) solutions can be observed in the optimal promotion policies. In both Cases, the junior cohorts are the first to experience promotion due to their relatively lower salaries. In Case (1), promotion occurs throughout the period of expansion since the goal levels of manpower increase more rapidly in grade two than in grade one. In contrast, in Case (2), we see that promotion occurs during the period of declining goal levels. To explain the Case (2) solution, we note that during the contraction period, the goal levels in grade two decrease at an almost negligible rate causing the marginal values of the junior cohorts in grade two to remain positive and greater than the corresponding marginal values in grade one, so that promotion occurs. Later, due to this promotion during the contraction period in Case (2), it is unnecessary to respond to the almost negligible increase in the goal levels of manpower in grade two in the expansion period and therefore, promotion does not occur.

From the examples of this Section, we can conclude the following.

1. When the goal levels of manpower are subject to rapid changes over time, the optimal policies advocate severe responses in terms of large amounts of separation or promotion.

2. In contrast, when the goal levels of manpower change slowly, we observe stable (small) responses in the optimal policies.

In fact, since both Cases (1) and(2) begin and end with identical goal levels of manpower in both grades one and two, we can conclude that compared to goals that change slowly over time, rapidly changing goal levels cause unstable behavior (a large amount of separation followed by a large amount of promotion) throughout the organization. The particular cohort group that experiences separation or promotion is determined by the interaction between the time and direction of the changing goal levels, the salary versus effectiveness relationship as   defined by the organization, and the initial levels of manpower in each grade.

6. CONCLUSION

In this paper, we have assumed that changes in the economy or technology can be translated into changes in the future goal levels of manpower. We have examined the effects on the careers of cohorts in response to the interaction between goal levels of manpower corresponding to a period of contraction followed by a period of expansion, and (1) the effectiveness versus salary relationship, (2) the initial size of the workforce, and (3) the rate of change in the goal levels of manpower. A more detailed analysis of the results presented in this paper can be found in [3].

In Section 3, we demonstrate that the success or failure of a cohort's career in response to changing goal levels of manpower depends on the functional representations of salary and effectiveness as defined by the organization. For organizations in which effectiveness dominates salary considerations, we find that cohorts of senior organizational age experience the more successful careers. Naturally, the converse is true, as well.

In Section 4, we present results that illustrate the
relationship between changing goal levels of manpower and the
initial size of the workforce. We show that if the initial level of
manpower in a grade is relatively large, then the optimal solution
advocates the movement of personnel out of the grade and either into
the next grade (promotion-out) or out of the organization
(separation). In addition, little movement occurs into the grade,
(limited promotion-in). If the initial level of manpower in a grade
is relatively small, then the optimal policies advocate less
movement out of the grade, (less separation and promotion-out), and
more movement into the grade, (more promotion-in). Furthermore, the
converse holds for each of the conclusions. The specific cohort
groups that experience promotion or separation are dependent on the
interaction between the changing goal levels of manpower and the
salary versus effectiveness relationship.

Finally, in Section 5, we conclude that when the goal levels of
manpower are rapidly changing, the optimal separation and promotion
policies have strong responses that correspond to the direction and
time of the changing goal levels. When goal levels of manpower
change slowly, the optimal solutions advocate less extreme and more
stable optimal policies. Again, the particular cohort group that
experiences separation or promotion is dependent on the salary
versus effectiveness relationship and the initial level of manpower
in each grade.

REFERENCES

[1] Bartholomew, D.J., _Stochastic Models For Social Processes_, John
     Wiley & Sons, NY, 1967.
[2] Freeman, J.H., M.T. Hannan, "Growth and Decline Processes in
     Organizations," _American Sociological Review_, 40 (1975), pp.
     215-228.
[3] Gaimon, C., "Analysis of Career Response to Fluctuations in the
     Economy," W.P., GSIA, Carnegie-Mellon University, Pittsburgh,
     PA, May, 1981.
[4] Gaimon,C. ,G.L. Thompson, "A Distributed Parameter Cohort
     Personnel Planning Model," W.P.#44-80-81, GSIA, Carnegie-Mellon
     University, Pittsburgh, PA, March, 1981.
[5] Haas, E., R.H. Hall, N.J. Johnson, "The Size of the Supportive
     Component in Organizations: A Multi-Organizational Analysis,"
     Social Forces, 42(1963-4), pp.9-17.
[6] Hawley, A.H., M.Boland, W.Boland, "Population Size and
     Administration in Institutions of Higher Education," _American
     Sociological Review_, 30(1965), pp.252-255.
[7] Hendershoot, G.E., T.F.James, "Size and Growth as Determinants
     of Administrative-Production Ratios in Organizations," _American
     Sociological Review_, 37(1972), pp.149-153.

[8] Holdaway, E.A., T.A.Blowers, "Administrative Ratios and
     Organization Size: A Longitudinal Examination," American
     Sociological Review, 36(1971), pp.278-286.
[9] Konda, S.L., S.Stewman, "An Opportunity Labor Demand Model and
     Markovian Labor Supply Models: Comparative Tests in an
     Organization," American Sociological Review, 45(1980),
     pp.276-301.
[10] Meyer, M.W., "Size and the Structure of Organizations: A Causal
     Analysis," American Sociological Review, 37(1972), pp.434-441.
[11] Rushing, W.A., "The Effects of Industry Size and Division of
     Labor on Administration," Administrative Science Quarterly,
     12(1967-8), pp.273-295.
[12] Stewman, S., "Markov Renewal Models for Total Manpower System,"
     OMEGA, The International Journal of Management Science,
     6(1978), pp.341-351.
[13] Stewman, S., "The Aging of Work Organizations: Impact on
     Organization and Employment Practice," forthcoming in The
     Elderly of the Future, National Academy of Sciences.
[14] Terrien, F.W., D.L.Mills, "The Effect of Changing Size Upon the
     Internal Structure of Organizations," American Sociological
     Review, 20(1955), pp.11-13.

A GOAL-FOCUSING APPROACH TO MANPOWER PLANNING

IN GRADED ORGANIZATIONS

Michael J. Feure

Drexel University
Dept. of Management and Organizational Sciences
Philadelphia, Pa. 19104

ABSTRACT

A manpower planning framework is developed for organizations
in which staffing targets are ambiguous due to the absence of a
well-specified production function. A variation of goal-programming
is introduced to guide the selection of hiring and promotion rates,
which are control parameters in a familiar Markov transition model.
Unlike applications that treat well-defined but mutually inconsistent
grade size targets, the goal-focusing model addresses the problem of
numerous feasible solutions. Attention is focused on a subset of
policy combinations with desirable properties, such as stability
of promotion rates, which are evaluated with respect to implied
interim and long-range staffing levels. The model is illustrated
by simulation of five-year planning scenarios for a university
faculty.

1. INTRODUCTION

Hierarchical manpower planning models generally address one of
two conceptual problems: (1) Assuming internal promotion rates are
stationary or vary in a predictable fashion, what are the grade-
size distributional consequences of alternate hiring and termination
policies? (2) Inversely, what are the promotion, recruitment, and
termination policies required to attain a defined grade-size distribution of staff in a specified amount of time?

This paper introduces a manpower planning approach that, while
based upon a familiar Markovian characterization of hierarchical
personnel flows, departs from the assumptions and applicability of
typical projection and control type models. Instead, our concern
here shifts to organizational settings in which (1) internal rates

359

of mobility are not stationary but fluctuate as a function of
grade-size requirements, hiring patterns, and attrition behavior;
(2) grade-size targets cannot, or need not, always be specified
exactly, especially if ambiguous organizational output measures
rule out the specification of a production function in the conventional sense; and (3) properties of personnel mobility, such as
equality or stability of promotion opportunities, are conceived
as organizational objectives to be achieved subject to budgetary
and procedural constraints.

Section 2 begins with a single transition Markov model of
personnel flows in a graded hierarchy, which is then extended to
treat multiple period staffing. A goal programming variation is
then developed in Section 3, in order to focus attention on tradeoffs
among processes of mobility and staffing outcomes. In Section 4 the
model is applied to the evaluation of academic staffing scenarios in
a university, and conclusions are summarized.

2.   MODEL STRUCTURE

Consider an organization of $\underline{k}$ grades in which mobility follows
a pattern of sequential steps with no skipping of grades and no demo-
tions. Allowing for total staff growth or contraction as well as
hiring and attrition at all levels, the intertemporal realignment of
personnel in the various grades can be characterized by

$$n_i(t+1) = n_i(t)p_{ii}(t) + n_{i-1}(t)p_{i-1,i}(t) + r_i(t)R(t) \qquad [1]$$

and $\quad n_i(t) \quad = n_i(t)p_{ii}(t) + n_i(t)p_{i,i+1}(t) + n_i(t)q_i(t) \qquad [2]$

where $n_i(t)$ denotes the population of grade i at time t; $p_{ij}(t)$ is
the rate of transition between grades i and j during period t, defined as the proportion of persons in grade i at time t that is in
grade j=i+1 at time t+1; $r_i(t)$ is the share of total recruits R(t)
entering grade i between time t and t+1, so that $\sum_{i=1}^{k} r_i(t) = 1$,
and it is understood that new recruits are all treated as though
they enter together at the beginning of period t+1; and $q_i(t)$ is
the rate of population attrition from grade i between time t and
t+1.

Equation [1] accounts for the inflow of personnel and defines
the next-period grade size in three terms--retentions, promotions
in from the grade below, and external recruitment into the grade,
respectively. Equation [2] accounts for outflows of personnel by
defining the current population in grade i in terms of expected
retentions, promotions out to the grade above, and number of quits.

To allow for multiple-period transitions the model adopts the following timing convention. Transition periods correspond to defined calendar periods, usually years or budget cycles. While individuals can be hired, promoted, or terminated at various times, a period is formally defined by the half-open interval $t = [t, t+1)$. Thus, the personnel distribution observed at time t remains in effect during the entire period up to but not including time t+1. A total of T interim transition periods are accounted for, beginning at point t=1 and ending with the final accounting point at time T+1.

Considering first a single transition, we subtract [2] from [1] to obtain

$$n_i(t+1) - n_i(t) = n_{i-1}(t)p_{i-1,i}(t) - n_i(t)p_{i,i+1}(t)$$
$$+ r_i(t)R(t) - n_i(t)q_i(t) \qquad [3]$$

where the change in population in grade i between point t and point t+1 is accounted for in terms of net internal mobility--promotions into the grade less promotions out--plus net external exchange--new recruits to the grade less the number terminating. (The concept of net exchange is explored in Feuer and Schinnar 1978.) A matrix account is obtained by first rearranging terms in [3] and writing

$$n_{i-1}(t)p_{i-1,i}(t) - n_i(t)p_{i,i+1}(t) + r_i(t)R(t) =$$
$$n_i(t+1) - n_i(t)s_i(t) \qquad [4]$$

where $s_i(t) = 1-q_i(t)$ denotes the survival rate in grade i during period t. This equation can now be extended to reflect the realignment of personnel in the entire organization, by writing

$$N(t)p(t) = g(t+1) \qquad [5]$$

where
$$N(t) = \begin{bmatrix} -n_1(t) & 0 & \cdot & \cdot & \cdot & 0 & r_1(t) \\ n_1(t) & -n_2(t) & \cdot & & & \cdot & r_2(t) \\ 0 & n_2(t) & \cdot & \cdot & & \cdot & \cdot \\ \cdot & & \cdot & \cdot & \cdot & \cdot & \cdot \\ \cdot & & & \cdot & \cdot & 0 & \cdot \\ \cdot & & & & \cdot & -n_{k-1}(t) & r_{k-1}(t) \\ 0 & \cdot & \cdot & \cdot & 0 & n_{k-1}(t) & r_k(t) \end{bmatrix}$$

$$p(t) = \begin{bmatrix} p_{12}(t) \\ p_{23}(t) \\ \cdot \\ \cdot \\ \cdot \\ p_{k-1,k}(t) \\ R(t) \end{bmatrix} \quad \text{and} \quad g(t+1) = \begin{bmatrix} n_1(t+1) - n_1(t)s_1(t) \\ n_2(t+1) - n_2(t)s_2(t) \\ \cdot \\ \cdot \\ \cdot \\ n_k(t+1) - n_k(t)s_k(t) \end{bmatrix}$$

For a single transition, [5] lends itself to the analysis of non-stationary promotion opportunities, by solution of

$$p(t) = [N(t)]^{-1}g(t+1) \qquad\qquad\qquad\qquad [6]$$

(for $N(t)$ nonsingular) and to their sensitivity under varying conditions of recruitment, attrition, and grade-size targets. (A formal treatment of sensitivity analysis in this context is found in Feuer and Schinnar, 1978, and Feuer, 1980.)

Our concern here, however, is with multiple-period mobility. We proceed by extending [5], viz:

$$
\begin{aligned}
N(1)p(1) &= g(2)\\[4pt]
N(2)p(2) &= g(3)\\
&\;\;\vdots\\
N(T)p(T) &= g(T+1)
\end{aligned}
\qquad\qquad [7]
$$

where a total of T transitions, spanning T+1 time periods (from t=1 to t=T+1), are accounted for simultaneously. Note that when interim and final grade sizes $n_1(t)$, t=2,...,T+1, are given, [7] becomes a sequence of single transitions, in which case periodic promotion opportunities may be computed by iterative application of [6].

To allow for variation in interim grade stocks, however, we rewrite [7] as a general accounting system in matrix form, i.e.,

$$N(t)p(t) + S(t)n(t) - n(t+1) = \theta \quad (t = 1, 2, \ldots, T) \qquad [8]$$

where $\quad S(t) = \begin{bmatrix} s_1(t) & 0 & . & . & . & 0 \\ 0 & s_2(t) & & & & \\ . & & . & & & \\ . & & & . & & \\ . & & & & . & \\ 0 & & & & & s_k(t) \end{bmatrix}$

$\qquad n(t) = \begin{bmatrix} n_1(t) \\ n_2(t) \\ . \\ . \\ . \\ n_k(t) \end{bmatrix}$

and θ is the k-order column vector of zeros.

While [8] is generally nonlinear, we shall treat both attrition and recruitment parametrically, allowing for a linear reformulation by substituting

$$f_{ij}(t) = n_i(t)p_{ij}(t) \tag{9}$$

where $f_{ij}(t)$ denotes the flow--in number of persons--from grade i to grade j during transition t, and by introducing a new term

$B(t) = \begin{bmatrix} -1 & 0 & . & . & . & 0 & r_1(t) \\ 1 & -1 & 0 & . & . & . & r_2(t) \\ 0 & 1 & -1 & . & & & . \\ . & . & . & . & & & . \\ . & . & . & . & . & & . \\ . & & & & . & -1 & r_{k-1}(t) \\ 0 & . & . & . & 0 & 1 & r_k(t) \end{bmatrix}$

which is a matrix similar in structure to N(t) but with ones in the place of grade stock parameters n(t). The linear system, which becomes the basis for our goal-programming model, closely resembles the conservation-of-flows model of Grinold and Marshall (1977), and is illustrated in Figure 1 for a hypothetical 3-grade organization.

3. GOAL FOCUSING

If interim grade stocks can be specified as numerical targets, [8] becomes a simultaneous system of $\underline{kT}$ equations in $\underline{kT}$ unknowns, as shown in Figure 1. However, a solution with strictly nonnegative promotions and hiring may not exist, under a given recruitment/attrition combination, if grade targets in any period are incompatible. This corresponds to the problem of "attainability" discussed by Bartholomew (1973) and Feuer and Schinnar (1978). If the organization's objectives can be defined as the attainment of specified grade targets $\underline{as}$ $\underline{closely}$ $\underline{as}$ $\underline{possible}$, the model is directly amenable to a goal programming formulation (see also Charnes and Cooper, 1977 and Charnes, Cooper, et al, 1975) in which promotion policies for each period in the planning horizon are chosen in order to minimize deviations from stipulated grade-size targets.

In many organizations, however, it may be impossible or undesirable to specify grade targets as objectives, for reasons pertaining

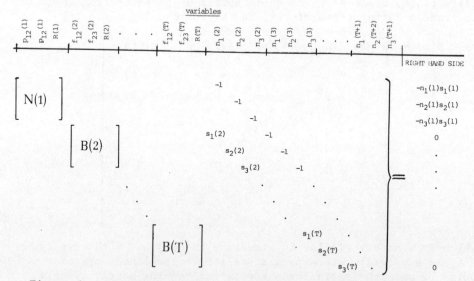

Figure 1. Linear Multiple Period Model of Personnel Flow
In a Three Grade Hierarchical Organization

to the organization's characteristics and to its external environment. First, typically in public-sector organizations, the absence
of a production function that relates output to required manpower
inputs rules out a precise numerical specification of periodic
grade targets in advance. While this problem is most acute if output measures are ambiguous--such as in universities--there is a
large class of organizations in which periodic manpower requirements can at best be specified only within certain ranges. And
even in firms where output is clearly defined, unpredictable external labor-supply conditions may weaken substantially the usefulness
of advance planning that is based on precise grade-size targets.
For these firms, too, it may be preferable to think of periodic
grade distribution of staff in terms of desirable ranges, allowing
for fluctuation as a function of exogenous changes and/or the
specification of changing internal objectives.

Second, even if long-range labor requirements can be predicted
accurately, management must take into account the short and medium-
term implications of policies executed to achieve them. For example, if productivity is linked to expectations of promotion, then
stability of advancement opportunities may be viewed as a goal
of the transition process. In certain organizations the concentration of individuals in various ranks is important, suggesting that
a desirable transition process maintain certain specifiable inter-
grade ratios. Similarly, one's tenure in a particular grade, especially the lower ones, frequently influences organizational
loyalty and other, productivity-related, variables, so that advancement opportunities should be structured not only with respect to
stability but with respect to grade-specific duration as well.
Finally, the assurance of equal opportunities for advancement of
various population groups necessitates an explicit accounting of
the processes by which certain internal labor force distributions
are attained. If affirmative action is defined in certain cases
as preferential hiring and promotions policies, these must be
incorporated as goals along with the desirable grade size distribution of staff.

To accomodate these sorts of concerns we make use of [8] in a
framework suggested by Charnes, Cooper, et al (1979), whose analysis of household production outputs bears resemblance to characteristics of the manpower problems raised here. In their model, house-
holds transform input resources, i.e., private or public consumption
goods and services, into outputs such as life expectancy and educational achievement. In theory, the household utility function
would select certain efficient input combinations from the larger
production possibility set. In the absence of a fully-specified
utility function, however, the authors develop "goal artifacts" that
represent certain desirable output ratios, and incorporate these as
goal statements in a goal programming framework. For example, desired
preference relations may be of the variety "2 weeks vacation for every

25 weeks of employment," constraints which, in combination with expressions to establish upper and lower bounds and other solution properties, are employed to "focus" in on particular impact combinations.

In a similar fashion, we note first that by treating interim
grade stocks as variables, our model consists of $\underline{kT}$ equations in
$\underline{2kT}$ unknowns, suggesting the existence of a wide range of efficient
solutions; and second, a well-specified and operational utility
function (or production function) may not exist. Following the
goal-focusing approach, then, we proceed by introducing a series
of goal artifacts that reflect properties of desirable interim and
final staffing distributions and of the structure of advancement
opportunities; by including also a payroll budget constraint the
model enables an evaluation of tradeoffs among feasible promotions/
hiring policies.

Goal artifacts and constraints are grouped in six sets of
linear equations, as follows:

$$N(1)p(1) - n(2) = -S(1)n(1) \qquad\qquad\qquad [10.1]$$

$$B(t)f(t) + S(t)n(t) - n(t+1) = \theta \ , \ t = 2, 3, \ldots, T \qquad [10.2]$$

$$f(t) - n(t) \ \leq \ \theta, \qquad t = 2, 3, \ldots, T \qquad\qquad [10.3]$$

$$\phi[n(t)] - \delta g^+ + \delta g^- = \phi*(t), \ t = 2, 3, \ldots, T+1 \qquad [10.4]$$

$$\psi[p(1);f(t)] - \delta p^+ + \delta p^- = \psi*(t), \ t = 2, 3, \ldots, T \quad [10.5]$$

$$c(t) - \delta b^+ + \delta b^- = c*(t), \ t = 2, 3, \ldots, T+1 \qquad [10.6]$$

The first three sets of equations and inequalities are constraints that reflect the structure of intertemporal staff mobility.
We assume initial grade stocks $n(1)$ are given, so that [10.1] is
linear in promotion rates $p(1)$ and has attrition accounted for on
the right hand side. For transition periods $t = 2, \ldots, T$, represented in [10.2], grade stocks are treated as unknowns with given
coefficients of survival. To effect the linear substitution of
person-flows for rates of flow we let $f(t)$ denote the k-order column
vector of flow rates $f_{ij}(t)$. Rates of promotion are constrained by
[10.3] to the range $p_{ij}(t) \leq 1$.

The equations in [10.4] denote goal artifacts for grade size
distribution of staff during the various periods. The form of ϕ
will depend on definitions of desirable grade distribution of staff.
For earlier periods, those for which planners may have more precise
information, precise targets may sometimes be entered. In general,
however, ϕ will reflect certain desirable properties of the grade
distribution, such as inter-grade ratios of personnel. An example,
as illustrated in section 4 below, is the desire in universities to

maintain a "tenure fraction," i.e., a proportion of total faculty
with tenure that reflects certain educational standards; by introducing these goal artifacts, the range of transition patterns nar-
rows towards solutions with particular properties. Note that the
goal artifact statements include so-called "deviational variables,"
δ^+, δ^-, which allow for incompatibility across goal artifacts and
the derivation of compromise solutions in accordance with a rank-
ordering reflected in the goal functional (discussed below).

Similarly, [10.5] are goal artifacts for advancement opportunities. Promotion rates may be difficult to specify, while their
properties are frequently the subject of planning and discretion.
As suggested earlier, stability of promotion opportunities is a
reasonable objective, from the perspective of management as well
as labor, and will be illustrated in our empirical example. Note
that the specification of stable promotion rates as an objective
departs from many sociological applications in which stationarity
is assumed or tested (e.g., Stewman, 1975). Deviation variables
allow for examination of tradeoffs among goals defined for properties
of advancement and other organizational goals.

Finally, [10.6] denotes a budget constraint, for which we introduce a vector of grade specific wages $w(t) = [w_1(t)\ w_2(t)\ \dots\ w_k(t)]$
so that $c(t) = w(t)n(t)$ gives the total payroll expenditure associated with the grade size distribution of personnel in period t. By
including the target payroll level $c^*(t)$ we allow for budget decisions
to be viewed as exogenous constraints on the manpower planning process.
Note, however, that unlike typical economic treatments of a budget
constraint, ours is not a constraining upper bound, but rather allows
for examination of costs of certain manpower objectives in terms of
budget overruns. In this manner, manpower plans can be evaluated
with respect to direct dollar costs—necessitating, for example,
deficit financing in certain periods—as well as with respect to
tradeoffs in personnel outcomes.

Given these goal artifacts, or constraints, we introduce a
goal functional (see Ijiri, 1965 or Charnes and Cooper, 1977), or
objective function, whose general form is

$$\min \quad \partial_1 M_{v1} \delta_1^+ + \partial_2 M_{v2} \delta_2^+ + \partial_3 M_{v3} \delta_3^+$$
$$+ \partial_4 M_{v4} \delta_1^- + \partial_5 M_{v5} \delta_2^- + \partial_6 M_{v6} \delta_3^- \qquad [11]$$

In this expression M_{vi} are "preemptive priority factors" that rank-
order whole sets of goals, ∂_i are weights for intra-group
ranking, and δ_i^+, δ_i^- are column vectors of deviation variables
for the various categories of goals (grade stocks, promotion flows,
and the payroll). To illustrate, consider an example in which a
certain property of the interim grade distribution is valued more

highly than a desirable range of promotion rates, but both goals
are secondary in importance to the budget constraint. The goal
functional of [11] would then have the form

$$\min \quad M_1 \delta_b^+ + M_2 [\, \partial_g \delta_g^+ + \partial_g \delta_g^- + \partial_p \delta_p^+ + \partial_p \delta_p^- \,] \quad [12]$$

where $\partial_g > \partial_p$, and $M_1 >>> M_2$ denotes the preemptive priority of

attaining category 1 first and then pursuing category 2. We turn
now to an illustration of the model as applied to faculty manpower
planning in a large university.

4.   EMPIRICAL ILLUSTRATION

     In this section we consider two simulated five-year personnel
plans at a major university. Table 1 shows the sample data, consisting of the 1978 distribution of academic staff in the three ranks of
assistant, associate, and full professor; the average recruitment and
attrition rate for the 1972-1978 period; and average annual salaries
by rank in 1978.

     We proceed by writing goal artifacts for interim grade stocks,
promotion opportunities, and the payroll budget. (While the structural constraints in [10.1], [10.2], and [10.3] are included in the
actual simulation their form in the example is omitted for brevity.)
Let us suppose it is desirable to maintain a 70 percent tenure
policy, which reflects the past decade at our sample university.
We write the goal artifact in terms of a ratio of tenured to non-
tenured faculty, i.e.,

$$-2.3n_1(t) + n_2(t) + n_3(t) - d_1^+(t) + d_1^-(t) = 0 \qquad [13]$$

$$t = 1979, 1980, \ldots, 1983$$

where subscripts 1, 2, 3 denote assistant, associate, and full professor ranks, respectively, $d_1^+(t)$ and $d_1^-(t)$ are deviation variables
for over and under attainment of the goal, respectively, and the
2.3:1 ratio corresponds to 70 percent tenured faculty.

     The goal of stability in promotion opportunities is written in
two sets of constraints, for flows into tenure (from assistant to
associate) and within tenure (from associate to full), respectively.
The general form of these goal statements is

$$f_{12}(t) - f_{12}(t+1) - d_{12}^+(t) + d_{12}^-(t) = 0 \qquad [14.1]$$

$$\text{and} \quad f_{23}(t) - f_{23}(t+1) - d_{23}^+(t) + d_{23}^-(t) = 0 \qquad [14.2]$$

where $d_{ij}^+(t)$ and $d_{ij}^-(t)$ are deviation variables for fluctuation

Table 1.   Sample Data from University
            Faculty Personnel Files

	Assistant Professor	Associate Professor	Full Professor
1978 Population	117	98	222
Attrition rate, average, 1972–78	.114	.015	.033
Hiring rate, average, 1972–78	.84	.06	.10
Average Salary	$16,107	$21,597	$32,705

in flows between grades i and j between periods t and t+1. Note that for each goal three points in time (or two transition periods) are covered. In our example, these constraints are entered as

$$117 p_{12}(1) - f_{12}(2) - d_{12}^{+}(1) + d_{12}^{-}(1) = 0$$

$$f_{12}(2) - f_{12}(3) - d_{12}^{+}(2) + d_{12}^{-}(2) = 0$$

$$\vdots$$

[15.1]

$$f_{12}(4) - f_{12}(5) - d_{12}^{+}(4) + d_{12}^{-}(4) = 0$$

and

$$98 p_{23}(1) - f_{23}(1) - d_{23}^{+}(1) + d_{23}^{-}(1) = 0$$

$$f_{23}(2) - f_{23}(3) - d_{23}^{+}(2) + d_{23}^{-}(2) = 0$$

$$\vdots$$

[15.2]

$$f_{23}(4) - f_{23}(5) - d_{23}^{+}(4) + d_{23}^{-}(4) = 0$$

where transition periods are denoted 1 though 5 for brevity.

Finally, we assume the desirability of constant payroll expenditures, which is incorporated as a goal statement of the form

$$16{,}107 n_1(t) + 21{,}597 n_2(t) + 32{,}705 n_3(t)$$

$$-d_3{}^{+}(t) + d_3{}^{-}(t) = \$11{,}261{,}535 \qquad [16]$$

In this fashion we pursue a distribution of personnel in the three ranks that attains as closely as possible the total payroll in 1978, shown on the right hand side of [16].

In the first simulation we investigate a scenario in which all goals are weighted equally, i.e., with the goal functional

$$\min \quad \sum_{t=2}^{6} \; d_1{}^{+}(t) + d_1{}^{-}(t)$$

$$+ \quad \sum_{u=1}^{5} \; d_{12}{}^{+}(u) + d_{12}{}^{-}(u) \;\; + \sum_{u=1}^{5} d_{23}{}^{+}(u) + d_{23}{}^{-}(u) \qquad [17]$$

$$+ \quad \sum_{t=2}^{6} \; d_3{}^{+}(t)$$

where pairs of transition periods (t, t+1) are indexed u = 1, ..., 5. Note also that underdeviation from the budget ceiling is not undesirable, so that $d_3{}^{-}(t)$ terms are omitted.

The results of this simulation are presented in Table 2 and demonstrate that all goals cannot be attained simultaneously. The solution that is "cheapest," in terms of total deviations, necessitates a gradual increase of flows into tenure, starting with the second transition (1979–1980); and instability of flows within tenure (an increase of 9) in the same period. While the constant payroll expenditure constraint is satisfied exactly, the desired tenure fraction is exceeded in the first year, and falls slightly short in the following years. We note also that upper grade promotions and tenure appointments undergo a freeze in the early years, while in the last year a positive level is restored.

We consider now the costs of stipulating a preemptive priority for the attainment of the desired tenure fraction. Assuming again that this objective may be incompatible with a stable promotions policy and/or with the desirable level of payroll expenditure, we minimize a goal functional of deviation variables, weighted so that the tenure fraction goals are attained as closely as possible first, followed by goals for stability and the budget. The goal functional is given by

$$\min \quad M_1 \quad [d_1^+(t)]$$

$$+ \quad M_2 \left\{ \sum_{t=2}^{6} d_1^+(t) + \sum_{t=2}^{6} d_1^-(t) \right.$$

$$[18]$$

$$+ \quad \sum_{u=1}^{5} [d_{12}^+(u) + d_{12}^-(u)]$$

$$+ \quad \sum_{u=1}^{5} [d_{23}^+(u) + d_{23}^-(u)] + \sum_{t=2}^{6} d_3^+(t) \left. \right\}$$

where all goals except the 1979 tenure ratio are grouped in the second priority class.

The effects of this policy are presented in Table 3, where we see a serious budget deficit incurred as a result of rigidly enforcing the tenure policy. The cost of attaining a distribution with precisely 70 percent of the total faculty tenured is approximately

Table 2. Results of Five-Year Plan under
Equally Weighted Preferences

A. Promotion Opportunities

	Assistant to Associate		Associate to Full	
	Flow	Rate*	Flow	Rate
1978-79	0	0	0	0
1979-80	0	0	0	0
1980-81	3	.02	9	.09
1981-82	4	.03	9	.10
1983-83	4	.03	9	.10

B. Grade Size Distribution

	Assistant	Associate	Full	Tenure Fraction	Budget Deficit
1978 (given)	117	98	222	73.2%	0
1979	126	98	217	71.5%	0
1980	135	98	213	69.7%	0
1981	135	92	217	69.6%	0
1982	134	87	221	69.7%	0
1983	133	81	225	69.7%	0

C. Hiring of New Faculty[†]

	Assistant (84%)	Associate (6%)	Full (10%)	Total (100%)
1978-79	23	2	2	27
1979-80	24	2	2	28
1980-81	18	1	3	22
1981-82	18	1	3	22
1982-83	18	1	3	22

*/ Rounded to two significant digits.   †/ Rows sums equal total.

Table 3.   Results of Five-Year Plan under Preemptive
Priority for Attainment of Tenure Fraction

A. Promotion Opportunities

	Assistant to Associate		Associate to Full	
	Flow	Rate*	Flow	Rate
1978-79	0	0	0	0
1979-80	0	0	0	0
1980-81	3	.02	9	.09
1981-82	4	.03	9	.10
1983-83	4	.03	9	.10

B. Grade Size Distribution

	Assistant	Associate	Full	Tenure Fraction	Budget Deficit
1978 (given)	117	98	222	73.2%	0
1979	138	98	219	70%	$250,000
1980	135	98	213	69.7%	0
1981	135	92	217	69.6%	0
1982	134	87	222	69.7%	0
1983	133	81	225	69.7%	0

C. Hiring of New Faculty[†]

	Assistant (84%)	Associate (6%)	Full (10%)	Total (100%)
1978-79	34	3	4	41
1979-80	12	1	2	15
1980-81	18	1	3	22
1981-82	18	1	3	22
1982-83	18	1	3	22

*/ Rounded to two significant digits.   †/ Rows sums equal total.

$250,000, explained primarily by the transfer of 12 additional
assistant professors, a solution which minimizes deviations from
all other goals. The fact that the "cheapest" solution involves
a significant dollar expense, rather than deviations from other
goals which may be less expensive in dollars, points to a difficulty
of including incommensurable goals in the same priority group. (See
Ijiri, 1965.) The drastic increase in recruitment, from 27 to 41 in
the 1978-79 hiring period may also be unreasonable from an organizational implementation perspective, especially since recruitment of
faculty requires considerable advance preparation and physical ac-
comodations (office space and the like) that may not be immediately
available. It is interesting, however, that many of the other
five year goals can be achieved through a one-time budget deficit,
an investment that may seem worthwhile in the long run.

The purpose of these simulations has been to illustrate the
usefulness of the model in focusing internal promotions policies
towards the attainment of certain objectives. In a real planning
environment one would expect a further refinement of plans to emerge
from simulations reflecting various goal artifact definitions and
preference rankings. As discussed in Feuer (1980) a wide range of
planning problems can be explored, including, for example, EEO,
wage determination, and mandatory retirement.

REFERENCES

Bartholomew, D.J., 1973, "Stochastic Models for Social Processes,"
    John Wiley and Sons, London.
Charnes, A. and W.W. Cooper, 1977, "Goal Programming and Multiple
    Objective Optimization," Eur. J. of O.R., 1.
Charnes, A., W.W. Cooper, K. Lewis, and R. Niehaus, 1975, A Multi-
    Objective Model for Planning Equal Employment Opportunities,
    in "Multiobjective Programming," M. Zeleny, ed., Springer-Verlag.
Charnes, A., W.W. Cooper, A.P. Schinnar, and N. Terleckyj, 1979, "A
    Goal Focusing Approach to Analysis of Tradeoffs among Household
    Production Outputs," Consortium Research Paper No. 8, Proceedings
    of the American Statistical Association.
Feuer, M.J., 1980, "Academic Staffing in Universities: Planning and
    Regulation of Advancement Opportunities in a Hierarchical Organ-
    ization," unpublished Ph.D. dissertation, School of Public and
    Urban Policy, University of Pennsylvania.
Feuer, M.J., and A.P. Schinnar, 1978, "Exchange of Advancement Op-
    portunities in Hierarchical Organizations," Fels Discussion
    Paper No. 136, School of Public and Urban Policy, University
    of Pennsylvania.
Grinold, R. and K. Marshall, 1977, "Manpower Planning Models," North-
    Holland, Amsterdam.
Hopkins, D., 1974, "Faculty Early Retirement Programs," Oper. Res.,
    22, 3.
Ijiri, Y., 1965, "Management Goals and Accounting for Control," North-
    Holland, Amsterdam.
Stewman, S., 1975, "Two Markov Models on Open System Occupational
    Mobility: Underlying Conceptualizations and Empirical Tests,"
    Amer. Soc. Rev., 40, June.

ON THE MOVEMENT OF ECONOMISTS AMONG ACADEMIC

INSTITUTIONS IN THE UNITED STATES

Martin J. Beckmann

Technical University Munich,Germany
Brown University, Providence, USA

1. A standard model of organizational mobility considers personnel in a given organization as upward mobile.
A career consists of a number of promotions after which
the individual may stay in the organization until the
normal age of retirement. This is the stay-on system.
Under this system the individual need not but may voluntarily transfer to another organization in order to
obtain an opportunity for further promotion.Such transfers are subject to a cost, but may be taken repeatedly. In equilibrium some persons choose to stay, some
to move after their final promotion in an organization.
This equilibrium condition determines the salary
spans that are needed to sustain a market with horizontal mobility between organization at all levels and
some vertical mobility in organizations [1].

The academic world possesses an additional feature in
the fact that a well defined system of ranking exists
not only within but also between organizations. Thus a
transfer between organizations need not occur at the
same individual rank but may involve a compensating
change in both the rank of the institution and the
rank in the institution. It is also possible that none
or only one rank is changed in such a transfer.The
mobility of academic personnel thus generates some interesting theoretical questions concerning salary
structure and personnel flow between organizations of
various ranks involving personnel of various ranks. The
purpose of this paper is to explore some of these

relationships that operate in the market for academic
economists. We restrict ourselves to the flow relationships while ignoring the salary aspects. Nor do we dwell
on the implications for the transfer of knowledge that
follow from this mobility of personnel.

2.   The theoretical model that underlies the following
analysis considers ranks r = 1,...4 of instructor,
assistant professor, associate professor, and full professor   and grades i = 1,...4 defined by the type of
institution: " chairmens' group " [2] , PhD granting
universities, MA granting universities and BA granting
colleges. At the instructor level and at the assistant
professor level an " up or out " system operates. At the
associate and full professor levels it is "stay-on".
Flows i.e. transfers are possible only between certain
adjacent ranks and between adjacent types of institutions.

In figure 1 we look at the aggregate of all institutions of one type viz. BA granting colleges

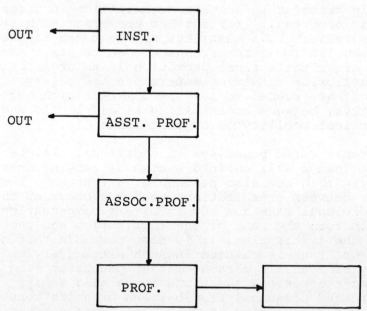

Fig.1 Flow Chart for BA Institutions

Here it is assumed that tenured personnel who have demonstrated superior qualification will have gained
promotion to full professor before being considered for
tenured slots in higher ranking institutions.Instruc-
tors and assistant professors who fail to be promoted
within the period set by the contract have no alternative but to quit the academic field. It is not ruled
out that they may stay on as life-time assistant professors, even though this is frowned upon by the AAUP.

Turn next to institutions of rank 2.

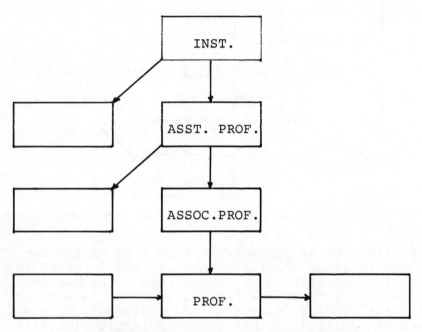

Fig.2 Flow Chart for MA Institutions

Instructors failing to get promoted to assistant professors have the opportunity of being hired as assistant professors in BA granting institutions, and the
same goes for assistant professors failing to reach
rank of associate professor in the MA type of institution. Full professors having tenure have no incentive
to move to lower ranking institutions — there is no
pay incentive either. Thus at the full professor level
once more mobility occurs only in the direction of
higher ranking institutions.

We consider next rank 3 institutions.

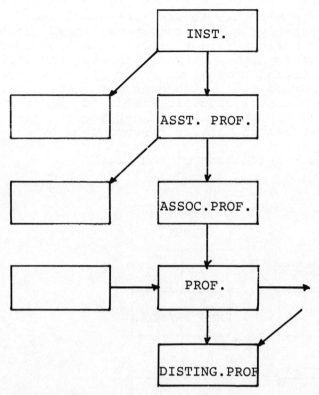

Fig.3 Flow Chart for Other PhD Granting Institutions

The direction of movement is once more up in rank
and down in grade of organization at the level of instructor and assistant professor, and to higher ranking institutions at the level of full professor. An
additional rank of "distinguished professor" or named
chairs etc. exists in this grade of institution.This
rank is filled either by promotion within the institution or from professors of the highest grade of institution, the "chairmens' group".

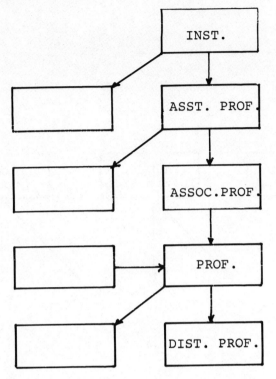

Figure 4.   Flow Chart for Chairmens' Group

The flow patterns for the chairmens' group presents nothing new.

After retirement there is an additional flow of personnel in a downward direction among institutions: lower ranking institutions are willing to hire professors from higher ranking ones beyond the ages of normal retirement. This movement is not considered here. Nor is the fact that there is some out movement to the non-academic world at all levels in all institutions and also a small amount of hiring from the outside at all levels. Our concern is with the mixing of personnel within the academic system. Entry into the instructor level or sometimes assistant professor level is by graduation from MA or PhD granting institution or from the chairmens' group.

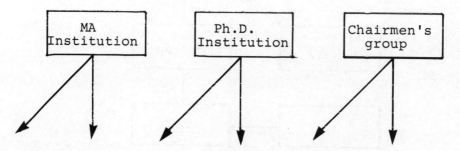

Figure 5.   Flow Chart for Graduates

3.   Putting the pieces together we have the following
composite flow diagram (fig.6). This diagram shows
first of all the set of possible career patterns.De-
pending on initial conditions certain intermediate or
final states can be reached, others cannot. In the diagram as shown, every final position can be reached
from every initial one, but not every intermediate position can be reached. Thus the assistant and associate
professor levels in higher ranking institutions are
inaccessible to graduates of MA institutions.

Restrictions on attainable final positions result when
a limit is imposed on the total number of allowed
transitions. Thus with five moves allowed an MA from
an MA granting institution can at most become a professor at an MA granting institute; with 6 moves at
most a professor in a PhD granting university. A person starting with a PhD from the chairmens' group has
the chance of becoming distinguished professor in the
same group in only 5 moves.

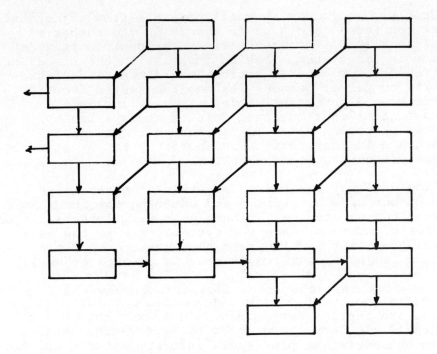

Fig.6 System Flow Chart

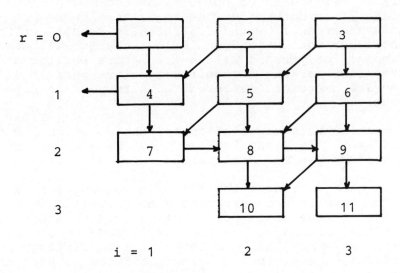

Fig.7 Simplified Flow Chart

Secondly the diagram shows all possible flow adjustments
that can accomodate a given change in the number of
positions (or of occupancy times for positions) at any
point in the system. Thus an increase in the number of
professorships ceteris paribus would mean a decrease
in the number of persons leaving the system from instructor and assistant professor positions in the
lowest grade institutions. This is accompanied by
higher promotion rates to tenured positions in at least
the grade 1 institutions but possibly in the other
types of institutions as well.

In the same way the possible result of an increase in
the number of PhDs granted and entering the academic
system can be studied.Ultimately it will force a larger
number of PhDs to leave the system again at the assistant professor level unless the number of positions
or the length of service in them is changed as well.

4.   Suppose that the total flow through any position is
specified per unit of time,which flows between positions are thereby determined ? To answer this question
we shall simplify the diagram by aggregating instructors and assistant professors into non-tenured and the
associated and full professors into the tenured positions. Moreover we shall aggregate the two lower grades of institutions into a single sector.
Let the number of positions in each rank for each
grade of institution be given and also the number of
years a position is occupied by an incumbent on the
average. This determines the annual inflow and an
equal annual outflow under stationary conditions.There
is no substitution between inflowing and outflowing
personnel for persons will enter before serving and
leave after serving in their respective positions.
(They are therefore different with respect to experience).

To avoid cumbersome notation let the positions be numbered as in fig.7 and let the flows be numbered by the
pairs of origin and destination.

Let us also introduce the following assumption. It is
assumed that tenured promotions to positions 8 and 9
are made on the basis of normal retirement and that
additional hiring at the same level takes place for
senior personnel that have left for more distinguished
positions, the exception being positions 7 which are
necessarily filled from the rank below.

The number of equations equals that of unknowns. Are the flows uniquely determined ?

We arrange the flow equations in a natural order beginning with the unique inflow to 6.

$$x_{36} = a_6$$

$$x_{35} = a_3 - a_6$$

$$x_{25} = a_5 - (a_3 - a_6) = a_5 + a_6 - a_3$$

$$x_{24} = a_2 - x_{25} = a_2 - a_5 + a_3 - a_6$$

$$x_{14} = a_4 - x_{24} = a_4 - a_2 + a_5 - a_3 + a_6$$

$$x_{1o} = a_1 - a_4 + a_2 - a_5 + a_3 - a_6 = a_1 + a_2 + a_3 - a_4 - a_5 - a_6$$

$$x_{69} = a_9$$

$$x_{68} = a_6 - a_9$$

$$x_{58} = a_8 - a_6 + a_9$$

$$x_{47} = a_7 - x_{57} = a_7 - (a_5 - a_8 + a_6 - a_9) + a_{1o} + a_{11}$$

$$= a_7 - a_5 + a_8 - a_6 + a_9 = a_7 + a_8 + a_9 - a_5 - a_6$$

$$x_{40} = a_4 - x_{47} = a_4 + a_5 + a_6 - a_7 - a_8 - a_9 - a_{1o} - a_{11}$$

$$x_{9,1o} = a_{11} \qquad x_{89} = a_{11} + x_{9,1o} \qquad x_{8,1o} + x_{9,1o} = a_{1o}$$

$$x_{78} = x_{8,1o} + x_{89} = a_{1o} + a_{11}$$

since

$$x_{89} = x_{78} - x_{9,1o} = a_{11} + x_{9,1o} = a_{11} + (a_{1o} - x_{8,1o})$$

Comparing the second and fourth parts of the equation

$$x_{78} = a_{11} + a_{1o}$$

Finally let

$$x_{9,1o} = \mu. \text{ Then}$$

$$x_{89} = a_{11} + \mu$$

$$x_{8,1o} = a_{1o} - \mu$$

All flows except $x_{8,9}$, $x_{8,1o}$ and $x_{9,1o}$ are thus uniquely determined, provided the position flows a_i satisfy the appropriate inequality constraints:

Table 1. Numbers of positions

$$a_6 \leqq a_3 \leqq a_5 + a_6$$

$$a_5 + a_6 \leqq a_2 + a_3 \leqq a_4 + a_5 + a_6$$

$$a_4 + a_5 + a_6 \leqq a_1 + a_2 + a_3$$

$$a_9 \leqq a_6 \leqq a_8 + a_9$$

$$a_5 + a_6 \leqq a_7 + a_8 + a_9 + a_{10} + a_{11}$$

$$a_7 + a_8 + a_9 + a_{10} + a_{11} \leqq a_4 + a_5 + a_6$$

$$a_8 \leqq a_{10} + a_{11}$$

One degree of freedom exists in the flows between positions 8,9 and 1o and this permits choosing anyone of them e.g. $x_{9,1o}$ in the range

$$a_8 - a_{11} \leqq x_{9,1o} \leqq \,^9 1_o$$

On the other hand when the rank "distinguished professorship" is collapsed into the professorial rank, then the equations for x_{78} and x_{89} are modified while the variables $x_{8,1o}$, $x_{9,1o}$, $x_{9,11}$ drop out

$$x_{78} = x_{89} = \nu = \text{exogen.}$$

It is assumed here that additional senior level recruiting is induced by either the creation of a new chair or senior personnel leaving the academic field. The equations for all x_{ij}, $i \leqq 6$ remain unchanged.

5. Table 1 lists the number of positions for economists at different ranks in 4 grades of institutions [2] .

r	i	1	2	3	4	
O		153	183	90	131	OTHER
1		258	356	121	128	INSTRUCTORS
2		403	238	332	34o	ASST.PROFESSORS
3		327	212	287	192	ASSOC.PROFESSORS
4		290	264	468	663	PROFESSORS

CHAIRMENS'OTHER MA   BA

GROUP   PhD INST.INST.

INSTITUTIONS

Table 2. Average Lengths of Stay

Assume the following average lengths of stay τ in various ranks

r	τ
0	2
1	2
2	8
3	8
4	20

Table 3. Flows through Positions

This generates the following flows

129	178	61	64
50	30	42	45
41	27	36	24
15	13	23	33

In fig.8 these flows have been allocated according to the schema of fig.7. Inspection of fig.8 shows that reality is more complex then the simple theoretical schema of figures 1-7. The calculated flows are compromises. It turns out that personnel will leave the system not only from non-tenured positions in the lowest grade of institutions but from better placed positions as well. This makes the internal flows non-unique. An alternative interpretation of these outflows is that system growth has occured which would absorb at least part of what in a steady state are unavoidable outflows.

An alternative estimate of steady state flows is obtained by using the actual movement of persons into position during 1978/9 from source [2]. These contain also the movements between universities in a given class and should therefore be larger then those of figure 8. The best interpretation of fig.9 is that it shows representative institutions blown up to the size of their respective class.

Actual in-and-outflows are given in figure 1o but not by source or destination, since these were not reported in [2].

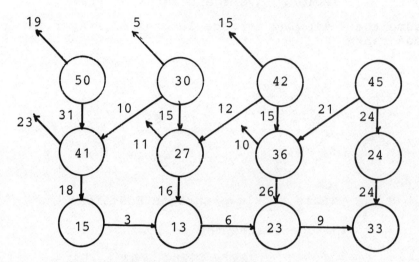

Fig.8 Allocation of Flows

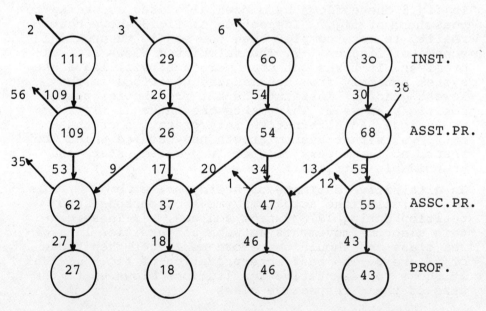

Fig.9 Allocation of Reported Through Flows

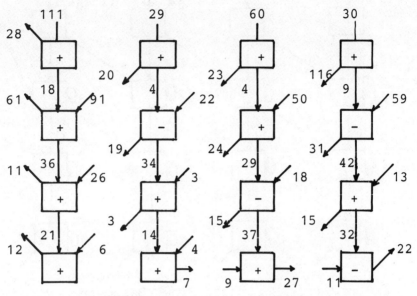

Figure 10.   Reported Actual Flows (Without Retirements)

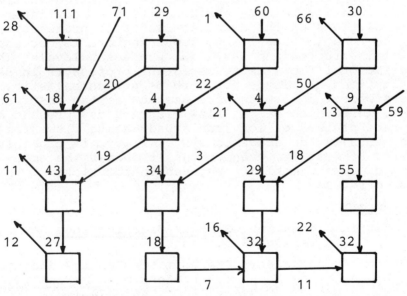

Figure 11.   Estimates of Flows by Origin and Destination
            (No Intrasectoral Flows)

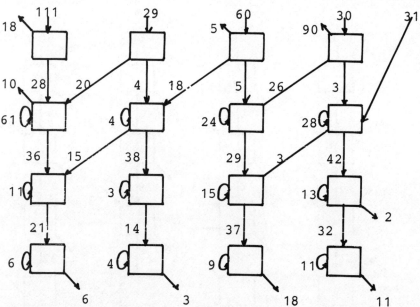

Figure 12.   Estimated Flows Based on Maximal
Intrasectoral Flows

If reported hiring and leavings are allocated as far
as possible to the same sector and rank, fig.12 results. Here intersectoral flows are minimized. Notice
that this system can be sustained without any intersectoral flows at the associate and full professor level, but that movement into lower sectors is necessary
at the instructor and assistant professor levels. The
intersectoral flows among assistant professors in BA
institutions are large but perhaps not unrealistic.No
intersectoral movements were assumed at the instructor
level. While none of the flow diagrams may present an
accurate picture of long run steady state flows for
economists in the academic system,fig.12 is probably
closest to reality, even though it denies the occurence of upward mobility between institutions at the
tenured levels.

References:

[1] Martin J.Beckmann: Rank in Organizations,New York,
     Springer Verlag,1978,chapter 57-71.

[2] Anne Friedlaender: The Committee on the Status of
     Women in the Economics Profession. American Eco-
     nomic Review 70,2(May 1980), 466-474,Table 1:
     Distribution of Full Time Faculty by Type of In-
     stitution Academic Year 1978/9.

NA

SECTION 8

CONCEPTUAL ISSUES

The conceptual issues covered by the Symposium extend into
a number of dimensions. Some of these issues are addressed in
papers which could not be included in their entirety because of
page limitations. One page abstracts and author addresses are
provided for those wishing the full texts.

ESTIMATING PROGRAM IMPACTS ON HUMAN RESOURCES

Robert E. Boynton

Defense Resources Management Education Center
Naval Postgraduate School
Monterey, California 93940

ABSTRACT

This paper proposes a theoretical framework for a human resources impact statement to fill a gap in the existing consideration given to human resources. At the macro-level, government
considers human resource effects in attempting to integrate full
employment goals with major economic policies. At the organization level, human resource effects are considered in developing
manpower plans to meet specific operating requirements. The human
resource impact statement (HRIS) would provide organization decision makers with organizationally relevant information to structure
and direct consideration of human resource effects when choosing
among alternative methods for implementing programs and policies.
  The HRIS framework proposes two principal dimensions - EMPLOY-
MENT OPPORTUNITIES and DEVELOPMENT OPPORTUNITIES. These dimensions
would be examined through four facets of impacts - PRODUCTION,
DISTRIBUTION, STABILIZATION, and GROWTH.
  Policy guidance developed within this theoretical framework
would enable managers to consider appropriate human resource
effects, rather than evaluating only budgetary or other effects,
in choosing among implementation alternatives. Desired human
resource impacts are then more directly achieved with the consequent reduction of present problems involved in attempting to
mitigate effects of decisions made on other bases.

OCCUPATIONAL WELFARE AS AN ASPECT OF QUALITY OF WORKING LIFE

*David Bargal, and **Boas Shamir

*School of Social Work - **Department of Sciology
Hebrew University of Jerusalem
Jerusalem, Israel

The Quality of Working Life movement (QWL) has been associated in the social sciences with two main streams of thought and practice. In the U.S. the main line has been 'Job Enrichment' or 'Job Redesign', with its emphasis on the job content and its motivational potential. In Western Europe the mainstream has followed socio-technical principles, which combined similar job content considerations with consideration of the social organization of the production unit. Both approaches have demonstrated their validity and utility in certain cases, but accumulating evidence has also shown some of their limitations. Perhaps the greatest limitations stem from the fact that in many production and service organizations the technological and economic considerations simply do not enable the levels of autonomy, variety, meaningfulness and significance of operative level jobs to be significantly increased.

It is against this background that we should consider the re-emergence of occupational welfare as a major organizational phenomenon. What is occupational welfare? What are the job's components, how do employees gain from its functioning and what is its contribution to organizational effectiveness, and to quality of working life? We begin by defining the area of occupational welfare, delineating briefly the background of its development in recent times. We proceed to the description of domains and methods through wich occupational welfare contributes to the work life, as revealed by our studies in Israel.

392

TECHNOLOGY SELECTION CONSIDERING MACRO-ECONOMICAL AND SOCIAL FACTORS

V.Dogan Sorguc

Technical University of Istanbul
İTÜ Faculty of Civil Engineering
Istanbul (Taşkışla) - Turkey

ABSTRACT

It is a known fact that the results of the competition between the nations in the world are evaluated according to the stages of development i.e. the maximum social change realised. This change is a function of macro-economical performance. Therefore technology selection which is the basis of macro-economical performance requires careful consideration of the related system with all its (economical and social) factors.

The paper presents the (mathematical) model of the economical development with its factors and discusses them for technology selection considering social facts (factors); the most important of them being the unemployment. Thus especially referring to the problems of developing countries as partly experienced in Turkey the paper gives the landmarks of a sound strategy of development based on the performance of macro-economical production heavily influenced by the selection of optimum technology.

It also shows how-far technology selection mainly entrusted to engineers is an interdisciplinary problem. It is thus shedding light on needs for (setting up and managing) interdisciplinary teams in handling this problem and is at the same time a challange for engineering education.

TECHNICAL INNOVATION IN CONSTRUCTION MANAGEMENT

A CRITICAL APPRAISAL OF DEVELOPMENT IN PROJECT PLANNING AND CONTROL

David Arditi

Middle East Technical University
Department of Civil Engineering
Ankara-Turkey

ABSTRACT

Research and development activity, purchase and flow of knowledge, talent level and distribution, economic and market structure, investment and availability of financing are the factors which affect the speed at which an innovation spreads. The use and diffusion of modern planning and control techniques such as CPM and PERT are examined in this paper within the above mentioned analytical framework. The similarities and discrepencies between industrialized and developing countries are emphasized.

KINETICS OF TECHNOLOGICAL CHANGE:

SOME THEORETICAL BASIS

Amar Dev Amar

Department of Finance & Quantitative Methods
Montclair State College
Upper Montclair, New Jersey   07043

ABSTRACT

This work provides a new theoretical and operational definition of technology both at macro and micro levels. Seven constructs of each level of technology are provided. Three types of
forces that alter technology are introduced and kinetically
modeled. Consequence of these forces on technology constructs is
presented and denoted 'technological change'. Two additional
measures of change, viz. rate of change of change, and rate of
change of rate of change of change are operationalized. A classification of organizations according to change states is given.
Although only theoretical bases are provided, there is enough
substance for further research and experimentation.

ON A MULTISTAGE GAME OF MANPOWER PLANNING:

PERSONNEL MANAGEMENT VS. FRUSTRATED EMPLOYEES

Alexander Mehlmann

Department of Operations Research
University of Technology
Vienna, A-1040

ABSTRACT

We consider a discrete-time pull-type manpower system, where
the management's decisions, by generating frustration, can change
both the requirements and the flows of personnel between grades.
An adaptive optimization problem is formulated, which is based on
a multistage game situation between personnel management and
frustrated employees. The personnel management is the dynamic
leader. It is assumed that the frustrated employees know the
leader's strategy and make their decisions (leave or stay) based
on future career prospects. This reaction behaviour is known to
the leader who optimizes his choice. A simple dynamic programming
argument will solve the problem.

> Men, working too hard in rooms that are too big,
>   Reducing to figures
> What is the matter, what is to be done.
>   W.H.Auden "The Managers"

THE HUMAN SIDE OF PROJECT-TEAMS: SOCIAL START-UP OF TEAM WORK

P.L.R.M. van Hooft

Berenschot Management Consultants
Churchill-laan 11, 3527 GV Utrecht

ABSTRACT

   Use is being made, at steadily increasing rates within
organizations, of teams to accomplish specified tasks. This
paper discusses the social start-up of project teams following
the process through the team's life-cycle. Particular attention
is placed on the use of a team building conference where the
principal makes his expectations manifest and the team explores
the job to be done. During these discussions there can be well-
planned interventions to assess (with the group) the ways in
which the decisions are to be made, the place of each individual
in the group, motivation and satisfaction of the group members
and so on.

PARTICIPANTS

NATO COUNTRIES

BELGIUM

Rauol GOBEL
Generale Bankmaatschappij
Directie Social Zaken (021B)
Warandeberg 3
1000 Bruxelles, Belgium

Raymond GORBITZ
UNIPAC
Avenue D'Auderghem 49
B-1040 Bruxelles, Belgium

Andre MEERS
Generale Bankmaatschappij
Directie Social Zaken (021B)
Warandeberg 3
1000 Bruxelles, Belgium

Paul VERHAEGEN
Université Catholique de
  Louvain
Schoonzichtlaan 21
B-3009 Herent, Belgium

CANADA

Alain MARTEL
Faculté des Sciences
 de L'Administration
Université Laval
Quebec, Canada G1K 7P4

Pierre MARTEL
Personnel Policy Branch
Treasury Board Secretariat
Ottawa, Canada K1A OR5

Wilson L. PRICE
Faculté des Sciences
 de L'Administration
Université Laval
Quebec, Canada G1K 7P4

LTCOL D. J. SLIMMAN
National Defence Headquarters
ATTN: Director Personnel
 Development Studies
Ottawa, Canada K1A OK2

FRANCE

Pierre MIRET
Eurequip
19, Rue Yves du Manoir
92420 Vaucresson
France

NETHERLANDS

P.M. BAGCHUS
Dept. of Industrial Engineering
Eindhoven University of
 Technology (P.B. 513)
5600 MB Eindhoven
Netherlands

Paul L.R.M. VAN HOOFT
Berenschot Management Consul-
 tants
Churchill-laan 11
3257 GV Utrecht
Netherlands

LCOL H.J.D. WANDERS (RNLAF)
Directie Personeel KLu,
 Afderling Gedragswetenschappen
2509 LM The Hague
Spul 47
Netherlands

TURKEY

Sinasi AKSOY
Department of Political Science
 and Public Administration
Middle East Technical University
Ankara, Turkey

David ARDITI
Department of Civil Engineering
Middle East Technical Univer-
 sity
Ankara, Turkey

V. Dogan SORGUC
Technical University of Istanbul
Istanbul, Turkey

UNITED KINGDOM

Paul BLYTON
Department of Business Admini-
 stration and Accountancy
University of Wales Institute
 for Science and Technology
Cardiff, Wales, U.K.

Lisl KLEIN
Tavistock Institute of Human
 Relations
120 Belsize Lane
London NW3 5BA, U.K.

Richard PEARSON
Institute of Manpower Studies
University of Sussex
Mantell Building
Falmer, Brighton BN1 9RF, U.K.

Gillian URSELL
Trinity and All Saints College
Brownberrie Lane, Horsforth
Leeds L518 5HD, U.K.

UNITED STATES OF AMERICA

Amar Dev AMAR
Montclair State College
Upper Montclair, NJ 07043
USA

Donald M. ATWATER
13750 Raywood Drive
Los Angeles, CA 90079
USA

Robert E. BOYNTON
Defense Resources Management
 Education Center
Naval Postgraduate School
Monterey, CA 93940
USA

Edward S. BRES, III
Office of the Assistant
 Secretary of the Navy
(Manpower, Reserve Affairs &
 Logistics)
Washington, D.C. 20350
USA

Michael J. FEUER
College of Business
 and Administration
Drexel University
Philadelphia, PA 19104
USA

Cheryl GAIMON
Graduate School of Industrial
 Administration
Carnegie-Mellon University
Pittsburgh, PA 15213
USA

Barbara A. GUTEK
5001 Murietta Avenue
Sherman Oaks, CA 91423
USA

George KOZMETSKY
Graduate School of Business
GSB 1.104
University of Texas at Austin
Austin, TX 78712
USA

Marty KAPLIN
Weatherhead School of Management
Case Western Reserve University
Cleveland, OH 44106
USA

Richard J. NIEHAUS
Office of Assistant Secretary
 of the Navy
(Manpower, Reserve Affairs &
 Logistics)
Washington, D.C. 20350
USA

William A. PASMORE
Weatherhead School of Management
Case Western Reserve University
Cleveland, OH 44106
USA

Murray ROWE
Navy Personnel Research and
 Development Center
San Diego, CA 92152
USA

Arie P. SCHINNAR
School of Public and Urban
 Policy
University of Pennsylvania
39th & Market Streets
Philadelphia, PA 19104
USA

John R. SCHMID
B-K Dynamics, Inc.
15825 Shady Grove Road
Rockville, MD 20850
USA

Andrew SCHOU
University of Central Florida
Box 25000
Orlando, FL   32816
USA

Carol T. SCHREIBER
General Electric Company
RM W2H1
2135 Easton Turnpike
Fairfield, CT   06431
USA

James A. SHERIDAN
American Telephone and
  Telegraph Company
1776 on the Green, RM 48-4A18
Morristown, NJ   07960
USA

Rami SHANI
Weatherhead School of Management
Case Western Reserve University
Cleveland, OH   44106
USA

James W. TWEEDDALE
Office of the Assistant
  Secretary of the Navy
(Manpower, Reserve Affairs &
  Logistics)
Washington, DC   20360
USA

Harry M. WEST III
Office of the Chief of Staff
  of the Army
RM 2C725, The Pentagon
Washington, D.C.   20310
USA

WEST GERMANY

Wolfgang ANDERS
Universität Hannover
Limmerstrasse 41
3000 Hannover 91
West Germany

Martin J. BECKMANN
Institut für Statistik und
  Unternehmensforschung
Technische Universität
  München
Barerstrasse 23
8000 Munich 2
West Germany

Jurgen GOHL
Dow Chemical GmbH
Werk Stade/Personalabteilung
Postfach 1120
2160 Stade
West Germany

Gert HARTMANN
International Institute of
  Management
Wissenschaftszentrum Berlin
Platz der Luftbrucke 1-3
D-1000 Berlin 42
West Germany

Volker KORNDOERFER
Institut für Arbeit und
  Organisation
Holzgartenstrasse 17
D7000 Stuttgart 1
West Germany

Siegfried LEHNBERG
Industrieanlagen-Betriebsgesell-
  schaft mbH
Einsteinstrasse
8012 Ottobrunn
West Germany

Rainer MARR
Hochschule der Bundeswehr
Werner-Heisenberg - Weg 39
8014 Neubiberg
West Germany

Dieter SADOWSKI
Universität Trier
FBIV-Betriebswirtschaftslehre
Postfach 3825
D5500 Trier
West Germany

Klaus ZOLLER
Hochschule der Bundeswehr
Holstenhofweg 85
2 Hamburg 70
West Germany

Gerhard O. MENSCH
International Institute of
  Management
Wissenschaftszentrum Berlin
Platz der Luftbrücke 42
D-1000 Berlin 42
West Germany

Wolfgang H. STAEHLE
Freie Universität Berlin
Institut für Unternehmensführung
Garystrasse 21
1 Berlin 33
West Germany

NON-NATO COUNTRIES

AUSTRIA

Alexander MEHLMANN
Department of Operations Research
University of Technology Vienna
Argentinierstrasse 8/4a
A-1040 Wien, Austria

IRELAND

Tevfik DALGIC
Faculty of Engineering and
  Systems Sciences
Trinty College
Dublin 2, Ireland

ISRAEL

David BARGAL
Paul Baerwald School of Social Work
The Hebrew University of Jerusalem
Jerusalem, Israel

JAPAN

Taro TANIMITSU
Centeral Research Laboratory
Mitsubishi Electric Company
80 Nakano, Minami-Shimizu
Amagasaki, Hygo
Japan 661

INDEX